T0342474

Smart and Sustainable Approaches for Optimizing Performance of Wireless Networks

Smart and Sustainable Approaches for Optimizing Performance of Wireless Networks

Real-time Applications

Edited by

Sherin Zafar
Jamia Hamdard
New Delhi
India

Mohd Abdul Ahad
Jamia Hamdard
New Delhi
India

Syed Imran Ali
University of Technology and Applied Sciences
Al Musannah
Sultanate of Oman

Deepa Mehta
Senior Data Scientist Great Learning

M. Afshar Alam
Jamia Hamdard
New Delhi
India

The right of Sherin Zafar, Mohd Abdul Ahad, Syed Imran Ali, Deepa Mehta, and M. Afshar Alam to be identified as the authors of the editorial material in this work has been asserted in accordance with law.

Registered Offices
John Wiley & Sons, Inc., 111 River Street, Hoboken, NJ 07030, USA
John Wiley & Sons Ltd, The Atrium, Southern Gate, Chichester, West Sussex, PO19 8SQ, UK

Editorial Office
The Atrium, Southern Gate, Chichester, West Sussex, PO19 8SQ, UK

For details of our global editorial offices, customer services, and more information about Wiley products visit us at www.wiley.com.

Wiley also publishes its books in a variety of electronic formats and by print-on-demand. Some content that appears in standard print versions of this book may not be available in other formats.

Library of Congress Cataloging-in-Publication Data

Names: Zafar, Sherin, editor. | Ahad, Mohd Abdul, editor. | Ali, Syed
    Imran, editor. | Mehta, Deepa, 1981- editor. | Alam, M. Afshar,
    editor.
Title: Smart and sustainable approaches for optimizing performance of
    wireless networks : real-time applications / edited by Sherin Zafar,
    Mohd Abdul Ahad, Syed Imran Ali, Deepa Mehta, M. Afshar Alam.
Description: Hoboken, NJ : Wiley, 2022. | Includes bibliographical
    references and index.
Identifiers: LCCN 2021040689 (print) | LCCN 2021040690 (ebook) | ISBN
    9781119682509 (cloth) | ISBN 9781119682523 (adobe pdf) | ISBN
    9781119682530 (epub)
Subjects: LCSH: Network performance (Telecommunication) | 5G mobile
    communication systems.
Classification: LCC TK5105.5956 .S63 2022 (print) | LCC TK5105.5956
    (ebook) | DDC 004.6–dc23
LC record available at https://lccn.loc.gov/2021040689
LC ebook record available at https://lccn.loc.gov/2021040690

Cover Design: Wiley
Cover Image: © Krunja/Shutterstock

Set in 9.5/12.5pt STIXTwoText by Straive, Chennai, India
Printed and bound by CPI Group (UK) Ltd, Croydon, CR0 4YY

C9781119682509_190122

Contents

About the Editors

Dr. Sherin Zafar is Assistant Professor Computer Science & Engineering in the School of Engineering Sciences & Technology, Jamia Hamdard, with a decade of successful experience in teaching and research management. She specializes in Wireless Networks, Soft Computing, Network Security, Data Visualization, Design Thinking, and Machine learning, etc. and has a great profile in Scopus, Mendeley, Google Scholar, Research Gate, and Publons. She has about a half century of papers published in Scopus, SCI, and peer reviewed journals and a double century of papers reviewed by the Editorial Board and Editor in Chief of many reputed and Scopus indexed journals. She has a published patent and one patent granted under her name, Co-Pi for the FIST project of DST and Pi for a completed project for Unnant Bharat Abhiyan. A strong believer in the power of positive thinking in the workplace, Dr Sherin regularly develops internship and career campaigns for students through Internshala and Epoch (Literary and Cultural Society) Groups and has guided a huge number of graduates, postgraduate, and PhD students. She has been session chair for 7+ international conferences, keynote speaker, resource person for 1000+ webinars and FDPs for renowned institutions, AICTE STTP, and for AICTE ATAL FDP. Sherin enjoys a good Netflix and Cricket binge but can also be found on long drives along country roads.

Dr. Mohd Abdul Ahad is currently working in the Department of Computer Science and Engineering, School of Engineering Sciences and Technology, Jamia Hamdard. He has a rich experience of more than 14 years in the field of Computer Science and Engineering. He obtained his PhD in the field of Big Data Architecture. His research areas include Big Data Architecture, Distributed Computing, IoT, and Sustainable Computing. He has published several research papers in various international journals of repute in Q1 and Q2 categories. His cumulative Impact factor for the last three years is 49.8 (as per Clarivate/JCR). He has chaired several sessions at international conferences of Springer, Elsevier, etc. He is a Certified Microsoft Innovative Educator and a certified

Google Educator. He is a life member of the Indian Society of Technical Education (ISTE), as well as an active Senior Member of IEEE. He is also on the review board of several prestigious journals such as *Journal of Networks and Computer Applications*, Elsevier, *Sustainable Cities and Society*, Elsevier, *Journal of Ambient Intelligence and Humanized Computing* (JAIHC), Springer, and *Computers in Biology*, Elsevier, etc.

Dr. Syed Imran Ali is currently working in the Department of Engineering, Computer Engineering Section, University of Technology and Applied Sciences Al-Musannah, Sultanate of Oman.

He has 16 years of academic and research experience and worked at various reputed institution in India and Oman. He obtained his B. Tech, M. Tech and PhD degree in the field of Computer Engineering.

Syed Imran Ali's research areas include Image Processing, Artificial Intelligence, Networks, and Security. He has published 19 research papers in various international journals of repute including IEEE and Scopus indexed journals.

He published a book on Artificial Intelligence in 2007 and another book on wireless sensor networks in 2021.

He is a life member of the Computer Society of India and he is also on the editorial review boards of several prestigious journals.

Dr. Deepa Mehta, Senior Data Scientist in Great Learning, has also been associated with KR Mangalam University for Research Industry Interactions & Incubation. She has a PhD in Computer Science Engineering from Manav Rachna International Institute of Research and Studies. Her research areas are Machine Learning and Artificial Intelligence. She is an expert in the area of Computer Networks and Machine Learning and has taught subjects in this area for 8+ years in reputed colleges and universities. She has authored more than 20 research papers in journals and national/international conferences. She has organized several successful workshops and Winter Schools in innovative and neoteric areas. She is also a certified trainer of Python.

Deepa Mehta's current research interests are on sustainable energy using the Artificial Intelligence and Machine Learning techniques. She maintains an upward learning curve by adapting to the new technologies and applying the new technologies to achieve the sustainable environment goal.

Prof. M. Afshar Alam is the Vice Chancellor, Jamia Hamdard and also Professor and Dean in SEST, Jamia Hamdard. He has a rich teaching and research experience of more than 24 years. He has guided more than 25 PhD students. He has published more than 130 research paper in various international journals and conferences of repute. Prof Afshar is also a member of several high-powered committees of Government of India Agencies such as DST, UGC, SERB, MHRD, etc. He has achieved a huge number of awards of national and international importance.

List of Contributors

Prashant Abbi
Department of Information Science and
Engineering
R V College of Engineering

Ayesha Hena Afzal
Department of Computer Science and
Engineering
FET
MRIIRS
Haryana
India

Shakil Ahmed
Department of Electrical and Computer
Engineering
Iowa State University
Ames
Iowa
USA

M. Afshar Alam
Department of Computer Science and
Engineering
School of Engineering Sciences and
Technology
Jamia Hamdard
New Delhi
India

Mohammed Amjad
Department of Computer Engineering
Jamia Millia Islamia
New Delhi
India

Khushi Arora
Department of Computer Science and
Engineering
R V College of Engineering

K.B. Ashwini
Department of Master's in Computer
Applications
R V College of Engineering

Banuroopa
Department of Computer Science
Karpagam Academy at Higher Education
Coimbatore
Tamilnadu

Bharat Bhushan
School of Engineering and Technology
Sharda University
Noida
Uttar Pradesh
India

Satrupa Biswas
Department of Information
Communication and Technology
Manipal University
MAHE

Siddhartha Sankar Biswas
Department of Computer Science and
Engineering
School of Engineering Sciences and
Technology
Jamia Hamdard
New Delhi
India

Md Mudassir Chaudhary
Department of Computer Science and
Engineering
School of Engineering Sciences and
Technology
Jamia Hamdard
New Delhi
India

V. Chayapathy
Department of Electrical & Electronics
Engineering
R V College of Engineering

Mehajabeen Fatima
SIRT Bhopal
Madhya Pradesh
India

Ashu Gautam
MRIIRS Faridabad
Haryana
India

Aju Mathew George
Department of Civil Engineering
Amal Jyorhi College of Engineering
Kanjirapally
Kerala
India

Majumder Fazle Haider
Department of Electrical and Electronic
Engineering
BRAC University
Bangladesh

Nida Iftekhar
Department of Computer Science and
Engineering
School of Engineering Sciences and
Technology
Jamia Hamdard
New Delhi
India

Christina Jane
Department of ECE
Mar Ephraem College of Engineering
Elavuvilai
Tamil Nadu
India

Lijo Joseph
Department of EEE
Amal Jyothi College of Engineering
Kanjirapally
Kerala
India

Maheswari K
Department of Mathematics
Kumaraguru College of Technology
Coimbatore
Tamilnadu

Chalapathiraju Kanumuri
RV College of Engineering
Bengaluru
Karnataka
India
and
S.R.K.R Engineering College
Bhimavaram
Andhra Pradesh
India

Samia Khan
Department of Computer Science and
Engineering
School of Engineering Sciences and
Technology
Jamia Hamdard
New Delhi
India

Afreem Khursjeed
IIIT Bhopal
Madhya Pradesh
India

Avinash Kumar
School of Engineering and Technology
Sharda University
Noida
Uttar Pradesh
India

S.N. Kumar
Department of EEE
Amal Jyothi College of Engineering
Kanjirapally
Kerala
India

Praveen Kumar Gupta
Department of Biotechnology
R V College of Engineering
Bengaluru
India

Gunjan Madaan
Deloitte Consulting India Private Limited
Bengaluru
India

CH. Renu Madhavi
S.R.K.R. Engineering College
Bhimavaram
Andhra Pradesh
India

Rashima Mahajan
MRIIRS Faridabad
Haryana
India

Ayasha Malik
Noida Institute of Engineering Technology
Greater Noida
Noida
Uttar Pradesh
India

Deepa Mehta
Department of EEE
GD Goenka University
Sohna
Haryana
New Delhi
India

Mohankumar
Department of Computer Science
Karpagam Academy at Higher Education
Coimbatore
Tamilnadu

Md Tabrez Nafis
Department of Computer Science and
Engineering
School of Engineering Sciences and
Technology
Jamia Hamdard
New Delhi
India

Nishitha R. P.
Department of EEE
Amal Jyothi College of Engineering
Kanjirapally
Kerala
India

Ashiqur Rahman Rahul
Department of Electrical and Electronic
Engineering
BRAC University
Bangladesh

Danish Raza Rizvi
Department of Computer Science and
Engineering
Jamia Millia Islamia
New Delhi
India

Akhil Sabuj
Department of EEE
Amal Jyothi College of Engineering
Kanjirapally
Kerala
India

Saifur Rahman Sabuj
Department of Electrical and Electronic
Engineering
BRAC University
Bangladesh
and
Department of Electronics and Control
Engineering
Hanbat National University
Korea

Neha Sharma
Department of DEEE
GD Goenka University
Sohna
Haryana
India

Safdar Tanweer
Department of Computer Science and
Engineering
School of Engineering Sciences and
Technology
Jamia Hamdard
New Delhi
India

M.J. Vidya
Department of Electronics and
Instrumentation Engineering
R V College of Engineering

Sana Zeba
Department of Computer Engineering
Jamia Hamdard
New Delhi
India

Editors:

Mohd Abdul Ahad
Department of Computer Science and
Engineering
School of Engineering Sciences and
Technology
Jamia Hamdard
New Delhi
India

Syed Imran Ali
University of Technology and Applied
Sciences
Al Musannah
Sultanate of Oman

Sherin Zafar
Department of Computer Science and
Engineering
School of Engineering Sciences and
Technology
Jamia Hamdard
New Delhi
India

1

Analysis and Clustering of Sensor Recorded Data to Determine Sensors Consuming the Least Energy

Prashant Abbi[1], Khushi Arora[2], Praveen Kumar Gupta[3], K.B. Ashwini[4], V. Chayapathy[5], and MJ. Vidya[6]

[1]*Department of Information Science and Engineering, R V College of Engineering, Bengaluru, India*
[2]*Department of Computer Science and Engineering, R V College of Engineering, Bengaluru, India*
[3]*Department of Biotechnology, R V College of Engineering, Bengaluru, India*
[4]*Department of Master's in Computer Applications, R V College of Engineering, Bengaluru, India*
[5]*Department of Electrical & Electronics Engineering, R V College of Engineering, Bengaluru, India*
[6]*Department of Electronics and Instrumentation Engineering, R V College of Engineering, Bengaluru, India*

1.1   Introduction

Wireless sensor networks (WSNs) [1] are wireless networks that configure themselves and are used to observe environmental or physical circumstances, such as pollutants, motion, vibration, pressure, temperature, sound, etc. Through the network, the data is returned to a principal position. The data is observed and analyzed at this station. The station behaves as an interface between the network and its users. The essential information can be retrieved from the station by raising queries and consolidating and assembling the required results. Usually, a WSN comprises a sizable number of sensor nodes, sometimes as high as hundreds of thousands of nodes. The mode of communication between the sensor nodes is typically through radio signals [2].

Currently, there are numerous areas where these networks can be applied, assisted by the vast open research on WSNs [3]. A few of these application areas, among others, are monitoring, tracking, automation, surveillance, military applications, and agriculture. In all such applications, one of the main objectives for the design of the sensor node and network is to keep the WSN working, functional, and efficient for as long as possible [4]. Thus, the way the network is formed is a key element. The topology is usually defined on the basis of the context and environment of the network application.

WSNs allow new areas of implementation, and need unconventional prototypes for protocol design, due to various restrictions [5]. Due to prerequisites such as the low complexity of devices along with low energy consumption and a long lifetime of the network, a suitable balance between signal and data processing capacity and communication must be found and established. During the last decade, this has motivated huge research activities, industrial investments, and standardization processes in this area [6–11].

Smart and Sustainable Approaches for Optimizing Performance of Wireless Networks: Real-time Applications, First Edition.
Edited by Sherin Zafar, Mohd Abdul Ahad, Syed Imran Ali, Deepa Mehta, and M. Afshar Alam.
© 2022 John Wiley & Sons Ltd. Published 2022 by John Wiley & Sons Ltd.

1.2   The Working of WSNs and Sensor Nodes

Sensor nodes have sensory devices that allow data sensing, while transceivers help them communicate [12]. When a stimulus is detected, sensor nodes, known as sources, generate data packets, and the corresponding sensor information is typically recorded via the accessible "gateways" in the topology. They then transfer them using the network to one or more special nodes, known as sinks or base stations. If a high transmission power is used on the node that is transmitting, direct communication would be possible. However, in largely established networks, lots of energy would be used, and high transmission power alone would not be adequate to reach the sink. This complication can be controlled by multihop communication, i.e. sensors acting as both data routers and data generators [13].

A WSN is assembled using computing and sensing devices, power components, and radio transceivers. In a WSN, the individual nodes have inherently restrained resources, i.e. the processing speed, communication bandwidth, and storage capacity are limited [14]. Once the sensor nodes are set, they are supposed to organize themselves into a suitable network infrastructure, usually along with multihop communication. This is followed by the onboard sensors collecting the required information. Wireless sensor devices also perform specified commands or provide sensing samples as a response to queries raised by a control site. The mode of operation of the sensor nodes might be either event driven or continuous. The Global Positioning System (GPS) and the algorithms used for native positioning could be used to procure information about positioning and location [15]. Wireless Sensor and Actuator Networks are essentially wireless sensor devices that are fitted with actuators to "act" on specific given conditions.

The wireless sensor network's design complexity and intricacy is dependent on the requirements of the application, the utilization or consumption power, the number of nodes, the sensor lifespan, the data that has to be collected, its timing, the environment, and the context, geography, and location of usage of the sensor [16–21].

Cable Mode Transition (CMT) [22] is a proposed algorithm for the determination of the least number of sensors and the optimum architectures needed to remain active to sustain the k-coverage [6, 23, 24] of a terrain along with the k-connectivity [25–29] of the network. Essentially, based only on native information, it assigns phases of idleness for the cable sensors without impacting the connectivity and coverage requirements of the network. Another proposition for WSNs is a network structure for data collection that is delay-aware [30–33]. The objective of this proposal is to reduce delays to the maximum extent possible in the data collection processes in WSNs. This results in extension of the lifespan of the network. A third proposition considers the relay nodes to alleviate the geometric insufficiencies of the network, and uses algorithms based on Particle Swarm Optimization (PSO) to detect the ideal sink position in relation to the relay nodes to overcome the lifespan challenge.

Another proposal suggests a geometry-based solution for finding the ideal sink location to maximize the lifespan of the network. Usually, homogeneous sensor nodes have been considered in the research on WSNs. However, researchers now are focused more on leveraging the use of sensor nodes that are dissimilar in terms of their energy. These are known as heterogeneous WSNs. The provision of fault tolerance with elevated network correspondence in heterogeneous WSNs using a deployment of relay nodes where sensor nodes have different radii of transmission is a major problem in research, which still needs to be, and is being, addressed. Rapid advancements in technology and new network architectures based on heterogeneous devices broaden the scope of possible applications for WSNs and eliminate the present-day limitations considerably.

1.3 Classification of WSNs

WSNs are often classified into two broad categories – distributed and centralized techniques. Within a distributed system, nodes are self-governing and therefore the link is merely between neighboring nodes, whereas in a centralized system, one device or appliance controls the network formation. WSNs can also be classified into two categories according to the type of their deployed sensors: static and mobile. Static WSNs are denoted simply as WSNs, while mobile WSNs are abbreviated as MWSNs. This chapter focuses on centralized and distributed networks.

Centralized networks [34, 35]: These are appropriate for those networks wherein the power capacity of processing relies totally on a singular appliance. Thus, this single appliance is responsible not just for the regulation and coordination of the sensed information but also for subjecting it to a series of actions in order to achieve a required result. The data that is sensed is forwarded and provided to the required sink node. The foremost benefits of this method are:

- Inside the network, roaming is allowed.
- Better application design in terms of the placement of nodes, application awareness, etc. is made possible through context information availability.
- Network coverage analysis is simplified.
- Energy management is more efficient in centralized schemes.
- Centralized networks receive instructions from a singular appliance.

This centralized node is answerable to the network for delivering operational services like node localization, event detection, and effective routing of traffic. An appropriate topology for this technique would be a star topology. Centralized networks are often classified based on the algorithm for processing the information. A few of these categories are mentioned below:

- Single sink: Essentially, the objective of the developmental plan is to find a method of reducing the forwarding time for routing the knowledge toward a singular sink. However, a point to be noted is that there is a lack of redundancy in these systems.
- Multisink [36]: When the assigned tasks are diverted among several nodes, multiple sinks are employed. Coverage, efficient distribution of traffic flows, network density and lifetime, redundancy, and possible energy utilization are a few possible areas where Multisink is applied.
- Multiple task devices: According to prominent research works, the utilization of additional network appliances is often responsible for performing selected activities inside the network, like controlling the node movements and defining target nodes, while also having knowledge of the environment that informs the definition of a route in order to optimize the general WSN application performance.

Based on the dynamics of the node roles, a further classification can also be made, between defined operational networks (DON) [36–39], hierarchical networks [40–44], and static networks [45–47].

- DON: The action or behavior of the node can be defined during the processing function of the network. The application begins with the successful detection of an event by the node. Thus, the node forwards its data to the required sink node.
- Hierarchical networks: Sensors define priorities specific to their role within the network. A lower precedence is seen in nodes that forward traffic, compared with fully functional nodes, which can sense, consolidate, and forward relevant information and data. The network management execution is carried out in a hierarchical way and is clearly designated to support the roles.
- Static networks: In these networks, the nodes are usually positioned strategically and in suitable places prior to the launch of the application. The main aim here is to provide a much better and enhanced data collection and processing performance. The network is partitioned into independent clusters. The centroids corresponding to these newly formed clusters are set in place with the positioning of the sinks. Node positions are set based on some measures. The assembled formation is based on Particle Swarm Optimization (PSO).

Distributed networks: In these networks, each node manages the information: decisions are taken locally and restricted to its neighborhood, known as single-hop neighbors. The foremost characteristics of this type of network are listed below:

- The appliances are autonomous.
- The information is shared by each node to its neighborhood.
- There is no requirement for interconnection devices, such as routers, bridges, etc.
- These networks are appropriate for applications that are distributed, like self-organized systems, multiagent systems, etc.
- Targeting harsh environments is made possible, due to their flexibility
- Information is specifically forwarded to at least one node.

Distributed techniques are applied when the appliance needs to maintain essential properties, such as the quantity of connections, memory, or energy conservation, or when information processing is ineffective. These techniques have some noteworthy characteristics:

1. Independence: This is observed in those instances where a user is the only one who has the liberty to choose where and when the knowledge could be stored, altered, or removed. The knowledge saved is completely independent and does not depend on any other devices. Device data is instrumental in supporting important decisions.
2. Reduced information management: Information is shared by each node to its neighbors, so that additional interconnecting devices are not required, and the management of information is made simpler compared to other network types.
3. Scalability: Depending on the appliance, more nodes can be added to the network and the architecture can be scaled without changing the overall performance of the network, which suggests that the modifications made do not affect the network as a whole.

Networks that are centralized are easily divided between those that operate with a single sink and those that use multisink environments. These networks may be classified further, based on the application in different categories, as hierarchical networks or application-based networks, and can be analyzed by topology.

Hierarchical networks: A notable research paper proposes the Adaptive Epidemic Tree Overlay Service (AETOS) [48]. This approach is based on using replacement agents to develop and maintain robust tree topologies on demand. They react to changes in the environment by rewiring their connections. Interactions between the agents and thus across the appliances are managed with the help of an AETOS proxy, with another native agent. The AETOS proposal focuses on virtual tree topologies built on fundamental network infrastructures [49–54].

The tree overlay is influenced by the node, and so is the way that the native software agents interact with each other to ensure that the topology is robust against failure. An optimal number of child nodes should be retained for each node in order to control the processing cost, which is not in the purview of the AETOS. The behavior of every agent in AETOS is cut-throat, which suggests continuous optimization of its location within the tree by selecting links with sturdier agents, which causes substitutions in the closeness criteria. This in turn suggests a high energy consumption. Every node within the tree, not including the leaves, is presumed to have a spread of child nodes. This demonstrates native adaptation and structurally fluid nodes. They coordinate and collectively use an auto-organization scheme based in AETOS. The researchers failed to consider energy constraints or a multi-link environment. There is no restructuring for the main node in the tree.

Application-based networks [55–61]: These networks are often further sub classified into event-based networks and routing-based networks. In an event-based network, the starting point for execution or formation is events, while in a routing-based network, the focus is more on finding the most appropriate route. The choice of route is informed by a number of different metrics that evaluate the nodes for their energy level, distance, number of hops, number of visited nodes, etc.

Networks compared by topology [62–68]: The foremost common topologies utilized in sensor networks were considered. A number of metrics such as energy consumption, reliability, and latency are presented as emergent properties of this approach. Other topologies where the node behavior affects the execution of the network were also analyzed.

1.3.1 Benefits and Drawbacks of Centralized Techniques

The chief benefit of centralized approaches is as follows: they are controlled by a singular central appliance which knows the entire neighboring domain and thus the locations of all the stationed appliances. Usually, the event source is known, and thus the knowledge is shipped to a specified sink or target. There are not many reception or transmission disputes since the central appliance manages each node.

The routing is simple to compute and thus the simplest paths are often chosen, keeping in consideration the entire environment. The optimum positions of the nodes and sink, and thus the number of hops, are often computed considering metrics like the space between nodes, the number of nodes within the neighboring domain, and the energy level of each node [69].

One of the disadvantages of this sort of technique is immense energy consumption. Since the nodes have to transfer information, they need to know to which node they are delivering the message. Normally, GPS or other such approaches are implemented on each node to determine their locations, which suggests plenty of processing and energy loss. These approaches normally assume faultless behavior of nodes within the network, but this does not consider memory constraints, which suggest a loss of data. Also, interference and obstacles are not taken into consideration.

In these approaches, the implementation of reconfiguration is simple, but more network resources are required, along with an excessive energy cost. The network does not support a high density of nodes, due to the huge quantity of knowledge produced within the network. Network interactivity is not assured, since during a period of quiet activity for the application it is normal to decide to connect nodes that are active instead of the whole network. Robustness and verification depend on the appliances.

The responsibility for repairing a failure lies with the central device. The rectification of a failure is difficult for a couple of nodes, because full rectification is necessary. The central appliance needs all of the data that is in the neighboring domain for the repair. The network is broken when the central device fails.

1.3.2 Benefits and Drawbacks of Distributed Techniques

The distributed approach is typically used when the system has to handle huge amounts of data, and having some expendability and some verification of the knowledge is convenient. In these approaches, the benefits and drawbacks are defined based on the appliances, the assets, and thus the neighboring domain.

The chief benefits are as follows: The knowledge is native. Therefore, a node only keeps information pertaining to its neighborhood, which may consist of one-hop, or at maximum two-hop, neighbors. The algorithms in distributed WSN systems are considered to be scalable. Reconfiguration takes place natively only on the afflicted parts. Since nodes are self-governing, substitutions are made by each node and are consistent with its place or its work. The prime concerns and information that is accessible within the network are interpreted by every node. Even when the death of a node occurs, the network still remains operational, and thus execution is not affected so much. Distributed techniques allow noisy environments to be handled, including obstacles. Every node reduces energy consumption. Normally, routing starts when an occurrence is detected or when a target is set. This suggests that before the procedure starts, there is no unnecessary energy depletion [70–75].

A few drawbacks of the distributed approach are as follows: Since nodes only have native information, there is no guarantee of connectivity of the entire network. Bottlenecks can arise when multichip transmission is used and there is only a single sink node. A lot of energy is required for the mobility of nodes. When only one sink node is present, the network ceases to function effectively, rendering it useless in that situation.

Some factors can impact the execution of distributed and centralized networks, like the number of hops to reach to a target or a specified appliance, the flow, the volume of retransfers, the volume of linking, and thus the number of appliances [76–78].

1.4 Security Issues

Security issues in WSNs necessarily depend upon understanding what is to be protected. A research paper [79] identified and defined a number of major goals in the field of security in WSNs, the four most important of which are Confidentiality, Authentication, Integrity, and Availability. The potential to hide messages containing confidential data from a passive or indirect attacker as they are communicated through the network is known as Confidentiality [80]. Authentication determines whether the data has truly come from the node that is declared to be the source. Integrity refers to the capability to verify that the data has not been meddled or interfered with while it has been on the network. The question of whether a node has the power to use the network and its resources, and whether they are available for propogating the data, is known as Availability.

A further security goal of Freshness means ensuring that the data received by the receiver is recent, and no attacker can replay old data. This goal is crucial when the nodes of the wireless sensor network use shared keys for data interaction, where a possible adversary could initiate a replay attack or assault using the earlier key while the latest key is being refreshed and communicated to all or any of the nodes within the network [81]. To satisfy the goal of Freshness, certain mechanisms need to be added to every data packet.

In addition to establishing a base set of security standards for wireless sensor networks, the main possible security assaults in the networks are identified in [82].

Routing loop attacks target the knowledge exchanged between nodes. When an attacker changes and replays the routing information, false error messages are generated. Routing loops increase node to node latency and attract or repel the network traffic [83].

Selective forwarding attacks affect the network traffic when the network believes that each of the nodes within the network is a dependable forwarder of messages. In a selective forwarding attack, the malicious or infected nodes drop some of the messages instead of than forwarding them. Once a malicious or infected node reduces the number of messages that it forwards, latency is reduced, and the nearby nodes are deceived into believing that they are on a shorter route. The two factors that the efficiency of this assault depends on are: (i) primarily, the site of the infected node – the nearer it is to the base station, the more traffic it will lure [84]; and (ii) secondarily, the amount of data it drops – when a selective forwarder forwards less data and drops more, its energy level is retained, which enables it to continue to mislead the neighboring nodes.

In sinkhole attacks, the attacker lures the traffic to an infected node [85]. Only if the infected node is located where it can attract most of the data, which is likely to be near to the sink or base station, or by the infected node itself pretending to be the base station, can a sinkhole attack be made [86]. One explanation for why attackers use sinkhole attacks is to make selective forwarding possible, to attract the traffic toward an infected node. WSNs where all the data flows toward one base station are more susceptible to this sort of attack.

Sybil attacks are attacks in which a node in the WSN creates multiple illegitimate identities, either by fabricating them or by stealing the identities of legitimate nodes [87]. These attacks are often used against routing algorithms and topology maintenance; they reduce the effectiveness of fault tolerance schemes like distributed storage and disparity.

1.4.1 Layer- or Level-based Security

The application layer: Data is gathered and managed at this level, therefore it is crucial to ensure the authenticity of the data. A resilient aggregation scheme has been presented in research; it can be implemented in a cluster-based network in which a cluster leader acts as an assembler [88]. However, this system can only be implemented if the aggregating node is within range of all the source nodes and there is no interceding aggregator between the aggregating node and source nodes. To prove the validity of the aggregation and the reliability of the data, cryptographic techniques are used by the cluster leaders.

The network layer: This level is responsible for routing messages between cluster leaders and the base station, between cluster leaders, between nodes and cluster leaders, and between nodes [34, 35, 89].

The data link layer: This level is responsible for fault identification and rectification, and encoding the data. This level is susceptible to jamming and DoS attacks. Link layer encryption has been introduced by TinySec [90], which is dependent on regulation strategy. However, an attack can still be staged if the attacker has better energy efficiency. A few protocols have anti-jamming properties that have proven to be viable corrective measures at this level [91].

The physical layer: This level focuses on the transfer connection between forwarding and collecting nodes. Signal strengths, frequency types, information rates, etc. are all considered within this level [92].

1.5 Energy Consumption Issues

To compute the lifespan of a WSN, energy consumption can prove to be a very valuable component, since sensor nodes usually operate on batteries [93]. Occasionally, energy optimization can prove to be much more complicated in WSNs, since it involves not only lowering of the consumption of energy but also elongating the lifespan of the network to the maximum possible [93]. The optimization is often performed by using energy awareness in all of the essential parts of design and operation. This also ensures that energy awareness is built into blocks of interacting sensor nodes, and therefore the entire network, not only within the individual nodes [94].

A sensor node typically comprises four sub-systems:

- Computing subsystem: This architecture comprises a microprocessor (often abbreviated as MCU), which is responsible for the governance and implementation of conveyance protocols, and usually functions in multiple modes for power regulation purposes. As these functioning modes involve the utilization of power, the energy consumption rates of the various modes should be taken into consideration when observing the battery lifespan of each node [95].
- Communication subsystem: This subsystem comprises a short-range radio, which communicates with neighboring nodes and thus the surface world. It is crucial to have the radio totally shut down, instead of keeping it in a mode in which it is idle, when it is not transmitting or receiving, to increase efficiency and reduce power consumption [96].
- Sensing subsystem: Actuators and sensors that connect the node to the surface world form the basis of this subsystem [97]. Energy consumption is usually lowered by using low-power components, and by lowering execution speeds to save power (where higher speeds are not critical) [98].
- Power supply subsystem: Power is supplied to the node by the battery that is part of this subsystem. The power drawn from the battery has to be monitored frequently, due to the

fact that a high current drawn from the battery for an extended period of time would lead to its rapid discharge. Usually, the minimum energy consumption of the node exceeds the nominal rated current capacity of the battery that is used in the sensor node. The lifespan of the battery can be extended by reducing this excess load. This can also be achieved by turning off the sensors every once in a while [94].

To reduce the overall energy consumption of the WSN to the maximum extent possible, differing kinds of algorithms and protocols have been designed and studied. The lifespan of a WSN is often extended remarkably if the appliance layer, the operating system, and thus the protocols of the network are designed to be energy conscious [99]. These protocols and methods need to take into consideration knowledge of the hardware, and must be able to utilize the relevant features of the microprocessors and nodes that transmit and receive information to reduce energy consumption [100]. This might lead to a better range of design and implementation solutions for sensor nodes. Different types of WSNs are a direct consequence of different kinds of sensor nodes. This could also lead to synergy between various sorts of algorithms in the WSN arena [101].

1.6 Commonly Used Standards and Protocols for WSNs

Protocols are crucial parts of the blueprint for sensor networks, mainly concerned with the interactions between sensor nodes. A quick outline of key protocols and standards that are used for wireless interaction is given below.

The standards are split into layers and sublayers. These guide the decisions relating to performing tasks, such as when and how to collect, transfer, dispatch, and process the information on each appliance [102]. In the link layer, information is collected and processed, after which it is sent to a different sensor appliance. Performance management for data transfer and possible fault rectification are provided at this level. It is split into two sublayers: medium access control (MAC) and logical link control (LLC) [103].

The LLC sublayer functions as an intermediary between the MAC sublayer and networking tasks. It is responsible for flow and fault supervision, and for data transfer between appliances on a network. Some standards utilized at this sublevel are 802.11, 802.5, Fiber Distributed Data Interface (often abbreviated as FDDI), and 802.3/Ethernet [104].

The MAC sublayer is responsible for governing the approach to the neighboring domain, including data frame verification, fault correction on transfers, transfer rates, package transfers, management of flow, acknowledgment messages, and so on [105]. In the MAC sublayer, there is an immediate impact on how the node approaches the neighboring domain to access information about the paths. The protocols are divided into two general groups:

- Slot-based or slotted protocols: In this subgroup of protocols, time is partitioned into time gaps known as slots or frames. In these protocols, the node has a state such as Transmit, Receive, or Sleep. The co-occurrence of these time slots is defined for the governance these states. The co-occurrence and preservation costs affect the energy consumption and therefore the bandwidth. A number of examples are the Sensor Medium Access Control (S-MAC) protocol, time division multiple access (TDMA), IEEE 802.15.4, and the T-MAC protocol, among others.

- Sampling-based protocols: A unique feature of these protocols is that they are generally turned off for most of the time, and are turned on only during specific instances, unlike the slot-based protocols. They search for activity within the channel. In the case that an action is detected, they begin receiving data. When these actions are not detected, they are shut down in order to conserve energy. These types of protocols are generally flexible in nature, and communication is permitted to any other sensor in scope. Communication, however, is not always possible, due to a shortfall of integration. ALOHA, B-MAC, Wiseman, and the protocol used by the Chipcon CC2500 transceiver are some popular examples of sampling-based protocols [106–110].

1.6.1 Slotted Protocols

Slotted protocols are commonly used in WSNs. Some samples of slotted protocols are as follows:

1.6.1.1 Time Division Multiple Access

The time division multiple access (TDMA) protocol is straightforward. It uses the technique of frequency division multiple access (FDMA) to split the radio spectrum into individual frequency channels through the use of a duplex channel. TDMA hence divides these frequency carriers further into smaller time slices. A frame is then formed by a combination of these time intervals. Transmission is structured in frames with a duration of T_i. The length of the interval is defined by $T = T_i/N$. Information is transferred through a burst of bits, and each conversation employs just one slot. Thus, instead of handling only one conversation, each and every radio frequency (RF) carrier can transport multiple conversations [111–115].

1.6.1.2 ZigBee/802.15.4

The IEEE 802.15.4 has been a standard since the early 2000s. It was created to address the need for a sensor network protocol that consumes less energy and is efficient in a wireless personal area network (WPAN) [116]. It is flexible, and operates on a narrow bandwidth. This protocol is designed to support two kinds of topologies:

- Star: this topology is used to implement power networks that consume less energy.
- Peer-to-peer: this topology is primarily used to implement wide, yet precise, networks [117].

This protocol functions with three types of role. All of these roles have specific functionality as described below:

- A reduced function device (RFD) is restricted to the star topology. These nodes interact solely with the relevant coordinators of the network, instead of adopting the role of a network organizer. These are straightforward devices with restricted resources, interactions, and communication requirements. They communicate solely with FDDs.
- A device with the personal area network (PAN) role acts as a router, and accordingly manages the network load.
- A full function device (FFD) is equipped to perform any task, and is able to communicate with all nodes within the network. The FDD adopts the role of an organizer.

To ensure the security and protection of the frame, the IEEE 802.15.4 standard is able to provide eight sophisticated security levels for this purpose [118–120]. They include:

- unsecured;
- encryption only;
- authentication only; and
- encrypted and authenticated.

Each category mentioned is available in three variations, depending on the size of the MAC and whether or not it supports authentication. On enabling the unsecured level, neither message integrity nor data confidentiality is ensured. In the other cases, data encryption is generally performed with the help of a variant of the Advanced Encryption Standard (AES), and the authentication of messages is performed using the AES-CBC technique. Each packet can carry a particular service payload. Communication with the MAC entity is made possible by using specific primitives to select the security level and the other various parameters that are instrumental to conducting the necessary security procedures [121].

The protocol involves the emission of beacon frames by the network coordinator for WPAN discovery and the detection of devices [122]. Devices are designed to function in two ways:

- Non-beacon-enabled: An active scan is initiated to send the frames as requested by the command frame.
- Beacon-enabled: The frames are sent in a periodic fashion by the network coordinator. A passive scan is initiated to detect the network [119, 123, 124].

The prevailing standards give rise to low data transmission and connectivity service cycles. The foremost reason for promoting a replacement protocol (ZigBee/802.15.4) as a standard is to ensure interoperability between devices that are manufactured by different companies. The intermediate nodes and sinks are FFD devices, while the final devices or leaf nodes are RFD devices. The network coordinator is chosen based on having the absolute best average energy level, using the energy detection (ED) procedure. Using different types of environmental sensors, the nodes measure moisture, temperature, and luminosity [125].

The fraction of nodes that are connected in the network is analyzed by mobilizing the sink nodes. A reconfiguration is taken into consideration to analyze the scenarios and run simulations to evaluate the performance. However, the characteristic node features such as noise, message losses, and delays are not considered. Hence, it is increasingly difficult to evaluate if the implementations are economically viable and are optimum, since there is no assurance that the connectivity of the network is consistent [126, 127].

1.6.1.3   Sensor Medium Access Control

Sensor Medium Access Control (S-MAC) is based on slots. It accordingly defines stages enabling each sensor to save energy [128]. The following tasks are performed by the nodes:

- Each sensor node, when in its Sleep period, is allowed to turn itself off. Consequently, it sets its timer so as to be awakened after a particular time slice.
- The node thus wakes up once this scheduled timer ends.

- The appliance, and hence the users, are responsible for the time for which a node is set to stay awake.
- Neighboring nodes are integrated together [129].

S-MAC is one of the earliest MAC layer protocols introduced for sensor networks. However, this layer consists of a few energy limitations. S-MAC is predicated on the Request To Send/Clear To Send method (more commonly referred to as RTS/CTS). This is used as a solution to the hidden terminal problem. A timer is used to switch alternately between transmission and reception modes [130].

The aim of the S-MAC protocol is to perform well-defined activities while conserving energy by turning the nodes on and off alternately during the work cycle. The perception radios of the nodes are set to operate in a listening mode, for possible emergent events or requests [131].

To transmit information, nodes have to compete for the medium, and accordingly leverage broadcasting to change their tasks [132, 133]. A transmission, once started, cannot be interrupted until it finishes completely. In fact, each node maintains a schedule in order to keep its transmission times synchronized, to prevent excessive losses of messages. S-MAC features a well-defined set of rules of conduct for nodes that need to leave or enter the network. These rules ensure that better control and support is exercised for the nodes within the network.

Every node has a contrasting TRM. The receiver specifies the listen and sleep times, which are kept consistent with the synchronization schedule, so that losses of messages can be avoided.

The S-MAC protocol provides the ability to define the trade-off between latency and energy, according to traffic conditions.

Primarily, there are three node states that are implemented by the protocol:

(i) Active: During this state, the sensor node remains working. This particular state also results in an unnecessary waste of energy, as a consequence of its advantage of maintaining availability to reply to any request.

(ii) Proactive: In this state, the sensor node conserves energy, thereby ensuring greater efficiency, by operating only when needed, for a particular period of time. This type of sensor suspends all actions when it is not in the Active state.

(iii) Inactive or Monitor: In this state, the sensor node is made to wait for further instructions from the channel, while it is monitoring the environment. The information then is merely forwarded until it is explicitly requested.

1.7 Effects of Temperature and Humidity on the Energy of WSNs

At present, there is a requirement for highly efficient energy-conserving algorithms and protocols. These can be made possible by conducting extensive research on WSNs and associated applications. Thus, the application domain is restricted to simple data-led monitoring, analyzing, and reporting applications [134].

1.7.1 Effects of Temperature on Signal Strength

Usually, when temperature falls, signal strength – measured using the received signal strength indicator (RSSI) – rises, and vice versa. This indicates a negative dependence between temperature and signal strength. This can be confirmed by computing the Pearson coefficient of correlation to measure the degree of linear dependence between RSSI and temperature. It is observed that the correlation factor varies depending on the channel and its corresponding link. Some link–channel pairs have a significantly strong correlation, whereas in some others the correlation is a not very strong. The correlation between differences in standard RSSI and changes in temperature, which is manifested in correlations between individually monitored link–channel performances and temperature, is clear. Thus, there is a lot of advantage to be gained from exploiting frequency diversity. The Pearson coefficient of correlation is robust, thus confirming the understanding and theory that RSSI varies inversely with temperature.

Below 0 °C, the variation or change in average RSSI values with temperature is high. However, there is a considerable degradation in correlation when the temperature fluctuates near 0 °C. Nevertheless, the indirect correlation still holds. Some differences are seen when the correlations between RSSI and temperature are compared at two different transmission power levels. Various experiments across the temperature range have shown a considerable effect of this parameter on signal strength. A linear negative trend is noticed for all links, though there is some disparity regarding the extent of the impact (the regression coefficient). The parametric statistic for changes in the value of RSSI is approximately −0.127, meaning that a 10 °C increase in temperature results in a 1.3 dB decrease of RSSI. The determination coefficient is significantly high, indicating that the RSSI variation is generally dependent to a great extent on the deviation in temperature.

Generally, there are not any significant differences between various ranges with reference to the magnitude of the impact. The parametric statistic ranges from −0.09 to −0.13. There are various other factors that lead to large deviations when the temperature fluctuates around 0 °C, resulting in reductions in the coefficient of correlation and the coefficient of determination. This is validated by the fact that the coefficient of determination decreases considerably toward new lows during frost, when the temperature is around 0 °C. In each measurement range, the regression coefficients are statistically significant. Keeping the extent of the effect in context, the variation between two different power levels is not striking. Values of the coefficient of determination are comparatively high with lower transmission power, meaning that the lower the transmission power is, the larger the variation in the RSSI is.

1.7.2 Effects of Humidity on Signal Strength

A transparent relationship is observed between both absolute humidity (AH) and relative humidity (RH) and signal strength. Above 0 °C, the ratios rise and fall alongside RSSI, indicating a positive correlation, while that of absolute humidity, below 0 °C, indicates a negative correlation.

As previously, we computed the Pearson correlation coefficients and straightforward rectilinear regression for both RH and AH. The coefficient of determination within the regression model is extremely high for RH, which suggests that RSSI variation might be explained to a high degree by the variation in RH. As for the change in RSSI, the parametric statistic of RH is about 0.03, which indicates that an increase in RH of 10% increases RSSI approximately by 0.3 dB. Both RH and AH have high, almost equally strong, yet opposite, correlations with RSSI, at temperatures below 0 °C. Also, the coefficient of determination is somewhat significant for both RH and AH, although for RH it is smaller.

In the near-0 °C range, correlation is somewhat low for both RH and AH.

However, it still follows the other ranges. Because of the decreased correlation, it is obvious that the coefficient of determination is also low for both RH and AH. Therefore, the rectilinear regression models are not sufficient to explain everything. On average, for the case of the correlation of RSSI with RH, the differences between the levels are relatively minor. For the case of AH, they are slightly more negative and stronger with lower transmission power in cold temperatures, as are the values of the coefficients of determination and regression.

1.7.3 Temperature vs. Humidity

Though correlation may be able to correctly predict a possible causal relationship, it does not imply causation, and might be caused by another factor. A high correlation value between the observed weather variables and RSSI in hot as well as sub-zero temperatures might be partly described by a high degree of correlation between temperature and humidity. Relative humidity correlates strongly to temperature in hot temperatures, while absolute humidity does so in cold temperatures. The close relationship between the weather variables that were studied complicated the attempts to differentiate between the particular impacts of humidity and temperature on RSSI.

Therefore, the combined effect that was studied was predominantly the effect of one specific weather variable, with the other variable to be taken under consideration. A number of the results showed high collinearity between temperature and humidity. In hot temperatures, temperature and RH were highly collinear, while in cold temperatures below 0 °C, temperature and AH were collinear. Therefore, it is questionable to use them together. In contrast with the previous statements, temperature and AH are not collinear in hot temperatures, and temperature and RH are not collinear in cold temperatures below 0 °C.

1.8 Proposed Methodology

The proposed methodology (Figure 1.1) contains three phases: (i) **visualization;** (ii) **clustering the data** using an unsupervised learning algorithm – namely, the k-Means clustering algorithm; and (iii) **classifying the data** using a supervised classification algorithm – the k-Nearest Neighbors (KNN) algorithm.

The data for this analysis was collected from the **Intel Lab Dataset,** which contains readings recorded in the year of 2004, corresponding to the humidity and temperature conditions of the sensor environment, with recorded input voltage values switching every 31 seconds in 54 sensors, which amounts to a total of 2.3 million readings.

Since the dataset does not contain any labels and is not presented in a structured order, it needs to be pre-processed and presented through the use of both supervised and unsupervised learning algorithms.

Figure 1.1 Proposed methodology.

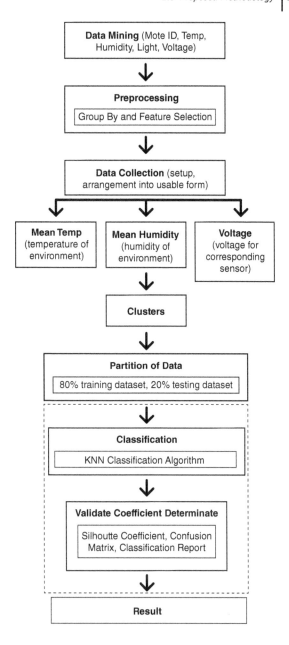

Unsupervised learning algorithms make use of a dataset without any labels. Accordingly, they attempt to find some organized structure within the available data. k-Means is one such widely used algorithm. It is used to discover latent "clusters" in the dataset, revealing new "features" by which to partition the data, so that a separate classifier can be trained on each partition.

Supervised classification algorithms need datasets with categorical dependent variables – "targets" – and sets of independent variables – "features" – to determine a relation between them. The training procedure for classification works on the training dataset to map the features to the required output variables, i.e. cluster numbers, thereby creating a relation to correctly identify and cluster the recorded data.

In this case, the training dataset contained 80% of the complete data, while 20% was set aside as a testing dataset (to determine the accuracy of the model).

The training phase ends with the generation of rules by quantifying the relationship between the outputs – the cluster numbers – and the given input features.

In the testing phase, the clusters are predicted based on the generated rules. These predicted values are then compared to the actual values (for the 20% of the data) to determine the accuracy of the model. Key metrics such as the standard R^2, F_1/F_β score, and confusion matrix are presented to give an accurate picture of the correctness of the model.

1.8.1 Information Gathering and Analysis

The data used in this chapter was collected from an Intel repository, consisting of the details of 54 MicaDot2 sensors installed within the Intel Berkeley laboratory between 28 February and 5 April 2004. The entire lab environment was planned precisely, with coordinates and setups for each of the 54 sensors within the lab. These sensors collected dynamic environment variables like the humidity and temperature of the surroundings, incident light, and sensor voltage values, and recorded these values every 31 seconds. This resulted in the recording of 2.3 million readings in total.

The dataset contained the following features:

- **Sdate**: Date (YYYY-MM-DD) on which the observation was recorded.
- **Stime**: Time (HH-MM-SS) at which the observation was recorded.
- **Epoch**: Monotonically increasing number from each mote.
- **Mote ID**: Unique ID of each sensor (1–54).
- **Temperature** (also named as Temp): Temperature of the sensor's environment.
- **Humidity**: Humidity of the sensor's environment.
- **Light**: Light intensity.
- **Voltage**: Output voltage.

The features that are part of this dataset include the humidity and temperature of the surroundings, the incident light, and the sensor voltage values over all of the 54 sensors. Temperature is measured in degrees Celsius. The value of Humidity varies from 0 to 100%. Light is measured in Lux (a unit used for measuring the intensity of incident light). Voltage is expressed in volts, with values between 2 and 3 V. This data is presented in a text file, and needs to be converted to a suitable form for operating and analyzing.

1.8.2 System Design and Implementation

An integral part of the pre-processing stage was to reduce the total number of observations and group them in order to streamline the data. Hence, the data was grouped by the Sdate and Mote ID features, aggregating the constituent entries into independent features.

This brought down the number of observations tremendously, and the aggregation resulted in cleaner data. The final dataset contained the following features:

- **Sdate**: Date (YYYY-MM-DD) on which the observation was recorded.
- **Mote ID**: Unique ID of each sensor (1–54).
- **Mean Temp**: Daily average temperature, per sensor.
- **Mean Humidity**: Daily average humidity, per sensor.
- **Mean Light**: Daily average light, per sensor.
- **Voltage (Mean Voltage)**: Daily average voltage, per sensor.
- **Max Voltage**: Daily maximum voltage, per sensor.
- **Count**: Total recorded observations, per sensor.

A heat map with Pearson correlation coefficients was plotted to check for redundant features, to visualize the relationships between them, and to select the important ones. Two types of correlation are possible: positive and negative. A positive correlation indicates that if a feature increases then the compared feature increases as well, and if the feature decreases then the compared feature decreases accordingly. Both features vary according to a fixed pattern, and they have a linear relationship. A negative correlation indicates that if the feature increases then the compared feature decreases, and vice versa.

From this heat map (Figure 1.2), it is clear that Temperature (Temp) is inversely correlated with Voltage and Humidity, so it can be inferred that hotter and drier environments correlate to more energy-efficient ones. To create clusters for the sensors, the Voltage had to be taken, along with Temperature (Temp), Humidity, and Light to identify an efficient sensor environment. However, Light was dropped since it had a low correlation with Voltage and would not significantly influence the cluster formation.

Hence, the final features were identified as Mean Humidity, Mean Temp, and Voltage (see Figure 1.3).

The Min-Max Scaler algorithm was used on the Temp and Humidity features for normalization. By doing so, the features were transformed such that their values were between 0 and 1, meaning that the maximum and minimum values of a feature were going to be 1 and 0 respectively. The main idea behind adopting this technique was that variables that

Figure 1.2 Correlation matrix of features.

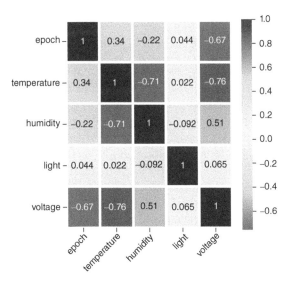

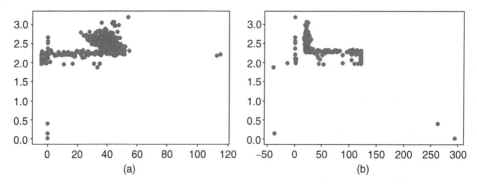

Figure 1.3 (a) Mean Humidity vs. Voltage; (b) Mean Temp (temperature) vs. Voltage.

are observed and measured do not contribute uniformly to model fitting and model function. Hence, to deal with such a problem, feature-wise normalization is usually used prior to model fitting, thereby preventing the creation of a bias.

On careful analysis, the possible algorithms for the clustering included Hierarchical Clustering, DBSCAN, and the k-Means clustering algorithm. In Hierarchical Clustering, the required time complexity is of order $O(n^2 \log n)$, and no step can be reversed, while DBSCAN cannot function accurately with clusters of similar density and data that have high dimensionality.

On the other hand, a few advantages of the k-Means algorithm were identified:

- it can be implemented on relatively large sets;
- its convergence can be guaranteed; and
- it can easily work on new examples.

The k-Means algorithm was hence best suited for this analysis.
The algorithms used in the model were:

1) the k-Means clustering algorithm;
2) the k-Elbow Visualizer; and
3) k-Nearest Neighbors (for classification).

The **k-Means** clustering algorithm is a widely used clustering technique attempting to find a certain number of clusters (k), as specified by the user, which are represented by their centroids. It is an unsupervised learning algorithm, meaning that it trains and learns from the data without receiving labels or assumptions. The number of clusters was determined using the k-Elbow Visualizer method.

The **k-Elbow Visualizer** utilizes the elbow method for suggesting the best possible number of clusters for k-Means clustering. The elbow method checks the k-Means clustering algorithm for various values of k, and computes a mean score for all clusters for every value of k.

Additionally, the accordant distortion score is computed; this is the value of the sum of square distances from each data point to the cluster center. When the metrics for every model are consolidated and plotted, it is possible to visually conclude the optimal value of k. The point of inflection in the plot is the best value of k, if the chart has the shape of an

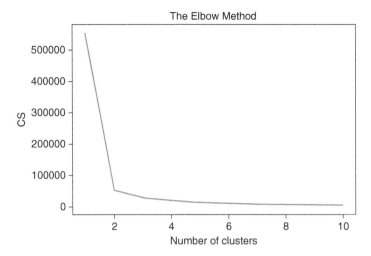

Figure 1.4   k-Elbow Visualizer.

arm. The case in which the arm is either up or down and there is a robust inflection point is an indication that the underlying model can be deemed the best fit at that time.

The value of k at the point after which the distortion/inertia starts decreasing in a linear fashion (the "elbow") gives the best possible number of clusters to form. Thus, for the given data, the optimum k value was 3 (see Figure 1.4). Hence, three clusters were formed containing the sensor readings.

On applying the model to the dataset, the clusters were formed and the corresponding labels were attached to each reading for ease of classification using the KNN classification algorithm.

k-Nearest Neighbors (KNN) is a supervised classification algorithm. This means that the training dataset is labeled, nonparametric, and lazy. The algorithm does not explicitly learn the model, but stores all the training data and uses the entire training dataset for classification. This indicates that the training process is very fast. The main idea behind KNN is that for an uncategorized observation, the nearest points (with minimum distance) are sought. These points define the category of the new observation by majority voting.

From Figure 1.5, there is a spike in accuracy for n_neighbors = 3. Hence, three points were taken as centers for the data, thereby confirming that clustering into three clusters was correct.

1.8.3   Testing and Evaluation

For k-Means clustering:

Inertia is defined as the mean squared distance between each observation and its closest centroid. Thus, a small value for inertia is preferred, showing the formation of tight clusters. The silhouette score is a metric that is calculated from the distances between data points in different clusters. The value of the silhouette score ranges from −1 to 1 (Figure 1.6).

For the k-Nearest Neighbors classification algorithm:

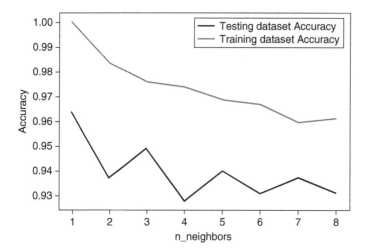

Figure 1.5 Accuracy vs. n_neighbors.

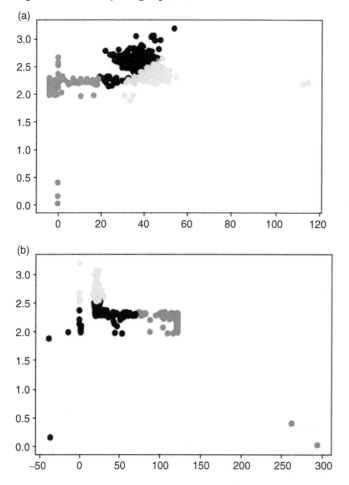

Figure 1.6 (a) Clustered plot of Mean Humidity vs. Voltage; (b) clustered plot of Mean Temp (temperature) vs. Voltage.

```
[[ 65   0   1]
 [  1  64   0]
 [  2   0 199]]
              precision    recall  f1-score   support

           0       0.96      0.98      0.97        66
           1       1.00      0.98      0.99        65
           2       0.99      0.99      0.99       201

    accuracy                           0.99       332
   macro avg       0.98      0.99      0.98       332
weighted avg       0.99      0.99      0.99       332
```

Figure 1.7 Confusion matrix (top); classification report (bottom).

A confusion matrix (Figure 1.7 top) was used to determine the accuracy of the KNN classifier. This is a summarized table used to define the overall performance and accuracy of a classification model for a testing dataset for which the corresponding true values are known.

In the classification report (Figure 1.7 bottom), **precision** is equal to the total number of true positives divided by the sum of true positives and false positives. Precision is the ability of the classifier not to label a negative sample as a positive value.

Recall is defined as the quotient of true positives divided by the sum of true positives and false negatives. The recall is intuitively the ability of the classifier to find all samples that are positive.

The **F_1 score** or **F_β score** (f1-score, F-beta score) can be obtained by calculating the weighted harmonic mean of the recall and precision metrics. An F_1 score or F_β score is between 1 and 0, with 0 indicating the worst and 1 indicating the best possible value.

Support refers to the number of occurrences of each class of the correct target values.

1.9 Conclusion

According to the Accuracy Report (Table 1.1) and Table 1.2, the sensor readings are accurately placed into three broad clusters. Out of the three, Cluster 2 contains readings depicting highly efficient sensors, followed by Cluster 1 and Cluster 3. As confirmed with the correlation heat map, the temperature in this cluster is the greatest and the humidity is the least, giving more insight about suitable sensor environments.

Table 1.1 Accuracy report – k-Means.

S.No.	Metric	Score
1.	Silhouette Average	0.9458
2.	Inertia Value	5.378

Table 1.2 Clusters – k-Means.

Cluster	Humidity (%)	Temperature (°C)	Voltage (V)
1.	37.721	23.499	2.501
2.	−2.528	121.161	2.158
3.	44.844	21.920	2.505

Since most sensor readings are grouped into these clusters toward the end of the recorded experiment, it remains to be seen whether sensors emitting more light and heat at lower voltages is a sign of them burning out, or gaining more efficiency. This gives rise to a scope for future exploratory data analysis and research in the domain concerned.

References

1 Jingcheng, Z. (2014). Wireless Sensor Networks. Linköping University, Campus Norrköping. 17 January 2014.

2 Pottie, G.J. (1998). Wireless sensor networks, Information Theory Workshop (Cat. No. 98EX131), Killarney, Ireland, pp. 139–140. doi: https://doi.org/10.1109/ITW.1998.706478.

3 Akkaya, K. and Younis, M. (2005). Surveys on routing protocols for wireless sensor networks. *Elsevier Journal of AdHoc Networks* 3 (3): 325–349. https://doi.org/10.1016/j.adhoc.2003.09.010.

4 Chiara, B., Andrea, C., Davide, D. et al. (2009). An overview on wireless sensor networks technology and evolution. *Sensors* 9 (9): 6869–6896. https://doi.org/10.3390/s90906869.

5 Iwendi, C. and Offor, J. (2012). Alternative protocol implementation for wireless sensor network nodes. *Telecommunication System Management* 2 (1): 1–7. https://doi.org/10.4172/2167-0919.1000106.

6 Akyildiz, I., Su, W., Sankarasubramaniam, Y., and Cayirci, E. (2002). A survey on sensor networks. *IEEE Communications Magazine* 40 (8): 102–114.

7 Tubaishat, M. and Madria, S. (2003). Sensor networks: an overview. *IEEE Potentials* 22 (2): 20–30.

8 Hac, A. (2003). *Wireless Sensor Network Designs*, 165–234. Etobicoke, Ontario: Wiley.

9 Raghavendra, C., Sivalingam, K., and Znati, T. (2004). *Wireless Sensor Networks*. New York: Springer.

10 Sohrabi, K., Gao, J., Ailawadhi, V. et al. (2000). Protocols for self-organization of a wireless sensor network. *IEEE Personal Communications* 7 (5): 16–27.

11 Culler, D., Estrin, D., and Srivastava, M. (2004). Overview of sensor networks. *IEEE Computer Society* 37: 41–49.

12 Chen, X. (2020). *Introduction, Randomly Deployed Wireless Sensor Networks*, 1–14. Elsevier ISBN 9780128196243.

13 Nilsson Plymoth, A. (2007). Wireless Multi Hop Access Networks and Protocols. Faculty of Engineering, LTH at Lund University, 2009.

14 Patil, H.K. and Chen, T.M. (2019). *Computer and Information Security Handbook (3 Edition)*. Morgan Kaufmann.

15 Mehbodniya, A. and Chitizadeh, J. (2005). An intelligent vertical handoff algorithm for next generation wireless networks. Second IFIP International Conference on Wireless and Optical Communications Networks. Dubai: WOCN. 244–249. doi: https://doi.org/10.1109/WOCN.2005.1436027.

16 Polastre, J., Hui, J., Levis, P.J. et al. (2005). A unifying link abstraction for wireless sensor networks. Proceedings of the 3rd International Conference on Embedded Networked Sensor Systems (SenSys '05). 76–89. New York: Association for Computing Machinery. doi: https://doi.org/10.1145/1098918.1098928

17 Klues, K., Hackmann, G., Chipara, O. et al. (2007). A component-based architecture for power-efficient media access control in wireless sensor networks. Proceedings of the 5th International Conference on Embedded Networked Sensor Systems (SenSys '07). 59–72. New York: Association for Computing Machinery. doi: https://doi.org/10.1145/1322263.1322270

18 Klues, K., Xing, G., and Chenyang, L. (2007). Link layer support for unified radio power management in wireless sensor networks. Proceedings of the 6th International Conference on Information Processing in Sensor Networks (IPSN '07). 460–469. New York: Association for Computing Machinery. doi: https://doi.org/10.1145/1236360.1236418

19 Ee, C.T., Fonseca, R., Kim, S. et al. (2006) Modular Network Layer for Sensornets. Proceedings of the 7th USENIX Symposium on Operating Systems Design and Implementation. pp. 249–262. Seattle: Rodrigo L C Fonseca.

20 Ye, W., Silva, F., and Heidemann, J. (2006). Ultra-low duty cycle MAC with scheduled channel polling. Proceedings of the 4th International Conference on Embedded Networked Sensor Systems (SenSys '06). 321–334. New York: Association for Computing Machinery. doi: https://doi.org/10.1145/1182807.1182839

21 Polastre, J., Hill, J., and Culler, D. (2004). Versatile low power media access for wireless sensor networks. Proceedings of the 2nd Internation Conference on Embedded Networked Sensor Systems (Sensys '04). 95–107. Baltimore, MD.

22 Chen, X. and Rowe, N. (2011). An energy-efficient communication scheme in wireless cable sensor networks. Proceedings of the IEEE International Conference on Communications (ICC 2011), Kyoto, Japan (5-9 June 2011) 1–6. Institute of Electrical and Electronics Engineers.

23 Bulusu, N., Estrin, D., Girod, L. et al. (2001). Scalable coordination for wireless sensor networks: self-configuring localization systems. Proceedings of the 6th IEEE International Symposium on Communication Theory and Application, Ambleside, Cumbria, UK: St. Martin's College, July 2001.

24 Huang, C.-F. and Tseng, Y.-C. (2003). The coverage problem in a wireless sensor network. Proceedings of the 2nd ACM International Conference on Wireless Sensor Networks and Applications (WSNA '03), San Diego, CA, USA (September 2003) 115–121. New York: Association for Computing Machinery. doi: https://doi.org/10.1145/941350.941367

25 Anjum, F. (2006). Location dependent key management using random key pre-distribution in sensor networks. *Proceedings of ACM WiSe*: 21–30.

26 Blackburn, S.R. and Gerke, S. (2009). Connectivity of the uniform random intersection graph. *Discrete Mathematics* 309 (16): 5130–5140.

27 Blackburn, S.R., Stinson, D.R., and Upadhyay, J. (2010). On the complexity of the herding attack and some related attacks on hash functions. Report 2010/030.

28 Chan, H., Perrig, A., and Song, D. (2003). Random key predistribution schemes for sensor networks. Proceedings of the IEEE 2003 Symposium on Security and Privacy (IEEE S&P 2003), Berkeley, CA, USA (11-14 May 2003) 13–18. Los Alamitos, CA, USA: IEEE Computer Society.

29 Di Pietro, R., Mancini, L.V., Mei, A. et al. (2008). Redoubtable sensor networks. *ACM Transactions on Information and System Security* 11 (3): 13:1–13:22.

30 Solis, I. and Obraczka, K. (2004). The impact of timing in data aggregation for sensor networks. *Proceedings of the IEEE International Conference Communication Paris* 6: 3640–3645.

31 Manjeshwar, A. and Agrawal, D.P. (2002). APTEEN: A hybrid protocol for efficient routing and comprehensive information retrieval in wireless sensor networks. Proceedings of the 16th International Symposium on Parallel Distribution Processes (IPDPS 2002). Fort Lauderdale, FL: APRPP. 195–204.

32 Fonseca, R., Gnawali, O., Jamieson, K. et al. (2006). The collection tree protocol (CTP). *TinyOS Enhancement Proposals (TEP)* 123: 1–6. http://www.tinyos.net/tinyos-2.1.0/doc/html/tep123.html.

33 Wang, W., Wang, Y., Li, X.-Y., Song, W.-Z. et al. (2006). Efficient interference- aware TDMA link scheduling for static wireless networks. Proceedings of the 12th Annual International Conference on Mobile Computing Networks (MobiCom '06). 262–275. Los Angeles, CA.

34 Mamei, M., Menezes, R., Tolksdorf, R. et al. (2006). Case studies for self-organization in computer science. *Journal of Systems Architecture* 52 (8–9): 443–460.

35 Schmeck, H., Müller-Schloer, C., Çakar, E. et al. (2010). Adaptivity and self-organization in organic computing systems. *ACM Transactions on Autonomous and Adaptive Systems* 5: 10.

36 Silva, R., Silva, J.Sá., Simek, M. et al. (2009). A new approach for multi-sink environments in WSNs. Proceedings of the IFIP/IEEE International Symposium on Integrated Network Management (IM '09). 109–112. New York: IEEE. June.

37 Cheng, S.-T. and Chang, T.-Y. (2012). An adaptive learning scheme for load balancing with zone partition in multi-sink wireless sensor network. *Expert Systems with Applications* 39 (10): 9427–9434.

38 Fortune, S. (1992). Voronoi diagrams and Delaunay triangulations. In: *Computing in Euclidean Geometry*, 1 Lecture Notes Series on Computing (eds. D.-Z. Du and F. Hwang), 193–233. Singapore: World Scientific.

39 Zhang, W. and Cao, G. (2004). DCTC: dynamic convoy tree-based collaboration for target tracking in sensor networks. *IEEE Transactions on Wireless Communications* 3 (5): 1689–1701.

40 IEEE Standard Association (2005), IEEE STD. 802.15.4: Wireless Personal Area Networks (PANs).

41 Silva, R., Leithardt, V.R.Q., Silva, J.S. et al. (2009). A comparison of approaches to node and service discovery in PAN wireless sensor networks. Proceedings of the 5th ACM

Symposium on QoS and Security for Wireless and Mobile Networks (Q2SWinet '09). 44–49. Tenerife, Spain: ACM.

42 Singh, M., Lal, N., Sethi, M. et al. (2010). A tree based routing protocol for mobile sensor networks (MSNs). *International Journal on Computer Science and Engineering* 1 (supplement 1): 55–60.

43 Heinzelman, W.B., Chandraasan, A.P., and Balakrishnan, H. (2002). An application-specific protocol architecture for wireless microsensor networks. *IEEE Transactions on Wireless Communications* 1 (4): 660–670.

44 Manjeshwar, A. and Agrawal, D.P. (2001). TEEN: a routing protocol for enhanced efficiency in wireless sensor network. Proceedings of the 15th International Parallel & Distributed Processing Symposium (IPDPS '01). 189, IEEE Computer Society.

45 Dandekar, D.R. and Deshmukh, P.R. (2013). Energy balancing multiple sink optimal deployment in multi-hop wireless sensor networks. Proceedings of the 3rd International Advance Computing Conference (IACC '13). 408–412, Ghaziabad, India: IEEE.

46 Venter, G. and Sobieszczanski-Sobieski, J. (2003). Particle swarm optimization. *AIAA Journal* 41 (8): 1583–1589.

47 Lin, Y. and Wu, Q. (2010). Energy-conserving dynamic routing in multi-sink heterogeneous sensor networks. Proceedings of the International Conference on Communications and Mobile Computing (CMC '10). April: 269–273. Shenzhen, China: IEEE.

48 Pournaras, E., Warnier, M., and Brazier, F.M.T. (2010). Adaptation strategies for self-management of tree overlay networks. Proceedings of the 11th IEEE/ACM International Conference on Grid Computing (GRID '10). October: 401–409. Brussels, Belgium: IEEE.

49 Lehsaini, M., Guyennet, H., and Feham, M. (2010). An efficient cluster-based self-organisation algorithm for wireless sensor networks. *International Journal of Sensor Networks* 7 (1–2): 85–94.

50 Förster, A., Murphy, L., Schiller, J. et al. (2008). An efficient implementation of reinforcement learning based routing on real WSN hardware. Proceedings of the 4th IEEE International Conference on Wireless and Mobile Computing, Networking and Communication (WiMob '08). October: 247–252, University of Lugano, Avignon, France: IEEE.

51 Colorni, M.D. and Maniezzo, V. (1991). Distributed optimization by ant colonies, in Proceedings of the First European Conference on Artificial Life (ECAL '91). 134–142. Paris, France: Elsevier.

52 Van Dyck, R.E. (2002). Detection performance in self-organized wireless sensor networks. Proceedings IEEE International Symposium on Information Theory. 13. Lausanne: IEEE. doi: https://doi.org/10.1109/ISIT.2002.1023285.

53 Saleem, K., Fisal, N., Hafizah, S. et al. (2009). Ant based self-organized routing protocol for wireless sensor networks. *International Journal of Communication Networks and Information Security* 1 (2): 42–46.

54 Fahmy, H.M.A. (2021). Simulators and Emulators for WSNs. In: *Concepts, Applications, Experimentation and Analysis of Wireless Sensor Networks. Signals and Communication Technology*. Cham: Springer https://doi.org/10.1007/978-3-030-58015-5_7.

55 Meng, M., Wu, X., Jeong, B.-S. et al. (2007). Energy efficient routing in multiple sink sensor networks. Proceedings of the 5th International Conference on Computational Science and Its Applications (ICCSA '07). August: 561–566, Kuala Lumpur, Malaysia: IEEE.

56 Wang, C. and Wu, W. (2009). A load-balance routing algorithm for multi-sink wireless sensor networks. Proceedings of the International Conference on Communication Software and Networks (ICCSN '09). February: 380–384. Macau, China.

57 Robert, S. and Wayne, K. (2015). Ghaps. In: *Algorithms, Chapter 4*, 4e. Princeton Editorial.

58 Guan Goh, H., Sim, M.L., and Ewe, H.T. (2006). Energy efficient routing for wireless sensor networks with grid topology. In: *Embedded and Ubiquitous Computing*, 4096 Lecture Notes in Computer Science, 834–843. Berlin: Springer.

59 Cuomo, F., Todorova, P., Cipollone, E. et al. (2008). Topology formation in IEEE 802.15.4: cluster-tree characterization. Proceedings of the 6th Annual IEEE International Conference on Pervasive Computing and Communications (PerCom '08). March: 276–281, Hong Kong: IEEE.

60 Heinzelman, W.R., Chandrakasan, A., and Balakrishnan, H. (2000). Energy-efficient communication protocol for wireless microsensor networks. The 33rd Annual Hawaii International Conference on System Sciences (HICSS '00). 2: Maui, Hawaii: HICSS.

61 Olascuaga-Cabrera, J.G., López-Mellado, E., Mendez-Vazquez, A. et al. (2011). A self-organization algorithm for robust networking of wireless devices. *IEEE Sensors Journal* 11 (3): 771–780.

62 Baker, J. and Ephremides, A. (1981). The architectural organization of a mobile radio network via a distributed algorithm. *IEEE Transactions on Communications* 29 (11): 1694–1701.

63 Swaszek, P.F. and Willett, P. (1995). Parley as an approach to distributed detection. *IEEE Transactions on Aerospace and Electronic Systems* 31 (1): 447–457.

64 Park, S., Shin, K., Abraham, A., and Han, S. (2007). Optimized self organized sensor networks. *Sensors* 7 (5): 730–742.

65 Chan, H. and Perrig, A. (2004). ACE: an emergent algorithm for highly uniform cluster formation. Proceedings of the European Workshop on Sensor Networks. January: 154–171, Berlin: Springer.

66 Shin, K., Abraham, A., and Han, S.Y. (2006). Self organizing sensor networks using intelligent clustering. Computational Science and Its Applications – ICCSA 2006. 3983 Lecture Notes in Computer Science, 40–49, Berlin: Springer.

67 Carlos-Mancilla, M., Olascuaga-Cabrera, J.G., López-Mellado, E. et al. (2013). Design and implementation of a robust wireless sensor network. Proceedings of the 23rd International Conference on Electronics, Communications and Computing (CONIELECOMP '13). March: 230–235, Cholula, Mexico.

68 Bandara, H.M.N.D. and Jayasumana, A.P. (2007). An enhanced top-down cluster and cluster tree formation algorithm for wireless sensor networks. Proceedings of the 2nd International Conference on Industrial and Information Systems (ICIIS '07) August: 565–570, Penadeniya, Sri Lanka: IEEE.

69 Ciancio, A., Pattem, S., Ortega, A. et al. (2006). Energy-efficient data representation and routing for wireless sensor networks based on a distributed wavelet compression algorithm. Proceedings of the 5th international conference on Information processing in sensor networks, IPSN '06. Association for Computing Machinery, 309–316. New York: IPSN. doi: https://doi.org/10.1145/1127777.1127824

70 Ramaswamy, L., Gedik, B., and Liu, L. (2005). A distributed approach to node clustering in decentralized peer-to-peer networks. *IEEE Transactions on Parallel and Distributed Systems* 16 (9): 814–829. https://doi.org/10.1109/TPDS.2005.101.

71 Fahmy, H.M.A. (2020). Energy management techniques for WSNs, Wireless Sensor Networks, Signals and Communication Technology. (1 Edition) 103–258. doi: https://doi.org/10.1007/978-3-030-29700-8

72 Ergin, M.O. (2018). Relative node position discovery in wireless sensor networks. January: 1–27. http://dx.doi.org/10.14279/depositonce-6625

73 Venugopal, K.R., Prakash, T.S., and Kumaraswamy, M. (2020). *An introduction to QoS in wireless sensor networks, QoS Routing Algorithms for Wireless Sensor Networks*, 1–21. Singapore: Springer https://doi.org/10.1007/978-981-15-2720-3.

74 Yick, J., Mukherjee, B., and Ghosal, D. (2008). Wireless sensor network survey. *Computer Networks* 52 (12): 2292–2330. https://doi.org/10.1016/j.comnet.2008.04.002.

75 Ergin, M.O. and Wolisz, A. (2013). Node sequence discovery in wireless sensor networks. International Conference on Distributed Computing in Sensor Systems. May: 394–401. Cambridge, MA. https://doi.org/10.1109/DCOSS.2013.24

76 Kassan, S., Gaber, J., and Lorenz, P. (2018). Game theory based distributed clustering approach to maximize wireless sensors network lifetime. *Journal of Network and Computer Applications* 123: 80–88. ⟨hal-02299974⟩.

77 Cudak, M.C., Mueller, B.D., and Kelton, J.R. (2000). Network protocol method, access point device and peripheral devices for providing for an efficient centrally coordinated peer-to-peer wireless communications. US Patent 6,058,106. 2 May, pp. 1–50.

78 Garudadri, H., Baheti, P.K., and Majumdar, S. (2014). Method and apparatus for distributed processing for wireless sensors. US Patent 8,917,798.

79 Chowdhury, M. and Fazlul Kader, M. (2013). Security issues in wireless sensor networks: a survey. *International Journal of Future Generation Communication and Networking* 6 (5): 97–116.

80 Deng, J., Han, R., and Mishra, S. (2002). INSENS: Intrusion-tolerant routing in wireless sensor networks, Technical Report CUCS-939-02, Department of Computer Science, University of Colorado at Boulder. November.

81 Karp, B. and Kung, H.T. (2000). GPSR: Greedy perimeter stateless routing for wireless networks. Proceedings of the 6th Annual International Conference on Mobile Computing and Networking. 243–254. ACM Press.

82 Zia, T. (2008). Security issues in wireless sensor networks. *International Journal of Communication* 2: 104–121.

83 Eschenauer, L. and Gligor, V.D. (2002). A key-management scheme for distributed sensor networks. Proceedings of the 9th ACM conference on Computer and communications security (CCS '02). 41–47. New York: Association for Computing Machinery. doi: https://doi.org/10.1145/586110.586117

84 Hadi, M.S., Lawey, A.Q., El-Gorashi, T.E.H. et al. (2018). Big data analytics for wireless and wired network design: a survey. *Computer Networks* 132: 180–199.

85 Kibirige, G. and Sanga, C. (2015). A survey on detection of sinkhole attack in wireless sensor network. *International Journal of Computer Science and Information Security* 13: 1–9.

86 Abdullah, M., Rahman, M., and Roy, M. (2015). Detecting sinkhole attacks in wireless sensor network using hop count. *International Journal of Computer Network and Information Security* 7: 50–56. https://doi.org/10.5815/ijcnis.2015.03.07.

87 Dhamodharan, U.S.R.K. and Vayanaperumal, R. (2015). Detecting and preventing Sybil attacks in wireless sensor networks using message authentication and passing method. *The Scientific World Journal*: 841267. 7 pages. doi: https://doi.org/10.1155/2015/841267.

88 Smith, I.G. (2012). The Internet of Things. 29–31. New Horizons. ISBN 9780955370793. https://www.internet-of-things-research.eu/pdf/IERC_Cluster_Book_2012_WEB.pdf.

89 Carlos-Mancilla, M., López-Mellado, E., and Siller, M. (2016). Wireless sensor networks formation: approaches and techniques. *Journal of Sensors*: 2081902. 18 pages. doi: https://doi.org/10.1155/2016/2081902.

90 Karlof, C., Shastry, N., and Wagner, D. (2004). TinySec: a link layer security architecture for wireless sensor networks. Proceedings of SenSys'04, Baltimore, Maryland, USA (3-5 November 2004). ACM Press.

91 Koubaa, A., Alves, M., and Tovar, E. (2006). A comprehensive simulation study of slotted CSMA/CA for IEEE 802.15.4 wireless sensor networks. IEEE International Workshop on Factory Communication Systems. 183–192. Torino, Italy. doi: https://doi.org/10.1109/WFCS.2006.1704149.

92 Shih, E., Cho, S.-H., Ickes, N.R. et al. (2001). Physical layer driven protocol and algorithm design for energy-efficient wireless sensor networks. Proceedings of the 7th annual international conference on Mobile computing and networking (MobiCom '01). 272–287. New York: Association for Computing Machinery. doi: https://doi.org/10.1145/381677.381703.

93 Zaman, N., Jung, L.T., and Yasin, M.M. (2016). Enhancing energy efficiency of wireless sensor network through the design of energy efficient routing protocol. *Journal of Sensors*: 9278701. 16 pages. doi:https://doi.org/10.1155/2016/9278701.

94 Medagliani, P., Leguay, J., Duda, A. et al. (2014). Bringing IP to low-power smart objects: the smart parking case in the CALIPSO project. In: *Internet of Things Applications – From Research and Innovation to Market Deployment* (eds. O. Vermesan and P. Friess), 287–313. Gistrup, Denmark: River Publishers.

95 Sethi, P. and Sarangi, S.R. (2017). Internet of things: architectures, protocols, and applications. *Journal of Electrical and Computer Engineering*: 9324035. 25 pages. doi:https://doi.org/10.1155/2017/9324035.

96 Sohraby, K., Minoli, D., and Znati, T. (2007). *Wireless Sensor Networks: Technology, Protocols, and Applications*. Hoboken, NJ: Wiley ISBN: 978-0-470-11275-5.

97 Lekshmy, V.M., Rekha, P., and Ramesh, M.V. (2016). Impact of algorithm complexity on energy utilization of wireless sensor nodes. 2016 International Conference on Advances in Computing, Communications and Informatics (ICACCI).

98 Olatinwo, S.O. and Joubert, T.H. (2019). Efficient energy resource utilization in a wireless sensor system for monitoring water quality. *Journal on Wireless Communications and Networking* 6: 5–19. https://doi.org/10.1186/s13638-018-1316-x.

99 Yadav, S. and Yadav, R.S. (2016). A review on energy efficient protocols in wireless sensor networks. *Wireless Networks* 22: 335–350.

100 Pour, N.K. (2015). Energy efficiency in wireless sensor networks. PhD thesis. University of Technology Sydney. https://export.arxiv.org/abs/1605.02393.

101 Nakas, C., Kandris, D., and Visvardis, G. (2020). Energy efficient routing in wireless sensor networks: a comprehensive survey. *Algorithms* 13 (3): 72.

102 Wang, C., Sohraby, K., Bo, L. et al. (2006). A survey of transport protocols for wireless sensor networks. *IEEE Network* 20 (3): 34–40. https://doi.org/10.1109/MNET.2006 .1637930.

103 García Villalba, L.J., Sandoval Orozco, A.L., Triviño Cabrera, A. et al. (2009). Routing protocols in wireless sensor networks. *Sensors* 9: 8399–8421.

104 Sharma, S and Jena, S.K. (2011). A survey on secure hierarchical routing protocols in wireless sensor networks. Proceedings of the 2011 International Conference on Communication, Computing & Security (ICCCS '11). 146–151. New York: Association for Computing Machinery. doi: https://doi.org/10.1145/1947940.1947972

105 Iannello, F., Simeone, O., and Spagnolini, U. (2012). Medium access control protocols for wireless sensor networks with energy harvesting. *IEEE Transactions on Communications* 60 (5): 1381–1389. https://doi.org/10.1109/TCOMM.2012.030712.110089.

106 Stojmenovic, I. (2006). Localized network layer protocols in wireless sensor networks based on optimizing cost over progress ratio. *IEEE Network* 20 (1): 21–27. https://doi .org/10.1109/MNET.2006.1580915.

107 Huang, P., Xiao, L., Soltani, S. et al. (2013). The evolution of MAC protocols in wireless sensor networks: a survey. *IEEE Communication Surveys and Tutorials* 15 (1): 101–120. https://doi.org/10.1109/SURV.2012.040412.00105.

108 Yu, S., Zhang, B., Li, C. et al. (2014). Routing protocols for wireless sensor networks with mobile sinks: a survey. *IEEE Communications Magazine* 52 (7): 150–157.

109 Liu, X. (2012). A survey on clustering routing protocols in wireless sensor networks. *Sensors* 12: 11113–11153.

110 Demirkol, I., Ersoy, C., and Alagoz, F. (2006). MAC protocols for wireless sensor networks: a survey. *IEEE Communications Magazine* 44 (4): 115–121. https://doi.org/10 .1109/MCOM.2006.1632658.

111 Ergen, S.C. and Varaiya, P. (2010). TDMA scheduling algorithms for wireless sensor networks. *Wireless Networks* 16: 985–997. https://doi.org/10.1007/s11276-009-0183-0.

112 Falconer, D.D., Adachi, F., and Gudmundson, B. (1995). Time division multiple access methods for wireless personal communications. *IEEE Communications Magazine* 33 (1): 50–57. https://doi.org/10.1109/35.339881.

113 Leatherbury, R.M., Johnson, R.E., and Beens, J.A. (2005). Time division multiple access over broadband modulation method and apparatus, US Patent 6,891,841.

114 Kobayashi, T. and Iwamoto, H. (1998) Time division multiple access radio communication system. US Patent 5,719,859.

115 Merakos, L., Jangi, S., and Li, F. (1996). Method and apparatus for providing mixed voice and data communication in a time division multiple access radio communication system. US Patent 5,521,925.

116 Zheng, J. and Lee M.J. (2004). Comprehensive Performance Study of IEEE 802. 15.4.

117 Kohvakka, M., Kuorilehto, M., Hännikäinen, M. et al. (2006). Performance analysis of IEEE 802.15.4 and ZigBee for large-scale wireless sensor network applications. 3rd ACM international Workshop on Performance Evaluation of Wireless Ad Hoc, Sensor and Ubiquitous Networks. Torremolinos, Spain. October 2006.

118 Yuan, L., Lin, F., and Lv, J. (2018). An improved asynchronous energy-saving mechanism for IEEE 802.15.5-based networks. *International Journal of Distributed Sensor Networks* 14 (9): 1–14. https://doi.org/10.1177/1550147718802184.

119 Buratti, C., Conti, A., Dardari, D. et al. (2009). An overview on wireless sensor networks technology and evolution. *Sensors* 9: 6869–6896.

120 Patil, H.K. and Szygenda, S.A. (2012). *Security for Wireless Sensor Networks Using Identity-Based Cryptography*, 1e, 134–172. Auerbach Publications https://doi.org/10.1201/b13080.

121 Mitchell, C. and He, C. (2005). Security Analysis and Improvements for IEEE 802.11 i, and Distributed System Security Symposium (NDSS '05, 2005 - Citeseer).

122 Tomić, I. and McCann, J.A. (2017). A survey of potential security issues in existing wireless sensor network protocols. *IEEE Internet of Things Journal* 4 (6): 1910–1923. https://doi.org/10.1109/JIOT.2017.2749883.

123 A Koubâa, Alves, M., and Tovar, E. (2005). IEEE 802.15.4 for wireless sensor networks: a technical overview. 1–21. http://hdl.handle.net/10400.22/3397

124 Meghji, M., Habibi, D., and Ahmad, I. (2012). Performance evaluation of 802.15.4 Medium Access Control during network association and synchronization for sensor networks. 4th International Conference on Ubiquitous and Future Networks (ICUFN), 27–33. Phuket. doi: https://doi.org/10.1109/ICUFN.2012.6261659.

125 Kuban, P.A. (2006). An architecture for the extension of fixed controller area networks to IEEE 802.15.4 wireless personal area networks. Thesis. University of Louisville.

126 Dahane, A. and Nasr-eddine, B. (2018). *Mobile, Wireless and Sensor Networks a Clustering Algorithm for Energy Efficiency and Safety*. Apple Academic Press.

127 Padhy, P. (2009). Autonomous energy efficient protocols and strategies for wireless sensor networks. University of Southampton, School of Electronics and Computer Science, Doctoral Thesis, 170 pp.

128 Yahya, B. and Ben-Othman, J. (2009). Towards a classification of energy aware MAC protocols for wireless sensor networks. *Wireless Communications and Mobile Computing* 9: 1572–1607. https://doi.org/10.1002/wcm.743.

129 Kafetzoglou, S. and Papavassiliou, S. (2011). Energy-efficient framework for data gathering in wireless sensor networks via the combination of sleeping MAC and data aggregation strategies. *International Journal of Sensor Networks* 10: 3–13.

130 Rajendran, V., Obraczka, K., and Garcia-Luna-Aceves, J.J. (2003). Energy-efficient collision-free medium access control for wireless sensor networks. Proceedings of the 1st international conference on Embedded networked sensor systems (SenSys '03). New York: Association for Computing Machinery. 181–192. doi:doi:https://doi.org/10.1145/958491.958513

131 Lu, G., Krishnamachari, B., and Raghavendra, C.S. (2004). An adaptive energy-efficient and low-latency MAC for data gathering in wireless sensor networks. Proceedings of the 18th International Parallel and Distributed Processing Symposium, 224–343. Santa Fe, NM, doi: https://doi.org/10.1109/IPDPS.2004.1303264.

132 Chilamkurti, N., Zeadally, S., Vasilakos, A. et al. (2009). Cross-layer support for energy efficient routing in wireless sensor networks. *Journal of Sensors*: 134165, 9 pages. doi:https://doi.org/10.1155/2009/134165.

133 Striccoli, D., Piro, G., and Boggia, G. (2019). Multicast and broadcast services over Mobile networks: a survey on standardized approaches and scientific outcomes. *IEEE Communication Surveys and Tutorials* 21: 1020–1063.

134 Luomala, J. and Hakala, I. (2015). Effects of temperature and humidity on radio signal strength in outdoor wireless sensor networks. Federated Conference on Computer Science and Information Systems (FedCSIS). 1247–1255. Lodz. doi: https://doi.org/10.15439/2015F241.

2

Impact of Artificial Intelligence in Designing of 5G

Maheswari K[1], Mohankumar[2], and Banuroopa[2]

[1]Department of Mathematics, Kumaraguru College of Technology, Coimbatore, Tamilnadu
[2]Department of Computer Science, Karpagam Academy of Higher Education, Coimbatore, Tamilnadu

2.1 5G – An Introduction

5G is the fifth-generation broadband cellular communication network which the telecommunication companies have been implementing since 2019 [1]. 5G is the digital cellular successor of 4G technologies which were introduced around 2010. 5G is the fifth-generation cellular communication network, which is the latest worldwide wireless standard after the 1G, 2G, 3G, and 4G networks. The 5G is a novel category of network technology which is fashioned toward linking almost all people and devices. The hype about 5G is the advanced, very high performance it facilitates. The 5G network has more speed than 4G networks; the speed it can support is 10 Gbit of data/second, which is 100 times faster than its predecessor.

The wireless devices in the 5G network will be connected to both the Internet and telephone network through radio waves. This is made possible by the antennae being incorporated into the devices which provides greater data download speeds. However, the existing 4G devices are not compatible with the 5G networks; they will not able to use this network because it requires specialized equipment. The 5G enabled devices will not only be used for mobile communication, but also can be used as laptops and desktop computers. These 5G networks are also opening up new avenues in data analytics and the Internet of Things (IoT) [2].

The peculiarity of 5G technology using radio waves for communication means these devices need different antennas for different ranges of communication. The latest 5G device will have smart antennae to serve this purpose. The existing infrastructure of 4G networks cannot be used for 5G technologies. As these devices use radio waves of high bandwidth, they cannot cover larger areas. 5G network is categorized into three, according to the service provided by the International Telecommunication Union (ITU) [3]:

1. Enhanced Mobile Broadband (eMBB) focuses on higher bandwidth requirements for the general population for viewing HD videos and playing games with virtual reality (VR) and augmented reality.
2. Ultra-reliable and Low-latency Communications (uRLLC) provides industries with fully dependable and reduced latency services.

Smart and Sustainable Approaches for Optimizing Performance of Wireless Networks: Real-time Applications, First Edition.
Edited by Sherin Zafar, Mohd Abdul Ahad, Syed Imran Ali, Deepa Mehta, and M. Afshar Alam.
© 2022 John Wiley & Sons Ltd. Published 2022 by John Wiley & Sons Ltd.

3. Massive Machine Type Communications (mMTC) is to develop smart city-like applications which require more of connections in a small area.

5G is the commencement of the upgrade of digitalization beginning with individual interest to society interconnection. Digitalization generates incredible chances for the mobile communication industry to possess and redefine mobile communication technologies. The spread of the possibilities for mobile networks augments the telecommunication network ecosystem. A number of conventional commercial enterprises, such as automobiles, healthcare industry, electrical energy enterprises, and city management systems partake in the building of this system. The features and a few of the applications of 5G mobile technology is listed below [4]:

- High-speed data download
- Almost nil latency
- Greater bandwidth
- Possibility of Wireless WWW
- Enabling of self driving cars, smart cities, smart agriculture, etc.
- Instant access to cloud services
- 3D video
- Playing games which have VR
- Connecting numerous IoT devices simultaneously

2.1.1 Industry Applications

Today's industries will benefit from implementation of 5G with reduction in costs, improved profits, increase in customer satisfaction, and many others. Some of these enterprises are listed below:

2.1.2 Healthcare

Use of 5G enabled devices in healthcare will facilitate physician and patients to be much associated than before. For example, the wearable devices of the patients could automatically monitor their heart rate, etc. and would alert a doctor to abnormal readings, thereby automatically alerting the hospital team.

2.1.3 Retail

The application of 5G technology in retail will create a novel customer experience. Stores of the future will not be like today's stores with their goods stacked in aisles. Future stores may allow the customers to add items to a virtual cart. They may also use 5G technology in stock management in real time. Stores may also go without cashiers entirely and will track the virtual cart without the need for the customer to stand in the conventional checkout line.

2.1.4 Agriculture

In the future, farms will be using auxiliary data and smaller amounts of chemicals. Capturing data with sensor devices positioned in open fields, will help recognize with accuracy as to which areas need watering, contain a disease, or need insect control management.

5G will make costs of wearable devices less, and easier to extend networks with a huge number of IoT devices. This makes it possible to continuously monitor the health and wellbeing of livestock. With this additional precise health data, breeders will diminish the utilization of medicines like hormonal injections, without compromising the wellbeing or the feeding of the animals.

2.1.5 Manufacturing

Factories will be completely altered as a result of the union of 5G, AI, and IoT. Further than prognostic safeguarding that assists in managing expenses and diminishing shutdown times, factories will also utilize 5G to manage and scrutinize engineering practices by means of an extraordinary measure of meticulousness.

The connectivity enhancement given by 5G will assist manufacturers to alter customary QA processes, reforming them by means of wireless devices and Artificial Intelligence (AI).

2.1.6 Logistics

Tracking of goods is costly, sluggish, and tricky with regards to distribution and logistics. 5G proffers the probability for superior communication between automobiles, as well as among the automobiles and the 5G infrastructure itself.

Supervising and navigating of fleets of vehicular traffic will be considerably easier with 5G. Navigation information for the drivers will be powered by means of an AR system that recognizes and marks impending danger, without distracting the driver's concentration from the road.

2.1.7 Sustainability of 5G Networks

By using 5G networks while implementing the scenarios, such as work-from-home, smart electricity grids, automated driving of vehicles, and futuristic precision farming, it has been found that the emissions of GHG can be reduced drastically [5]. GHG emissions play a critical role in climatic changes worldwide. The rise in temperature and melting of ice caps has been of huge concern for all us during the past few decades. Replacing the existing network infrastructure with fast, reliable, ultralow latency 5G cellular networks has high potential in curbing the emission of GHG. By 2030, when the 5G networks will become fully functional, the emission of GHG will also be drastically reduced, thereby helping the world toward more sustainable development.

2.1.8 Implementation of 5G

As the final vision of 5G is still being decided upon, it will probably become as vital in favor of connectivity as was the commencement of the Internet. However, the majority of the stimulating elements of 5G is that nobody can discern unerringly what the future holds.

The future of edge technology predominantly relies on 5G. Even though the complete deployment of 5G will take more time, it will be the main component in the evolution of several technologies like cloud-computing, which will pave the way to more distribution of computing environments. The investment in infrastructure for 5G networks could be $1 trillion before 2025 to support worldwide deployment of cellular, edge, and IoT systems in all facets in our lives [6].

It will empower more workloads to be executed efficiently and data to reside on edge devices. It will serve as a base for futuristic AI environments in which data-driven algorithms will drive almost all cloud-computing processes, sensor devices, and VR experience. In the same way, AI will be a strategic element in guaranteeing that 5G technologies are further enhanced.

2.1.9 Architecture of 5G Technology

According to US government, the number of 5G connections will increase to around 3 billion by 2025, as shown in the chart below (Figure 2.1). This is a major challenge for service providers worldwide to design scalable 5G networks which will also support the already existing lesser generation network technology users. Major network service providers like AT&T, Verizon, T-Mobile, and Sprint, have run out of bandwidth in crowded cities, which will affect the performance of the 4G LTE connection available [6]. 5G networks provide a solution by utilizing the millimeter wave (mm-Wave) bandwidth with higher frequency and wider bandwidth that is not yet utilized to its full potential. While designing the functionality of the 5G networks there is a need for conversion of 3G and 4G interoperability to 5G. The functional 5G network architecture is shown in Figure 2.2 [7].

To achieve low latency and high-speed data, 5G networks need small cells distributed densely throughout region.

Network slicing of 5G networks will provide different amounts of resources for various requirements through virtualization of network connections. This is usually done with the help of Software Defined Networks (SDN) and Network Function Virtualization (NFV). The SDN has materialized as a novel networking paradigm to reduce the dependency of

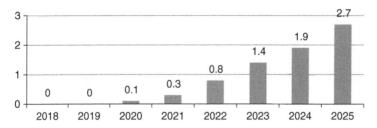

Figure 2.1 Growth of 5G connections worldwide.

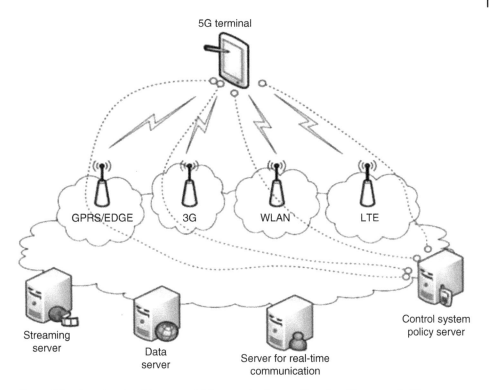

Figure 2.2 Functional architecture of 5G networks. Source: Based on Zhang, et al. [7].

hardware components by providing simple abstractions of the various hardware devices like Routers and Switches, and to describe their components, functionalities, and management of the devices logically with help of software. SDN creates a dynamic intelligent network for the users without the inconvenience the hardware needed to create it. In 5G networks, SDN and NFV can be used to provide network slicing capabilities to supply varied functions to end users. SDN and NFV make a single physical connection to a multiple virtual connection which can be shared to provide varied functionalities [8]. They use AI and Machine learning algorithms to provide this functionality. Many architectures are proposed and being tested by varied enterprises to implement network slicing through SDN for 5G networks (Figure 2.3).

2.2 5G and AI

Applying AI to an enormous quantity of data will be fast-tracked with high-speed, competent connectivity. For instance, in a smart city application, automated traffic analysis could correlate the traffic signals and implement any changes in the city's landscape. The security monitoring can become smart to involuntarily recognize likely security breaks or unsanctioned users.

5G will assist and facilitate AI deduction at the edge and will moreover collect and send data from devices to a vital cloud server for training or refining AI models. For instance,

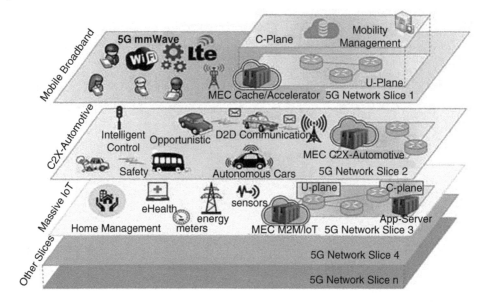

Figure 2.3 Network slicing of 5G networks.

vehicles which are connected could send real-time data about the current road conditions to the cloud server to improve the mapping process [9].

2.2.1 Gaming and Virtual Reality

5G promises further extra impressive experience for gaming. Due to high-speed data transfers in 5G, HD live streaming will be enhanced. The near zero latency of 5G will not tie down gaming to devices with elevated computing power, as cloud is used for processing, storage, and retrieval at the same time that the game is viewed and managed by a mobile device.

Applications that use VR depend on swift response and reaction times to offer a pragmatic capability will benefit from the low-latency feature of 5G networks. They will drive major innovation in all VR applications. The number of VR applications will probably increase and become more sophisticated as 5G technology and devices develop as the new normal.

As 5G edge-computing becomes more common, enterprises could radically increase their use of data and react to interpretation of that data more rapidly, repeatedly, and involuntarily.

In the future, AI will play a major role in the cloud-computing paradigm. Even now, many AI platform enterprises have invested in 5G-driven applications for their mobile communication, IoT, etc.

For the enhanced understanding of the impact of 5G on the online economy, let us reflect on this promising wireless architecture which will provide more value throughout the AI tool-chain:

- **Convergence of AI systems with edge technology:** Digital mobile technology, wireless LTE, and Wi-Fi technology are combined by 5G. 5G will empower every edge device to effortlessly switch between interior and wider exterior environments when implemented in cross-platform network interfaces. This 5G technology will combine the entire radio spectrum of channels and also could change the network interface device into a single chip powered by AI. These chips will be highly agile and maintain many connections through many radio access technologies. These 5G interface chips will use neural networks to provide low-cost and low-latency systems for many AI applications.

- **Real-time concurrency achievement of AI data lakes:** 5G technology will support many millions of devices within a small region and connect them all concurrently without any time lag. Such performance is higher than with 4G technology. This concurrency would allow enterprises to amass huge quantities of data unceasingly from many devices such as mobile phones, IoTs, and other 5G-enabled devices. 5G networks would begin by inundating data streams and lakes ubiquitously by means of renewed data coming from devices. Designers of the AI model will be able to create complicated data analysis from ML models intended for use in cases of IoT, mobile technology, engineering computerization in industry, smart city development, and numerous other areas.
- **Low-latency enhances fast and large volumes of data for AI:** The latency of 5G networks is much less than with 4G networks: 1–50 millisecond of latency. Therefore, 5G networks have much more rapid downloads and uploads than 4G: 20 gigabits per second to 5–12 megabits per second of 4G. This speed is 1000 times faster than with 4G networks. This feature of 5G is made possible by the ability of these networks to simultaneously transmit multiple DataStream between the base station and the devices. This produces greater bandwidth and transmission capacity for 5G. These performance merits will enable 5G networks to create AI DevOps with low latency in real-time situations for data consumption and preparation to model building, training, and serving. Furthermore, the 5G's combination of more rapid download speeds and lower latency would facilitate experts to gather, clean, and evaluate large amounts of data very much more rapidly than before.

2.3   AI and 5G

AI is a very important part in the infrastructure of 5G networks, which ensures that the complex 5G network can support huge workloads of applications. The latest published research shows that many wireless mobile operators across the world are planning to deploy AI to run the 5G and other networks.

The future AI application's effective execution depends on the performance of 5G networks. In order for this to be achieved, the 5G networks need to continuously evolve, heal, manage, and optimize on their own [10]. This sequentially depends on the concentration of ML and additional AI models for automating data traffic routing at the application level, assuring QoS, managing, and maintaining the performance, and if errors occur, perform root-cause analysis, etc., in a more scalable, predictable, rapid, and efficient manner than if done manually.

This capability, known as AIOps, would be important for the 5G networks to deliver on its guarantee of significantly quicker, dependable, and RF-efficient networks than previous wireless technologies. The capabilities of AIOps would have to be essential to the network virtualization and multi-cloud monitoring applications that monitor 5G network's associated applications in all aspects.

The quality-of service will become dynamic across 5G environments, and to meet this need AI-based models are used to provide end-to-end services. These AI-based controls will ensure that the infrastructure resources in 5G networks, such as RF channels and other interfaces, are shared on the go and with accuracy to upkeep varying quality-of-service necessities, traffic configurations, and application capacities.

AIOps controls will complement a 5G infrastructure potential known as "network slicing." Network slicing empowers 5G networks to execute numerous virtual networks over single physical connections [9]. AIOps tools will uphold this virtualized resource sharing competence. AIOps tooling will also provide prognostic and varying deliverance of divergent wireless QoS tiers for varied consumer categories and edge-device categories.

AIOps are used for improving 5G dynamic RF-channel allotment properties. As 5G networks have smaller cells than 4G, to reuse the same RF frequencies they must target base station antennas continuously at each edge device. 5G base stations, dynamically envisage and share the preeminent wireless route with every other device to guarantee quality of service, while endlessly facing the obstacles they come across in the way of walls and other solid objects. AI-based real-time data analytics are urgently required for executing these computations in real time through dynamically shifting wireless local loops.

Data management will be increasingly needed as more data is produced in the 5G networks and to analyze that data and to sustain it. This may lead to data lakes and automated machine learning AI repositories all over 5G networks. This much needed operational infrastructure in 5G networks ensures that the best-fit AI models are installed in real time. The complex public/private amalgamated environments, which would be a primary characteristic of 5G networks, line up with the data/model management infrastructure that will be deployed in cloud-to-edge structures. It is time for both network service providers and enterprise IT professionals to reconnoiter the vital role of AI in their 5G and edge-computing plans.

5G and AI are two of the utmost unruly concepts that humanity has encountered in years. Although each of them has exclusively transformed businesses and endowed novel practices, the amalgamation of 5G and AI is going to be revolutionized. Actually, this juncture is essential to realize the dream of the smart wireless edge which will have on-device data processing, cloud-on-edge, and 5G and will in combination create a permanent connectivity of smart devices and services.

AI and edge-cloud will use low-latency 5G networks to process substantial volumes of data to be processed closer to its source. Data processing nearer to the source via on-device AI is vital as it gives essential merits like confidentiality, personalization, and dependability, furthermore helping to scale intelligence. The intelligent wireless edge will enrich present applications but also empower novel applications.

By applying AI to the 5G networks and the devices will give way to new expeditious wireless communications, long-lasting batteries, and increased user capabilities. The key to harnessing such a powerful tool like AI is to concentrate on vital wireless challenges that are arduous to resolve with conventional processes, which are also a great fit for the machine learning. Profound wireless domain knowledge is a requisite to recognize and use AI's potential. AI will have a robust sway on numerous significant capacities of 5G network

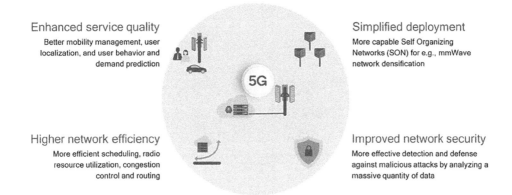

Figure 2.4 Intelligent 5G network management by AI.

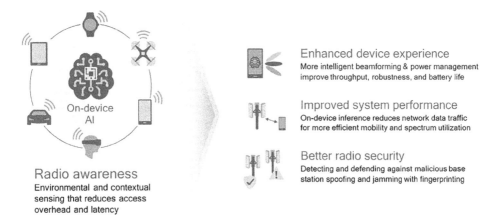

Figure 2.5 Improved 5G end-to-end system by On-device AI.

management, such as improved QoS, streamlined deployment, sophisticated network productivity, and enriched network security [10]. Figure 2.4 shows the intelligent 5G network management by AI. For instance, AI might be used to spot irregularities in network traffic, such as inundating or imitation, by perceiving infrequent spectrum usage.

Usage of an AI on-device to enhance a 5G end-to-end system has not been elaborated upon until now. Figure 2.5 it gives the details of the improved 5G end-to-end system. Awareness about the radio frequency waves will improve the use of AI rather than the machine learning algorithms. AI is the consummate instrument to create wisdom out of the complex RF signals around the device. Increased radio consciousness permits a variety of enhancements, such as enriched device capability, improved system performance, and upgraded security.

Flexible system solutions are made possible by the low latency and high capacity of 5G, which allows AI processing to be distributed among the device, edge cloud, and central cloud. This adaptable wireless edge architecture gives suitable tradeoffs to be made for

Figure 2.6 Personalized shopping through boundless AR.

various requirements. For instance, performance and financial compromises may assist in determining workloads to reach the necessary latency or calculate necessities on behalf of a specific application's need.

AI-enhanced practices will improve on scenarios such as personalized retail through limitless VR, and instinctive computer-generated aids through immensely reinforced voice User Interface, because of 5G networks and devices. In the future, a sample shopping and retail application may look like Figure 2.6. It shows the use of VR rendering for a very personalized shopping experience by viewing whatever is of interest to you in the shop's window display. This would make the shopping experience efficient, effective, and fun too. This is made possible by splitting the workload of AI-processes in both device and 5G networks.

2.3.1 Continuous Learning AI Model

AI should be moved from cloud-centric implementation to full distribution to expand its scale. Nowadays, partially distributed AI are implemented with the help of on-device AI inference, where the devices process and refine data before passing it on to the cloud plat-form for analysis. In the future, on-device AI will train the AI model on the device itself rather than from inference alone. This paradigm shift will allow an entirely disseminated AI with lifetime on-device learning as depicted in figure 2.7, which will provide the next level of personalization, privacy, and security [11].

This disseminated learning is made possible by the 5G enabled devices, which allow scaling of the training of the AI model beyond the cloud. To realize this, the initial phase is that an outstanding AI model is sent to all devices by central or edge cloud. In the next phase, all devices amass particular data and execute on-device training (see Figure 2.8). As enormous amount of training is computationally rigorous and has not been completed in the cloud platform until recently. But by performing minor training rounds on reduced datasets, the workload becomes further controllable. And above all, on-device AI abilities have been expanding exponentially, along with enhancements in algorithms and software.

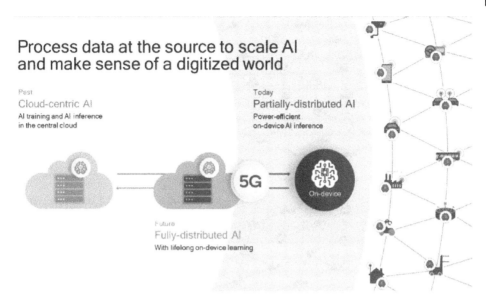

Figure 2.7 Entirely disseminated AI with lifetime learning.

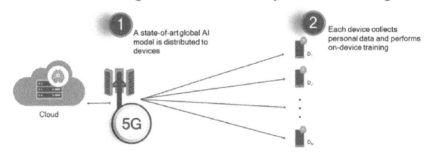

Figure 2.8 Distributed learning AI model over 5G Network (part 1 of 2).

On-device training has three extremely significant merits that will pilot the way to huge implementation of AI:

1. Scaling: By dissemination of processes over numerous devices, like masses of smart mobile phones, one could bind a substantial quantity of computational power.
2. Customization: The AI model innately customizes our own devices' training with our own data.

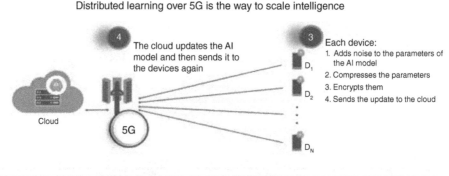

Figure 2.9 Distributed learning AI model over 5G Network (part 2 of 2).

3. Confidentiality: The raw data would not leave our device and enter the cloud. By on-device training our own data, we would be mining the significance of the data and so protect our confidentiality.

The subsequent steps are iterated for enhancing the state-of-the-art AI model. In order to achieve this, noise is added to the features of the AI model which obscures the data to preserve privacy, then the data is compressed and encrypted to pass on to the cloud. The enhanced AI model is sent to all the devices which have that particular application after the cloud processes the AI model from all the other devices (see Figure 2.9). This process iterates to the improvement of the AI model.

2.4   Challenges and Roadmap

AI displays outstanding prospects for cellular networks. But important challenges are yet to be mastered. The following are the significant challenges and a roadmap for viability of AI-enabled mobile networks for 5G and beyond 5G [12].

- **Issues in training the AI model.**
  System overheads and accessibility of training data would be the two vital hindrances associated with training AI models for cellular data networks. From a physical and MAC layer point of view, training a mobile device AI model is costly in terms of uplink control overheads. This is usually done with the back propagation algorithm. Therefore, minimizing the training overheads is an important issue for the viability of MAC layer-based AI models. Secondly, labeled training data is difficult to obtain because of the segregation of data across network protocol layers.
- **Performance Guarantee**
  Prediction of a worst-case scenario over a mobile cellular network is crucial for providing an acceptable Quality-of-service of the networks. Unlike conventional model-based approaches, which can predict system output distribution for a certain input distribution,

AI-based models will not be able to achieve this because of their non-linear characteristics, even though they perform well in live networks. The AI-Model cannot make available any worst-case scenario performance assurance. For seamless integration of AI into mobile cellular data networks, it is critical to assure a bearable and smooth dilapidation in a worst-case scenario.

- **Explainability Issues**
  As AI implements are frequently considered as black boxes, it is arduous to cultivate analytical models to assess their rightness and explicate their activity. This lack of explanation of this ability is a main disadvantage for instantaneous decision-making situations like vehicle-to-vehicle communications. Cellular networks and wireless criterions are premeditated built on mathematical models, hypothetical analysis, and channel dimensions. This methodology has been established for experts to fall back on to either hypothetical analysis or computer simulations to formalize communication system of cellular networks. It is desired by the AI models to have analogous echelons of explainability when deliberating for 5G networks.

- **Generalization Ambiguity**
  When the AI model is used for a communication task in cellular networks, it is often unclear as to whether the dataset used for training the model is sufficient for training that model for the given inputs encountered in reality. This is especially true in real-time scenarios where it is crucial to be cautious of sporadic happenings. Nevertheless, even though the learning model cannot know the data during training. it ought to still be capable of generalizing unconventional cases.

- **Interoperability Issues**
  Interoperability is an acute part in multifaceted cellular networks and this frees the consumers from dependency on vendors. The network's performance would decline if is there is any variation amidst AI modules from different vendors. Certain actions carried out by an AI-based component from one vendor could neutralize the activities carried out by another vendor whose network module may or may not be AI-based. This could cause unsolicited conflicts among the base stations and the performance of the overall network may be compromised.

2.4.1 Technical Issues

- **mm-Wave Communications:** The 5G networks function using millimeter waves in high radio frequency. The higher the frequency of the waves, the more it can be blocked by barriers, such as walls, furniture, and even persons. So, the challenge is to design components in the circuit and antennas to overcome this feature of the mm-Wave.

- **Device-to-Device Communication:** In 5G networks there are no restrictions on device-to-device communications. As there is no central controller to regulate the communication the resource allocation and inter-device communication, regulation and management is difficult. Another issue which arises is the security and privacy of the data in the end devices.

- **Backhaul network infrastructure:** The backhaul or backbone network infrastructure is the fiber optic cables, copper cables, or satellite communications which facilitate the transfer of huge amounts of data among the 5G devices. In many areas there is lack of adequate infrastructure such as fiber optic cables to support the future technology. This would become a bottle neck in widely implementing the 5G networks.

- **Security challenges:** 5G network would be implemented as an SDN with NFV. This slicing of networks virtually to provide various functionalities for various services provides the need for different security policies for each. Implementing varied security mechanisms on the same network is a challenge.
- **Electromotive force (EMF) levels:** The EMF radiation levels should be in the acceptable ranges when deploying t5G networks. As 5G networks would need more equipment and many closely placed cells for working efficiently, care should be taken to maintain the EMF radiation levels well within acceptable limits.

2.4.2 Technology Roadmap

As per the foregoing challenges, from a technology roadmap viewpoint, fresh training algorithms and neural network architectures should be explored to diminish convoluted training and amount of training required.

Additionally, understandable and explicable AI models are critical for gaining perceptions into their assessment building procedure. To exploit their toughness and to diminish ambiguity by oversimplification, an uncontested prerequisite could comprise of paralleling the AI model output alongside a well-understood hypothetical routine. Standards organizations, such as 3GPP, judiciously appraise the fundamental requirement of AI models.

The dawn of IoT devices along with smart phones and their small power necessities for devices at the edge of the network, the training of AI models could be divided between the edge and the cloud. The specification could contemplate innovative erudition use-cases based on disseminated learning, at edge devices. To conclude, the foreseeable cellular data networks would formulate innovative application situations that exploit parameters based on interface data and cross-layer data like application layer information, for working of the AI models. From a network design outlook, to take full advantage of ease of deployment, it is necessary to deliver uncontaminated crossing points, both inside and transverse to protocol stack layers, when adding feature contributions to AI models.

2.4.3 Deployment Roadmap

Given the emerging landscape of installing AI for 5G applications and the high levels of service sureties necessitated by multinational enterprises, it is a requisite to engage AI in a stage-by-stage routine. This would enable system developers to relate their lessons through initial AI deployment, and consequently hone their AI tools and testing procedures. Initial contemplation is of the time used by the AI models for implementing AI in cellular networks. It may be desirable initially for AI models to operate across longer time periods, such as in the order of minutes or hours, so that domain experts can supersede model endorsements, if needed. Foolproof procedures are necessary to curtail effects of surging faults owing to unanticipated AI model outputs.

Further sturdiness can be increased if the AI model adjusts its activities constructed on human professional's opinion. One such situation is where, on spotting a network anomaly, an AI model outputs a certain basic reason that seems flawed. The expert can offer feedback concerning the indecorous conclusion; the model can enhance and develop its decision-making process until it touches a point where its verdict a is the same as that from an expert.

2.5 Mathematical Models

Mathematical models lag behind technological growth, despite the significant growth in mathematical modeling literature. The two main areas of mathematical modeling of interest are the Optimization and Queuing models. Some mathematical models used for analysis and design of the 1G to 2G for channel allocation, loss, etc., are the Erlang B and Erlang C models, using Erlang models that were developed in 1917 for PSTN connections – developed for circuit switched and later used for packet switched. The Erlang B model is the M/M/m/m model, where in the queuing model, customers arrive at a queuing system with many servers but has zero waiting positions. It implies that if there are no servers available upon the arrival of the customer, then the customer returns back empty which is a great loss. For this reason, it is known as the loss system and the customers tend to undergo a blockage.

The important building block in computational queuing theory is given by the Erlang loss function:

$$Prob\,(m,x) = B\,(m,x) = \left(\frac{x}{n!}\right)\left(1 + x + \frac{x2}{2!} + \frac{x3}{3!} + \cdots + \frac{xm}{m!}\right)$$

where m is the number of servers for the probability of blockage or is the number of parallel (telephone lines) channels. $E = \lambda h$, λ is arrival rate and h is the average service time.

These models are not apt for wireless systems. Channel holding times no longer inevitably follow the exponential distribution. Poisson arrivals were most likely not suitable in a good number of cases. Plenty of research has been dedicated to finding more fitting arrival processes, services processes, and queuing models.

Some mathematical models are used for analysis and design of the 3G and 4G models. With 3G, determined by CDMA (even though CDMA was adopted in 2.5G also), the mathematical modeling efforts concentrated more on resource allocation. This led to further work on optimization – maximizing capacity, while allocating power and ensuring QoS is met. Most of the optimization problems were non-convex and most results were based on heuristics, some for which were too tightly bound to be considered [13]. For 4G, driven by OFDMA, mathematical modeling efforts were on optimization and game theory. State-of-the-art in mathematical modeling for Wireless Communications Queueing systems emerge in diverse forms in latter-day communication systems, including CRN, WSN, etc., which are unconventional. The need of the hour is proper queuing models in order to design an enhanced system. Likewise, composite optimization problems for resource allocations arise in current-day communication systems. The optimization problems are typically just manageable using heuristics, so nevertheless the need for good algorithm. With imminent implementation of 5G in the near future, the need to soon necessitate the fitting mathematical models should not be overlooked.

2.5.1 The Insights of Mathematical Modeling in 5G Networks

The digital economy has reached multiple stages in the social infrastructure and mobilizing in our day-to-day lives by the wireless systems. AI with that of the 5G wireless communications is perceived as one of the leading research fields in the next generation of wireless communication systems. The 5G standardization requirements has enabled AI's potential in the field of cellular networks [14]. Computational architecture is framed to distribute the power of networking functions closer to the users by certain machine learning techniques [15]. The challenges posed by AI in focusing on the challenges of 5G standards has enabled us to operate in a fully automated fashion that will be met with increased demand and with superior quality of experience. Here, we model a non-linear problem which requires optimization. So as to fulfill the diversified standards of 5G requirements, a model is so framed that it is operated in a fully automated fashion which serves as a superior quality of experience. Based on the level of supervision, the training stage of AI is divided into different levels, such as supervised learning, unsupervised learning, and reinforcement learning.

The concept of supervised learning comprises of the random vector x and its output y, and predicting the value of y which is obtained by x and its estimation is given as $P(y/x)$ with its particular properties of that distribution. On the other hand, unsupervised learning assesses the random vector x and learns its probability distribution. In the case of the reinforcement learning, an interaction with the environment takes place by obtaining the feedback loops between the learning systems and its experiences. The digital economy at its outset has taken on multiple shapes in mobilizing our daily life and social infrastructure in wireless systems. However, even though wireless networks using 5G networks are different from those of AI, it is seen that when both are joined together, they have rich potential for research into the concepts of machine learning algorithms that aim to distribute computer power, storage, and networking functions used to assist radio communication technologies in intelligent decision-making and in adaptive learning. It is also observed that the next generation technologies desire certain objective functions to minimize and maximize the process of decision-making. The problems faced in mobile and wireless networks are not all linear, which have been discussed in Bogale et al. [15]. Based on the level of supervision, the AI technique requires the following stages such as the supervised learning, unsupervised learning, and the reinforcement learning. These concepts of learning are invoked through the program where the machine experiences E based on the number of tasks T with the measure of performance P that becomes improved by the experience of E. In supervised learning, we have the random vector x with the vector y and predicting the value of y from x by estimating the probability $P(y/x)$. Aiming to learn the probability distribution $P(x)$ and its properties is done through unsupervised learning, whereas obtaining feedback loops between learning systems and its experiences by means of rewards are done through reinforcement learning.

To understand the difference between these three learning subcategories, it is made easier to understand by invoking the learning concept as: "A computer program is said to learn from experience E with respect to some class of tasks T and performance measure P, if its performance at tasks in T, as measured by P, improves with experience E" [16].

2.6 Conclusion

The era of using AI-based algorithms in wireless networks is the enormous generation of networks in exploring some of the successful cases in 5G technologies. These days, due to the availability of power and ubiquity of information, there is always the wide possibility of expanding the tools with AI-based models, where even a few milliseconds of these models in the 5G world makes a difference in latency. This latency can be enhanced with distance and through network links by applying and predicting a variety of AI tools.

References

1 Sauter, M. (2017). From GSM to LTE-Advanced pro and 5G: An Introduction to Mobile Networks and Mobile Broadband. Wiley.

2 Ghosh, A., Maeder, A., Baker, M. et al. (2019). 5G evolution: a view on 5G cellular technology beyond 3GPP release 15. *IEEE Access* 7: 127639–127651.

3 Parkvall, S., Dalhman, E., Furuskar, A. et al. (2017). NR: The new 5G radio access technology. *IEEE Communications Standards Magazine* 1 (4): 24–30.

4 Chen, S., Qin, F., Hu, Bo. et al. (2018). *User-Centric Ultra-Dense Networks for 5G*, Springer Science and Business Media LLC.

5 Bieser, J., Salieri, B., Hischier, R., et al. (2020). *Next generation mobile networks: Problem or opportunity for climate protection?* Zurich, St. Gallen: University of Zurich, Empa.

6 Barakabitze, I.A., Ahmad, A., Mijumbi, R. et al. (2020). 5G network slicing using SDN and NFV: a survey of taxonomy, architectures and future challenges. *Computer Networks* 167: 106984.

7 Zhang, W., Cai, W., Min, J. et al. (2020). 5G and AI technology application in the AMTC learning factory. *Procedia Manufacturing* 45: 66–71.

8 Shafin, R., Liu, L., Chandrasekhar, V. et al. (2020). Artificial intelligence-enabled cellular networks: A critical path to beyond-5G and 6G. *IEEE Wireless Communications* 27 (2): 212–217.

9 Zhang, S. (2019). An overview of network slicing for 5G. *IEEE Wireless Communications* 26 (3): 111–117.

10 You, X.H., Zhang, C., Tan, X.S. et al. (2019). AI for 5G: research directions and paradigms. *SCIENCE CHINA Information Sciences* 62 (2): 021301. doi:10.1007/s11432-018-9596-5.

11 Wu, S-Y. (2019). Key technology enablers of innovations in the AI and 5G era. *2019 IEEE International Electron Devices Meeting (IEDM). IEEE*.

12 Taheribakhsh, M., Jafari, A.H., Moazzamipeiro, M. et al. (2020). 5G implementation: Major issues and challenges. 25th International Computer Conference, Computer Society of Iran (CSICC). 1–5. 10.1109/CSICC49403.2020.9050110.

13 Su, R., Zhang, R., Venkatesan, Z. et al. (2019). Resource allocation for network slicing in 5G telecommunication networks: a survey of principles and models. *IEEE Network* 33 (6): 172–179.

14 Li, R., Zhao, Z., Zhou, X. et al. (2017). Intelligent 5G: when cellular networks meet artificial intelligence, *IEEE Wireless Communications*, 24 (5): 175–183. [Online]. Available at: http://www.rongpeng.info/files/Paper_wcm2016.pdf.

15 Bogale, T.E., Wang, X., and Le, L.B. (2018). Machine intelligence techniques for next-generation context-aware wireless networks. *ITU Special Issue: The impact of Artificial Intelligence (AI) on communication networks and services.* 1. [Online]. Available at: https://arxiv.org/pdf/1801.04223.pdfhttp://arxiv.org/abs/1801.04223.

16 Mitchell, T.M. *Machine Learning*, 1 ed. McGraw-Hill Science/Engineering/Math, 1997. [On-line]. Available at: https://www.cs.ubbcluj.ro/~gabis/ml/ml-books/McGrawHill-MachineLearning-TomMitchell.pdf

3

Sustainable Paradigm for Computing the Security of Wireless Internet of Things: Blockchain Technology

Sana Zeba, Mohammed Amjad, and Danish Raza Rizvi

Department of Computer Engineering, Jamia Millia Islamia University, New Delhi, India

3.1 Introduction

Advanced computerization, sustainable paradigms, and expeditious growth of the wireless Internet of Things (IoT) system functions with the use of sensors, actuators, and translators. In the recent era, the wireless IoT has become a rich field in the smart environment. It has rapidly increased the number of intelligent devices in the system and has effectively merged devices and the physical environment through the Internet, thus creating a smart system. The IoT network is an illustration of innovative automation and a rapid growth wireless system which exploits sensing, actuating, Artificial Intelligence, Blockchain Technology, and networking technologies to achieve the complete goal of the system. Firstly, the IoT network was invented in 1998 and developed in 1999 by Kevin Asthon [1]. The market of the IoT increased from $547 million in 2018 to $841 million in 2020, and according to a report by Gartner, by 2018, an estimated 23.14 billion devices were installed worldwide.

The IoT is well-defined as a framework or environment where intelligent devices or things are connected and communicate and transfer data between devices at anytime and anywhere. In the IoT, intelligent devices have circumscribed sources of power and limited storage capacity in the network. Its rapid growth, complexity, and heterogeneity of the objects of the system create many security concerns and issues related to the IoT system [2]. These objects might be sensors, radio frequency identifiers (RFID), actuators, mobile phones, and other smart devices. The IoT system is used in different applications or domains like Smart City, Smart Home, Smart Traffic Systems, Smart Parking Systems, Smart Enterprise, Smart Health Monitoring System, Smart Education Systems, etc. [3, 4]. Security is the biggest concern in all the applications of the IoT system, because of its open structure and nature (Figure 3.1).

Recently, many researchers have been working on Blockchain technology to try to find a cure for the threats which occur in IoT. The Blockchain technology perceived by Satoshi Nakamoto in 2008 [1, 5, 6] has a distributed immutable and public ledger of transactions. Resources or constraints of the IoT objects explain the working of smart IoT systems. The expeditious growth and regular assessment of the IoT environment increases security issues and problems of the system because of its wireless connectivity and cloud-based centralized architectures, etc. Security and threats can become a crucial problem in any wireless

Smart and Sustainable Approaches for Optimizing Performance of Wireless Networks: Real-time Applications, First Edition.
Edited by Sherin Zafar, Mohd Abdul Ahad, Syed Imran Ali, Deepa Mehta, and M. Afshar Alam.
© 2022 John Wiley & Sons Ltd. Published 2022 by John Wiley & Sons Ltd.

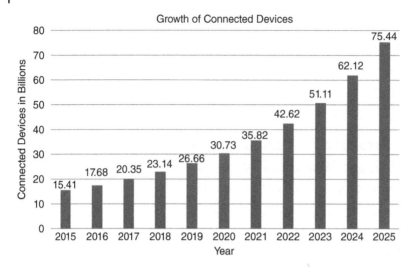

Figure 3.1 Growth of connected devices. Source: Statista. (trinamix.com/blog/blockchain).

networking system, such as the IoT, which hamper the computing power of that system and reduce its efficiency as well. Numerous security threats that occur in the IoT system are Signal or Radio Jamming, Spoofing, Malicious attacks, Denial of Service (DoS), Distributed Denial of Service (DDoS) attacks, Black Hole attacks, Sybil attacks, Data tampering, Tag cloning, etc. To address these problems, researchers are using many solutions like Blockchain technology, IoTA, etc. In the current IoT system, a scalability problem occurs due to its centralized nature where intelligent devices are identified, authenticated, and managed centrally in the system. In the current situation, Blockchain technology is the most recent, sustainable, and growing technology in the security area of IoT applications, because of its peer-to-peer and decentralized nature. The distributed behavior of Blockchain makes it compatible for integrating Blockchain with the IoT system. Hence, Blockchain provides the concept of distributed ledger, peer-to-peer network [7], and management and authentication mechanisms in the system which ensure the authenticity, privacy, and integrity as well as the security of the system. Thus, there is a need for Blockchain's integrated IoT application in the future.

There is also the need for a systematic study to recognize the need to overcome these security issues in IoT systems, with rapid evaluation of IoT. In this chapter, our goal is to perform a systematic review with summarized existing problems and solutions related to security in the Blockchain-based Internet of Things (BIoT) and the IoT system itself. We use research questions (RQ) to illustrate these problems. Through the contribution of RQ and literature surveys, we conclude with the actual security problems of the IoT and mechanisms of solutions with Blockchain.

The outline of this chapter is as follows: Section 3.2 describes the research background which discusses the wireless IoT system, Blockchain technology, and Integration of Blockchain with wireless IoT. Section 3.3 discusses related work in security and Blockchain. Section 3.4 explains research methodology in more detail. Section 3.5 compares various solutions. Section 3.6 discusses the research questions. Section 3.7 describes the future scope of this research, and Section 3.8 presents concluding comments.

3.2 Research Background

3.2.1 The Internet of Things

Every day, smart objects or devices increase in the IoT system. The IoT is the Internet working of items like smart vehicles, smart phones, or any smart devices which are embedded with sensors, actuators, software, electronics, and Internet connectivity [8]. Internet connectivity enables the things or objects, collects the data, and performs an operation accordingly. The architecture of the IoT uses different protocols for the purpose of routing the data, and management of the infrastructures, keys, intelligent devices, and security threats. Due to the open nature of the architecture of IoT, it is vulnerable to security threats and various attacks. Security threats are the main concern with this system. Here are some reasons for these concerns:

- It is not a secure environment because of the Internet networking of objects in the IoT. There are many opportunities for the execution of malware and many other threatening activities.
- In its system, things are communicated with each other in the network, which is why there are so many opportunities for encroaching on the privacy and integrity of data.
- The IoT system promotes versions of different technologies similar to Mobile Ad-hoc Network (MANET), Vehicular Ad Hoc Networks (VANET), Wireless Sensor Network (WSN), 2G, 3G, and 4G communication. Hence, it is also vulnerable to all the threats and attacks already suffered by these technologies.

3.2.1.1 Security Requirements in Wireless IoT

Security is the main concern in this system because of its open nature. Different security parameters are discussed below (Figure 3.2):

- **Integrity**: Preventing the unauthorized nodes from modifying the information.
- **Confidentiality**: Confidentiality means protecting the information from unauthorized nodes from leaks.

Figure 3.2 Security requirement in wireless IoT system.

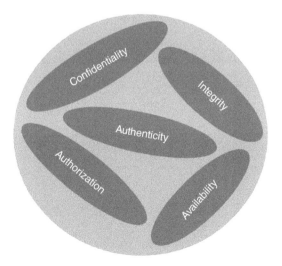

- **Availability**: Availability refers to information that is available for access when it is needed.
- **Authenticity**: Sensitive data or the system should not be accessed by unauthorized users.
- **Authorization**: Authorization refers to the process of giving permission to someone, or something, for doing anything.

3.2.1.2 Layered Architecture of Wireless IoT

In the IoT system, the operations of devices categorize the IoT system into different architectural layers [1, 10]. There are many options for the layered architecture of this system, such as 3 Layer IoT Architecture, 4 Layer IoT Architecture, 5 Layer IoT Architecture, etc. However, the literature review discusses the 4 Layer IoT Architecture, with its layers called the Perception Layer, the Middleware Layer, the Network Layer, and the Application Layer (Figure 3.3).

1. **The Sensor Layer** [9]: In the IoT system, the sensor layer is the topmost layer, because originally data or information are entered into the system through this layer. This layer is also known as the "Perception Layer" of the architecture. Acquisition of data from the physical environment through sensor devices and actuators is the function which is performed here. For the acquisition of data, different devices like RFID [9] reader, sensors, Global Positioning System (GPS), etc. are used in this layer.
2. **The Network Layer** [10]: The Network Layer of IoT plays the function of routing and transferring of the data across the network. The operations of this layer, such as routing, switching, Internet gateway, etc., are performed by using various latest technologies such as 3G, 4G, Zigbee, Wi-Fi, Bluetooth, etc.

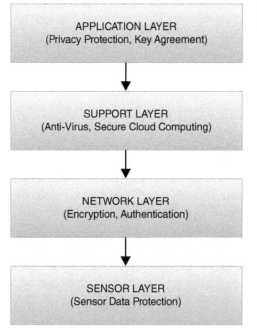

Figure 3.3 Wireless IoT system architecture [9, 11–13].

3. **The Support Layer** [9]: The Support Layer in the IoT system is responsible for processing and manipulating the data in the network. This layer acts as the border between the network and application layers. This layer is also known as the "Middleware layer" in the IoT system.

4. **The Application Layer** [10]: In the IoT system, the Application Layer is the furthermost significant layer. This layer ensures the integrity, authenticity, and confidentiality of the data. The main purpose of the IoT system is achieved by this last layer.

3.2.2 Blockchain Technology

The Blockchain concept acts as a database which is used for storage in the network in a decentralized manner. Blockchain is an arrangement of blocks, which store the complete list of records of transactions in the network [5]. Every block has one parent block and every first block is known as a genesis block in the Blockchain and has no parent block. Blockchain technology has peer-to-peer and distributed ledger concepts. The distributed ledger has three concepts known as block, chain, and transaction. The Block is the storage part which contains the transactions, hash values, and records, etc. The Chain is the linking part of Blockchain, which is used to connect the blocks and create the chain. Any valuable information which is circulated in the network is known as the Transactions. Overall, the Blockchain concept is mainly defined with the four terms which are [5]:

- **Peer-to-Peer Network**: Blockchain technology removes central dependencies of all nodes from any central party within the network.
- **Distributed and Open Ledger**: Each node in the network is validated individually and acts transparently.
- **Synchronization of ledgers copies**: Ledger copies are shared among all nodes in the network. Hence, synchronization of ledgers is required in the network when the new transactions occur, which are then validated and the new transactions are added into the validated ledgers of the transactions.
- **Mining Operation**: Mining operation of Blockchain adds past transactions in the public ledger.

3.2.2.1 Types of Blockchain

Blockchain is categorized on the basis of the access controlling power of the network [14]. There are four types of networks:

1. **Public Blockchain**: For access purposes, in Public Blockchain, there is no restrictions on the network. The Proof of Work (PoW) is the consensus algorithms of public Blockchain. The Proof of Stake (PoS) consensus and Delegated Proof of Stake (PoS) also occur in this Blockchain [15]. Examples of Public Blockchain are Monero, Litecoin, Dash, Bitcoin, Ethereum (Figure 3.4).

2. **Private Blockchain**: In this Blockchain, there is restriction on accessing the network. Nodes cannot join the network unless network administrators invite it. RAFT Consensus Algorithm (RAFT) and the Practical Byzantine fault tolerance (PBFT) consensus algorithms come under the Private Blockchain [16]. Examples of Private Blockchain are Multichain and MONAX.

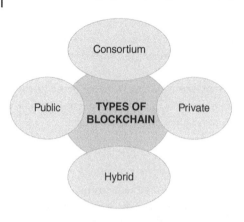

Figure 3.4 Types of blockchain.

3. **Consortium or Federated Blockchain**: In this Blockchain, the semi-decentralized concept is used. It is controlled and restricted by more than one organization. This type of Blockchain mainly uses government organizations like banks, etc. Examples of Consortium Blockchain are Corda, B3i (Insurance), EWF (Energy Web Foundation), and R3 (Bank).
4. **Hybrid Blockchain**: This type of Blockchain is a mixture of both the private and the public Blockchain. This Blockchain uses the features or characteristics of both types of Blockchain.

3.2.2.2 Integration of Blockchain with Wireless Internet of Things

In integrated Blockchain IoT applications, all the transactions with lots of information go through the Blockchain Network for storing immutable and persistent records of the data. The IoT system is vulnerable to security breaches in the system (Figure 3.5). Blockchain technology has the features of decentralizing, immutable, anonymity, etc., which merge with the IoT system and become a secure Blockchain-based IoT system. The different authoritative features of Blockchain technology are [5, 15, 17]:

- **Decentralized**: Blockchain technology make the system decentralized, which means there is no need for third centralized parties to validate any transaction in the system.

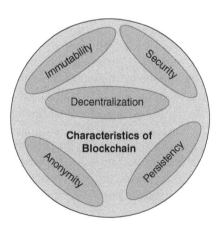

Figure 3.5 Key characteristics of blockchain.

- **Persistency**: Persistency is the quality which is determined to ensure or accomplish something. Due to this quality in Blockchain, the network can validate transactions quickly and identify invalid transaction immediately in the system.
- **Anonymity**: With an address, any user can interact directly with Blockchain. That is why Blockchain cannot assure privacy.
- **Immutability**: Blockchain technology offers the concept of distributed ledger and as this distributed ledger is immutable, it means that the majority of nodes verify data modification in the network. The immutable distributed ledger enhances the privacy and security of systems.
- **Security**: Blockchain uses the cryptographic technologies like public key, private key, and hash concept in the system, which increases the security automatically.

3.3 Related Work

In this section we present a summary of works in the previous literature related to the security problems of the IoT system, Blockchain Technology in IoT, and solutions to the IoT problems using Blockchain.

3.3.1 Security Issues in Wireless IoT System

Mohammed et al. [18] present a survey of the comprehensive approach which includes general DDoS attack motivations and specific reasons why attackers prefer the IoT devices to launch these attacks. They list different types of tools that are available as attacking devices to form a botnet and further tools are discussed which allow using bots to launch DDoS attacks and present a detailed and systematic classification of different types of DDoS attacks that take place on the cloud.

Akhtar et al. [19] present a study of privacy defense problems in the IoT system through a broad review of the state-of-threat, by mutually considering three major dimensions, namely the state-of-the-art principles of privacy laws, the IoT system architecture, and representative privacy enhancing technologies (PETs).

Zhou et al. [20] discuss "IoT Feature" concepts to better recognize the necessary details of new IoT threats and the challenges which focus on recent research. They also discuss security and privacy effects on the IoT system on the basis of different IoT features, including the threats and various existing solutions.

Liang et al. [21] have exposed a denial-of-service attack of an IoT. They discuss different attack tools, such as Qualify of Service (QoS) attacks, like the QoS attack tool such as Kali Linux, which is launched by using many methods and also comparing the different methods.

Frustaci et al. [22] have tried to monitor the IoT security scene by providing taxonomical monitoring from the viewpoint of the main layered architecture of the IoT system model. They have also introduced a new model, the Social Internet of things (SIoT), where the IoT system merges with different social networks, which allows devices and people to facilitate information and to interact.

Benzarti et al. [23] present a categorization of attacks from a variety of networks involved in the IoT system. This categorization discriminates common and specific attacks from each network and uses several criteria such as congestion, security attributes, and disturbance. Also, several existing security solutions are presented to expose the security requirements to protect IoT.

3.3.2   Solutions of Wireless IoT Security Problem

Lao et al. [24] give an overview of Blockchain-IoT architecture and also analyze protocols and their structures. They discuss various consensus protocols for Blockchain-IoT and make comparisons among various consensus algorithms of Blockchain. They also analyse the peer-to-peer traffic model. They have provided a traffic model for Blockchain based on its system for traffic distribution.

Satamraju et al. [25] discuss a model which uses the concept of Blockchain to lead to the security problem and authentication problem in the IoT system. They use various factors in Blockchain and IoT systems which effect the integration of both. They conclude with the hybrid model which is more powerful across distributed platforms.

Dabbagh et al. [26] analyze the bibliometric conference papers of Blockchain and related articles, and review papers that have been indexed in WoS from 2013 to 2018. They analyze these collected papers against five RQs. The results reveal some valuable insights, including yearly publications and citation trends, the hottest research areas, and the top-10 in influential papers, favorite publication venues, and most supportive funding bodies.

Varghese et al. [27] propose a system that forms a decentralized network offering transparency and security. It also guarantees authentication, synchronization, and data integrity. To overcome the limitations and issues in the IoT system, the authors propose an IoT system using Blockchain.

Hang et al. [14] propose a unified platform for IoT with Blockchain, which focuses on the integrity of sensitive data. They use another layer in its architecture known as the IoT-Blockchain Service layer which manages the Blockchain Network. Raspberry Pi devices are used to create the IoT Network, which submits the sensor data to the Blockchain Network. A Smart contract written in solidity language discusses the rules and logic of the Unified IoT-Blockchain platform.

Tang et al. [28] propose a decentralized passport for trust cross-platform using Blockchain technology. Arbitrary trust relations are established among each other using this Blockchain based platform.

Arbabi et al. [29] propose a decentralized based novel system using Blockchain technology. This system uses the smart contract concept in the implementation of the IoT system to overcome the high costs and centralized problem. An Ethereum Blockchain (BC) tool is used in the merged platform of IoT and Blockchain.

Ahram et al. [30] focus on breaking the ground by presenting and demonstrating the use of Blockchain in numerous industrial applications. A healthcare industry IoT application, Health chain, is formalized and developed on the foundation of Blockchain using International Business Machines (IBM) Blockchain inventiveness.

3.4   Research Methodology

In the field of research, there are two approaches for conducting a literature review, which is "Systematic Literature Review" and "Systematic Mapping Studies." In this chapter, the Systematic Literature Review are conducted with the help of different steps. Steps of Systematic Literature Review are shown below (Figure 3.6):

1. **Choose Databases**
   For conducting the literature review, different databases are used to collect the research articles. Different electronic databases used in the review are:
   - IEEE (Institute of Electrical and Electronic Engineers) Xplore Digital Library
   - Springer Link
   - ELSEVIER ScienceDirect
   - ACM (Association for Computing Machinery) Digital Library
2. **Research Questions**
   The main aim of this Systematic Literature Review is to identify the security problems in the IoT system and its solutions. Hence, the RQs are listed in Table 3.1 for this systematic study:
3. **Choose Time Range**
   For this research, the research articles were restricted to papers published from 2014 to 2020.
4. **Exclusion and Inclusion Measures for Selection**
   In the chapter, articles are examined on the basis of metrics, techniques, evaluation criteria, and results. The topics include basically focus on IoT and Blockchain in IoT. The major selection criteria are title and abstract. The inclusion and exclusion criteria are listed in Table 3.2 as follows:
5. **Search Strategies**
   On the basis of Research Question extracted Keywords for this Systematic Literature review (SLR), Table 3.3 represents the research keywords which are helpful for the SLR. The electronic site databases which are used provide sufficient articles for the SLR on the above research keywords like IoT and Blockchain:

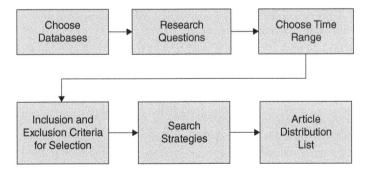

Figure 3.6   Steps of literature review [31, 32].

Table 3.1 Research questions of systematic study.

S. No.	Questions	Motivations
RQ1:	What security-related issues have been raised within the Sensor IoT System?	The aim of this RQ is to find out which different security-related issues and challenges of Sensor IoT systems are being faced.
RQ2:	What kinds of security solutions have been discussed to improve the sensor-based IoT system?	Aims to find the solution of RQ1 for improving the functioning the sensor-based IoT system.
RQ3:	What kinds of Blockchain solutions have been presented to improved security of the sensor-based IoT System?	The aim to find the solution of RQ1 with the help of Blockchain.

Table 3.2 Inclusion and exclusion criteria table.

Inclusion criteria	Exclusion criteria
Research articles related to IoT title	Articles in English language
Peer reviewed papers	Non-peer reviewed papers
Studies with title related to Blockchain	Articles from conference
Security concerns in IoT Environment	Books
Blockchain Technology in IoT for security	Articles on IoT and Blockchain without abstract or title
Blockchain used for attacks in IoT	Article, if available

Table 3.3 Keywords of research.

S. No.	Keywords
1	Internet of Things, Sensors of IoT
2	Blockchain Technology, Blockchain in IoT
3	Security in IoT
4	IoT Layered Architecture

- IEEE Xplore (www.ieeexplore.ieee.org/Xplore)
- Elsevier ScienceDirect (www.sceincedirect.com)
- SpringerLink (www.springerlink.com)
- ACM Digital Library (www.portal.acm.org/dl.cfm)

Paper title	Consensus algorithms/ approaches	Parameters considered/ solution	Blockchain types	IoT devices	Future work
BIFF: A Blockchain-based IoT forensics framework with identity privacy [4]	Modified Merkel Signature Scheme	Identification and privacy of evidence	Public Blockchain	Similar smart devices	Test the reliability with heterogeneous devices
A secured and authenticated IoT model using Blockchain architecture [25]	Binary Merkel Tree Scheme	Authentication problem	Permission less or private Blockchain	Smart devices	Improve Privacy and security
IoT Device management using Blockchain [27]	Proof of Work Concept	Authentication and data integrity	Ethereum Blockchain	Smart Phone and Raspberry Pi	Not mentioned
A Secured and Authenticated Internet of Things Model using Blockchain Architecture [10]	Any prototype of BC	Privacy and authenticity	Public Blockchain	Cameras, GPS Sensors	Not mentioned
Design and implementation of an Integrated IoT Blockchain platform for sensing data integrity [14]	Proof of Concept consensus algorithm	Sensing data integrity	Hyperledger Blockchain	Raspberry Pi device	Used other consensus algorithms for improving transaction rate
BlendCAC: A Blockchain enabled decentralized capability-based access control for IoTs [33]	Proof of Concept Prototype	Scalable and lightweight access control	Local private Blockchain	Raspberry Pi device	Still a long way to complete decentralized security and used in real type application
Blockchain-based solutions to security and privacy issues in the IoT [12]	Not discuss	Scalability, authentication, integrity, privacy	Public Blockchain (Bitcoin, Ethereum)	Smart IoT devices	Used GHOST Protocol and DAG model alternative to Blockchain
Blockchain-based secure time protection scheme in IoT [34]	Practical Byzantine Fault Tolerance (PBFT) consensus mechanism	Time synchronization of IoT	Public Blockchain	RFID, smart home appliances	Improve accuracy of time, reduce to offset
IoT-based smart security and home automation system [35]	Used standard IoT model	Motion detection	Not used	TI CC3200 launch pad	Lack in synchronization of alarm, so make more synchronized
A Denial-of-Service attack method for an IoT system [37]	Not used	Denial of Service attack	Not used	Arduino, Kali Linux	More attacks Study
Blockchain meets IoT: An architecture for scalable access management in IoT [38]	PoC prototype	Generic, scalable, and access control management system	Private Blockchain	Smart sensor device	Not mentioned
IoT application development: home security system [36]	Based on fundamental IoT architecture	Lower power system, detection, and authentication	Not used	Raspberry Pi device	Similar concept used for face detection using ML and AI

3.5 Comparison of Various Existing Solutions

Many earlier projected solutions are deliberated for the IoT applications with the help of different latest technologies. For example, Zhang et al. (2019) proposed a trust-based mechanism in smart manufacturing, which uses trust tax as the necessary information on transactions between any physical system, e.g., human, because of the lack of confidence and reliability [10]. Hang et al. (2019) worked on the integrity of the sensing data of its systems and for securing the integrity of this data they proposed the Blockchain-based platform of the IoT system [14]. For implementation using Raspberry Pi devices and Hyperledger fabrics, Xu et al. (2018) developed a BlendCAC scheme in which devices are self-controlled resources and there is no centralized monitoring of the devices [33]. This scheme is based on the private Blockchain. Yu et al. (2018) investigated the security concerns of Blockchain-based systems and developed a framework for dealing with privacy problems within the IoT system [12]. This chapter also discussed some solutions for the security of IoT using Blockchain and Ethereum tools. Fan and Wang (2019) assured the security related to the IoT applications during the time synchronization of the system [34]. The Blockchain scheme was used for the time verification and convenience synchronization of the distributed ledger in the network. Kishore Kodali (2016) proposed a Home automation system which sends an alarm to the owner when there is threat of any intruder activity [35]. This project used the micro controller TI-CC3200 and sends voice calls to alert the owner. Anvekar (2017 developed a model of the Home Security System, which focuses on detection of intruders, identification, and authentication of unknowns [36]. Table 3.4 lists the consensus protocols and security parameters, etc.

Although all research attempting to solve the issues is not new, there is still a deficiency in the critical and practical analysis of security issues of IoT applications using Blockchain technology. Most of the articles discuss the private or public Blockchain and few security issues, but there is a lack of a consensus mechanism analysis for implementing it.

3.6 Discussion of Research Questions

RQ1: What security-related issues have occurred within the sensor-based IoT System?
Overall, the systematic literature process identified the major security issues of IoT-based applications. The main issue of IoT systems is data integrity, privacy, authentication, different types of attack, and centralized monitoring of IoT systems.

RQ2: What kinds of security solutions have been discussed that will improve the wireless IoT System?
Many researchers have identified the security issues and provided solutions to this issue. They have used many technologies (DAG, IoTA, etc.), various approaches, and cryptographic algorithms (RSA, SHA, etc.) to find solutions for the security of IoT system.

RQ3: What kinds of Blockchain solutions have been presented to improved security of wireless IoT Systems?
Most articles have discussed the problems of the IoT system and discussed solutions to the related problems. Some researchers are found solutions to these security problems with

the help of Blockchain technology. These include some solutions to privacy problems and integrity problems in general.

Overall, most research has focused on only one security issue, and there is a lack of a practical approach. Many researchers have considered any one of the security issues including privacy, authentication, data integrity, etc. Security solutions of the IoT system have been projected without the use of the Blockchain technologies. There are also many researchers who propose solutions of IoT applications with the support of Blockchain. Security problems are discussed overall in detail and the researchers are approaching solutions by using Blockchain, but solutions to the problems have not been discussed in any detail.

3.7 Future Scope of Blockchain in IoT

IoT offers inclusive and comprehensive decisions to people's life styles by exchanging and converting IoT devices data. Blockchain has a decentralized network architecture which depends on researchers knowing how to use IoT. IoT security problems can be solved with autonomous technology such as the directed acyclic graph model. It is exciting to explore directed acyclic graph-based solutions for access control, integrity, privacy issues, and synchronization of IoT devices.

Another possible technology similar to Blockchain is IoTA (The Internet of Things Application), which theoretically uses a more scalable protocol Tangle than Blockchain. Tangle protocol of IoTA is fast and connects a greater number of people to the network.

Future work will provide a model for synchronized IoT applications with the help of Blockchain, with synchronized IoT devices that can deal with the different security parameters.

3.8 Conclusion

The goal of conducting SLR of the wireless IoT system is to provide an analysis of automated and smart IoT applications. Subsequently, the systematic literature review discusses how researchers have been attentive to the security concerns of the wireless IoT Network and move toward the latest technologies for solutions to the security concerns arising from it. Firstly, the main focus was on research solutions for the security which basically relates to privacy and integrity or attacks in the network. Secondly, the focus was on the Blockchain solutions of the researchers for the security problems of the wireless IoT system. Blockchain-based platforms were used in its implementation, which were dealing with any security parameters. This article elaborated on the survey of Blockchain technology for security in the wireless IoT and synchronized devices with it. We also explained different technologies liked DAG and IoTA architectures, as alternatives to the Blockchain decentralized architecture in IoT. Surely Blockchain acts as the liberator of security of the IoT and provides a model for synchronized wireless IoT application with the help of Blockchain, with synchronized IoT devices for solving the different potential security issues such as access control, authentication, and synchronization of its devices.

References

1 Zeba, S., Khan, D., and Ahmad, H. (2019). Survey on attacks in MANET based Internet of Things system. *IJSRD – International Journal for Scientific Research & Development*| 7 (7): |ISSN (online) 2321–0613. 64–69.

2 Guo, C. (2019). A brief analysis of privacy protection strategy for Blockchain-based Internet of Things system. *IJSRD – International Journal for Scientific Research & Development*| 7 (7): |ISSN (online) 2321–0613. 94–97.

3 Zeba, S. and Ahmad, H. (2019). Survey on attacks in MANET based Internet of Things system. *IJSRD – International Journal for Scientific Research & Development* 7 (6): |ISSN (online) 2321–0613. 96–102.

4 Le, D.P., Meng, H., Su, L. et al. (2019). BIFF: A Blockchain-based IoT forensics framework with identity privacy. IEEE Reg 10 Annu Int Conf Proceedings/TENCON. 2018-October: 2372–2377.

5 Zheng, Z., Xie, S., Dai, H. et al. (2017). An overview of Blockchain technology: architecture, consensus, and future trends. Proceedings of the 2017 IEEE 6th International Congress Big Data, BigData Congress 2017 June: 557–564.

6 Ren, Q., Man, K.L., Li, M. et al. (2019). Using Blockchain to enhance and optimize iot-based intelligent traffic system. Proceedings of the 2019 International Conference Platform Technology Services PlatCon 2019 2019: 1–4.

7 Lo, S., Liu, Y., Chia, S. et al. (2019). Analysis of Blockchain solutions for IoT : a systematic literature review. *IEEE Access* 7: 58822–58835.

8 Ahmad, N.M., Fatimah, S., Razak, A. et al. (2018). Improving identity management of cloud-based IotTapplications using Blockchain. 2018 International Conference Intelligent Advanced Systems. (i): 1–6.

9 Zeba, S., Khan, D., and Ahmad, H. (2019). Security threats and technologies in Internet of Things system. *IJSRD – International Journal for Scientific Research & Development.* 7 (7): |ISSN (online) 2321–0613. 64–69.

10 Zhang, Y., Xu, X., Liu, A. et al. (2019). Blockchain-based trust mechanism for IoT-based smart manufacturing system. *IEEE Transactions on Computational Social Systems* 6 (6): 1386–1394.

11 Fazal, K., Shehzad, H., Tasneem, A. et al. (2017). A systematic literature review on the security challenges of internet of things and their classification. *International Journal of Engineering Research and Technology* 5 (2): 40–48.

12 Yu, Y., Li ,Y., Tian, J. et al. (2018). Blockchain-based solutions to security and privacy issues in the internet of things. *IEEE Wireless Communications* 25 (6): 12–18.

13 Santhosh Krishna, B.V. and Gnanasekaran, T. (2017). A systematic study of security issues in Internet-of-Things (IoT). Proceedings of the International Conference IoT Soc Mobile, Anal Cloud, I-SMAC 2017 2017: 107–111.

14 Hang, L. and Kim, D.H. (2019). Design and implementation of an integrated IoT blockchain platform for sensing data integrity. *Sensors (Switzerland)* 19 (10): 1–26.

15 Park, J-H., Salim, M.M., Jo, J.H. et al. (2019). CIoT-net: a scalable cognitive IoT based smart city network architecture. *Human-Centric Computing and Information Sciences* 9; 1–20.

16 Yang, Y., Wu, L., Yin, G. et al. (2017). A survey on security and privacy issues in internet-of-things. *IEEE Internet ofThings Journal* 4662(c): 1–10. doi 10.1109/JIOT.2017.2694844

17 Zhou, L., Wang, L., Sun, Y. et al. (2018). BeeKeeper: a Blockchain-based IoT system with secure storage and homomorphic computation. *IEEE Access* 6 (8): 43472–43488.

18 Mohammed, M., Shailendra, S., Jong, R. et al. (2019). Distributed denial of service attacks and its defenses in IoT: a survey [internet]. *The Journal of Supercomputing*. Springer US. Available from: https://doi.org/10.1007/s11227-019-02945-z

19 Akhtar, M.M. and Rizvi, D.R. (2020). EAI endorsed transactions IoT-chain: Security of things for pervasive, sustainable and efficient computing using Blockchain. *Journal of the EAI Endorsed Transactions on Energy Web* 7 (30); 1–8. doi:10.4108/eai.13-7-2018.164628 09 2020 - 11 2020

20 Zhou, W., Zhang, Y., and Liu. P. (2019). The effect of IoT new features on security and privacy: New threats, existing solutions, and challenges yet to be solved. *Journal of the EEE Internet of Things* 6 (2): 1–11. doi: 10.1109/JIOT.2018.2847733

21 Liang. L., Zheng. K., Sheng, Q. et al. (2017). A denial of service attack method for IoT system in photovoltaic energy system. Lecture Notes Computer Science (including Subservice Lecture Notes Artificial Intelligence Lecture Notes Bioinformatics). 10394 LNCS: 613–622.

22 Frustaci, M., Pace, P., and Aloi, G. (2018). Evaluating critical security issues of the IoT world: present and future challenges. *IEEE Internet of Things Journal* 5 (4): 2483–2495.

23 Benzarti, S., Triki, B., and Korbaa, O. (2017). A survey on attacks in Internet of Things based networks. Proceedings of the 2017 International Conference on Engineering & MIS (ICEMIS)

24 Lao, L., Li, Z., Hou, S. et al. (2020). A survey of IoT applications in Blockchain systems: architecture, consensus, and traffic modeling. *ACM Computing Survey* 53 (1): 1–32.

25 Satamraju, K.P. and Malarkodi, B. (2019). A secured and authenticated Internet of Things model using blockchain architecture. Proceedings of the 2019 TEQIP – III Sponsered International Conference on Microwave Integrated Circuits, Photonics Wire Networks, IMICPW 2019. 2019: 19–23.

26 Dabbagh, M. and Sookhak, M. (2019). The evolution of belockchain: a bibliometric study. *IEEE Access*. 7: 19212–19221. doi: 10.1109/ACCESS.2019.2895646

27 Varghese, C. and Jose. J. (2019). IoT device management using Blockchain. *International Journal of Science Engineering Technology Research* 8 (3): 79–84.

28 Tang, B., Kang, H., Fan, J. et al. (2019). IoT passport: a Blockchain-based trust framework for collaborative Internet-of-Things. Proceedings of the ACM Symposium Access Control Model Technology – SACMAT, 2019: 83–92.

29 Arbabi, M.S. and Shajari, M. (2020). Decentralized and secure delivery network of Iot updated files based on Ethereum smart contracts and Blockchain technology. Proceedings of the 29th Annual International Conference Computer Science Software Engineering. CASCON 2019. 2020: 110–119.

30 Ahram, T., Sargolzaei, A., Sargolzaei, S. et al. (2016). Blockchain technology innovations. Proceedings of the 2017 IEEE Technology Engineering Managagment Society Conference. TEMSCON 2017. 137–141.

31 Aleisa, N. and Renaud, K. (2017). Privacy of the Internet of Things: a systematic literature review. Proceedings of the 50th Hawaii International Conference Systems Science. 2017: 5947–5956.

32 Witti, M. and Konstantas D. (2018). IoT and security-privacy concerns: a systematic mapping study. *International Journal of Network Security and Its Applications* 10 (6): 25–33.

33 Xu, R., Chen, Y., Blasch, E. et al. (2018). Blendcac: A blockchain-enabled decentralized capability-based access control for IoT. Proceedings of the IEEE 2018 International Congress Cybermatics. 2018 IEEE. 2018: 1027–1034.

34 Fan, K., Wang, S., Ren, Y. et al. (2019). Blockchain-based secure time protection scheme in IoT. *Journal of the IEEE Internet of Things* 6 (3): 4671–4679.

35 Kodali, R.K., Jain, V., Bose, S. et al. (2017). IoT based smart security and home automation system. Proceedings of the 2016 IEEE International Conference on Computer Communications Automation. ICCCA 2016. October: 1286–1289.

36 Anvekar, R.G. and Banakar, R.M. (2017). IoT application development: home security system. Proceedings of the 2017 IEEE Technological Innovation ICT Agricultural Rural Development TIAR 2017. January: 68–72.

37 Liang, L., Zheng, K., Sheng, Q. et al. (2016) A denial of service attack method for IoT System in photovoltaic energy system. Proceedings of the 8th International Conference on Information Technology in Medicine and Education 2: 613–622.

38 Novo, O. (2018). Blockchain meets IoT: an architecture for scalable access management in IoT. *Journal of the IEEE Internet of Things* 5(2): 1184–1195.

4

Cognitive IoT-Based Health Monitoring Scheme Using Non-Orthogonal Multiple Access

Ashiqur Rahman Rahul[1], Saifur Rahman Sabuj[1,2], Majumder Fazle Haider[1], and Shakil Ahmed[3]

[1]*Department of Electrical and Electronic Engineering, BRAC University, Bangladesh*
[2]*Department of Electronics and Control Engineering, Hanbat National University, Korea*
[3]*Department of Electrical and Computer Engineering, Iowa State University, Ames, Iowa, USA*

4.1 Introduction

The Internet of Things (IoT), which can be implemented in every sphere of life, is the most promising technology of fifth-generation (5G) and beyond 5G wireless communications. Wearable sensors, when deployed for healthcare services to sense physical data of the human body and transmit the measured data to the nearest gateway wirelessly facilitating IoT, form a network called the wireless body area network (WBAN). The concept of wireless technology integrated Internet-based healthcare services was first coined in 2000 by Laxminareayan and Istepanian [1]. The rapid advancement in the field of wireless technologies and wearable electronics in recent times transform WBAN into a hot topic in the fields of academia and industry, to carry out top-notch research to bring it to its full potential [2]. The main purpose of WBAN in healthcare services is the early detection of abnormalities through regular measuring of physiological data; thus, alerting the person to take necessary precautions and assisting the medical professionals to make sound decisions regarding the appropriate treatment to cure the disease as quickly as possible [3–5].

Due to the huge prospects of WBAN bringing revolutionary changes in telemedicine systems with real deployment of IoT, some different wireless standards have been proposed. IEEE wireless standard 802.15, defined as wireless personal area network (WPAN), is considered as a suitable technology to fulfill the purpose of e-health systems [4]. The IEEE 802.15 standard has several versions, such as 802.15.1 and 802.15.4. These are commonly known as Bluetooth and ZigBee respectively, which are widely used as low-power consumption, short ranged, and low data transfer enabled wireless technology for infotainment and healthcare services [6]. Later, IEEE originated 802.15.6, named as WBAN, particularly developed for telemedicine systems with improvement in data reliability, sensor life time, latency, and lower interference [7].

In the near future, a large number of sensors will be deployed on the body of the patient through WBANs, to manage healthcare services remotely and provide intuitive decisions of medical professionals more dynamically. Therefore, handling large amounts

Smart and Sustainable Approaches for Optimizing Performance of Wireless Networks: Real-time Applications, First Edition.
Edited by Sherin Zafar, Mohd Abdul Ahad, Syed Imran Ali, Deepa Mehta, and M. Afshar Alam.

of situation-awareness data in healthcare services for the future heterogeneous networks requires intelligent management of communication systems to ensure reliability, real-time feedback, and expected high-quality services by utilizing minimum resources as much as possible. To meet this challenge, cognitive IoT-based wireless systems in WBANs for healthcare services can play a crucial role [8–10]. When a communication system, where the nearby transceivers cooperatively participate to ease the communication toward the destination and interact intelligently with humans to convey the instructions properly, is known as a cognitive communication system. Some research workers investigated the potential of an IoT-cloud integrated communication framework for healthcare monitoring systems [11–13].

In this chapter, we focus on designing cognitive IoT-based wireless uplink communication systems by incorporating a non-orthogonal multiple access (NOMA) technique in WBANs for monitoring health conditions of categorized patients. The proposed system is intelligent enough to allocate high performance communication channels for exchanging information of the body nodes of intensive care unit (ICU) patients, with high priority communication (HRC) and moderate performance communication (MRC) channels for regular patients based on computation of the signal-to-interference-plus-noise (SINR) value. Theoretical analysis and simulation results of the performance metrics of such uplink communication validate the proposed optimized system to provide reliable data exchange applicable for healthcare services.

The remaining content of this chapter is structured accordingly: Section 4.2 deals with the related research work in this sector. Next, Section 4.3 describes the system model, which consists of network description, sensing, and transmission process, pathloss model, evaluation of the mathematical model, and optimization technique. Then, Section 4.4 illustrates the simulation outcomes through graphical representation, and finally, we conclude with Section 4.5 by discussing the best energy efficiency case with some suggestions for future scope of cognitive IoT.

4.2 Related Work

Several research works have been presented to highlight the prospects and challenges of WBAN and proposed optimal solutions to alleviate certain challenges. Inter-network interference is a common challenge that hinders the performance of WBANs when large numbers of wearable sensors act in a small dense area, where a power control approach with low complexity based on game theory has been proposed in [14] to mitigate inter-network interference in WBANs. The authors focused on optimizing transmission power while keeping the system throughput high in mitigating inter-network interference in WBANs. The authors in [15] introduce the implications and advantages of the human body communication physical layer (PHY) of the IEEE 802.15.6 standard over the other two PHY layer protocols, due to its high conductivity in the human body on the performance of WBAN for healthcare services.

Few researchers have made an attempt to implement the concept of energy harvesting in WBANs for telemedicine systems to notably prolong the lifetime of sensor nodes. In [16, 17], the authors analyze power management strategy in energy harvested WBANs.

A self-adaptive sensor has been proposed on a time division multiple access (TDMA) frame structure to enable lifetime operation of the wearable sensors. The author in [18] present a bandwidth allocation method for multiple WBANs along with beacon shifting and a super frame interleaving integrated scheduling algorithm, which can ensure efficient bandwidth utilization in WBANs.

A good number of research works, focusing on designing medium access control (MAC) layer protocols, have been presented. In [19–21], the authors highlight the significance of designing energy and delay-aware MAC protocols to extend the durability of sensor nodes in WBAN. The authors in [22] develop a SeDrip protocol incorporating Secure Hash Algorithm-1 based hash functions and Advanced Encryption Standard-based encryption techniques for highly secured and energy-efficient data dissemination in WBANs. In [23, 24], the authors demonstrate how the placement strategy of Body Node Coordinator (BNC) or hub in a WBAN can enhance the performance and lifetime of the body nodes notably. Moreover, the authors in [23] show that a suitable routing protocol plays a crucial role in prolonging the lifetime of body nodes and improves energy efficiency of a WBAN. The authors in [25, 26] introduce the impact of transmission delay of IEEE 802.15.6 carrier-sense multiple access with collision avoidance (CSMA/CA) on duty-cycles of WBANs and computed the overall delay in WBANs using the theory of probability. A tele-medicine MAC layer protocol under the consideration of IEEE 802.15.4 standard CSMA/CA enabled beacon mode has been proposed in [27], and the detailed analysis shows enhanced performance of the protocol in terms of reduced delays, reliability, and energy consumption. A hybrid MAC layer protocol has been proposed in [28], considering the advantages of both CSMA/CA and TDMA schemes simultaneously, in order to enhance energy efficiency and also prolong the lifetime of body nodes in WBANs. In [29–32], the authors emphasize different aspects of MAC layer protocols, such as optimization of duty cycle and different coding techniques to analyze the energy and delay tradeoff in WBANs. Considering adaptive multi-dimensional traffic load and class, an MAC layer protocol has been proposed in [33], which can provide better performance in terms of energy efficiency and avoidance of delays.

The authors in [34] discuss the potential security threats of WBANs through practical assessment and suggest implementing a forensic server and the use of hidden drones in the wireless network architecture to detect security threats. In [35], the authors demonstrate muscle strain sensor data measurement in a WBAN through prototype hardware development. The authors in [36] illustrate the deployment of magnetic induction-based wireless communication systems for WBAN instead of typical electromagnetic wave communication technologies and show that impedance matching can significantly improve the efficiency of magnetic induction-based WBAN.

The impacts of channel characterization on the performance of WBANs have been highlighted in a few research works. In [37, 38], the authors address the necessity of the Multi-Channel Broadcast (MCB) protocol to broadcast control signals from BNC to all the body nodes in WBANs. Additionally, by utilizing channel hopping sequences, the authors design a MCB protocol, which can ensure minimum broadcast delay in asymmetric duty cycling and is capable of broadcasting signal over multi-channels supported by the IEEE 802.15.6 standard. The authors in [39] introduce a channel allocation algorithm for typical WBANs using machine learning techniques. The proposed algorithm can allocate channels dynamically by considering traffic load and also provide optimum performance in terms of throughput. In [40, 41], the authors analyze the performance of WBANs by focusing on wireless channel modeling approaches. The authors in [40] highlight the comparison of different types of relay systems performance on channel modeling and computed performance metrics to provide good insights. The authors in [41] investigate the performance of WBANs, concentrating on the comparison of different channel modeling approaches for implanted and wearable sensors and calculate the bit error performance based on the movement of the human body.

On the other hand, the influence of gender and body shape on the performance of the body nodes in WBANs have been scrutinized in [42], the results show a higher pathloss and fading effect on males compared to females.

With respect to the above-mentioned cognitive approaches by various authors [8–10], in this chapter, we focus on the sensing scheme with CR to utilize the unoccupied spectrum. Later, with contrast to [16, 22], we derive the optimal power, which will be later utilized to maximize the energy efficiency and throughput for both IoT-based HRC and MRC devices for an uplink communication.

4.3 System Model and Implementation

4.3.1 Network Description

We have designed an uplink scenario for two types of devices concerning the health status of a patient with a cognitive IoT-based communication network, shown in Figure 4.1. Here, the distinguishing factor for the devices depends on the response of the medical staff toward the patient status. For instance, the HRC (high reliable communication) provides the medium or low latency with high-powered precise communication, for patients needing immediate care. Conversely, MRC (moderate reliable communication) devices have an average latency with low power, which deals with the conventional (non-urgent) responses toward regular patients. Moreover, in Figure 4.1, the HRC secondary transmitter (ST_H) and the MRC secondary transmitter (ST_M) are in communication with the secondary receiver (SR_x), which works as a BS in an uplink development. Additionally, PT_x is the primary transmitter and it does not fall under the regulations of HRC or MRC devices. Thus, in Figure 4.1, PT_x is shown as an example of signal interference. Furthermore, in this figure, P_H, P_M, and P_P are the powers necessary for transmission from ST_H, ST_M, and PT_x, respectively.

The HRC and MRC devices are utilizing NOMA methods within the same subcarrier (S_C) in an uplink communication [43]. The HRC and MRC devices use the same S_C for communication, distinguished by their power levels. Thus, the SR_x will perform successive interference cancellation (SIC) to decode the particular signal apart from other signals in the same S_C.

4.3.2 Sensing and Transmission Analysis

As sensing is an integral part of any cognitive device, our cognitive IoT system will also perform sensing operations in the radio spectrum to detect under-utilized channels. Both

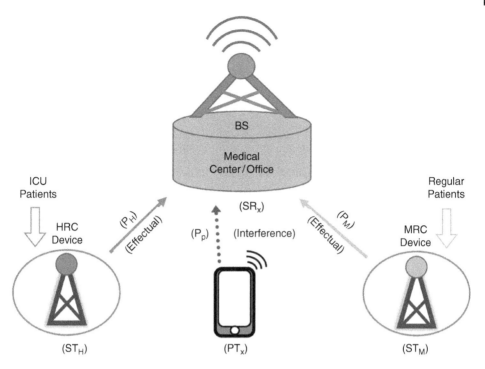

Figure 4.1 Communication linkage of a cognitive IoT-based health monitoring structure.

the secondary transmitters (i.e., ST_H and ST_M) perform the sensing process to identify any used or unused channels. This is established by comparing any received signal-to-noise ratio (SNR) from any device to a threshold SNR. In our model, if the received SNR (μ_P) of PT_x is lower than the threshold SNR ($\mu_{H/M}$) of any HRC or MRC devices, then it is considered as an effectual state. However, if the received SNR is higher, then PT_x is considered to be in a transmitting state, which is denoted as the interference state [44]:

$$z = \begin{cases} 0, when\ \mu_P < \mu_{H/M} \\ 1, when\ \mu_P \geq \mu_{H/M} \end{cases} \tag{4.1}$$

Here, z is an indicating factor for the active and inactive states of PT_x, where $z = 0$ indicates there is no interference ($\mu_P < \mu_{H/M}$) from the PT_x meaning effectual state, while HRC and MRC devices are transmitting and $z = 1$ indicates interference state ($\mu_P \geq \mu_{H/M}$) when PT_x is active in transmission.

4.3.3 Pathloss Model

Pathloss characterizes our system model as the attenuation of any signal strength, propagating through space between the transmitters (i.e., ST_H, ST_M, PT_x) and the receiver (i.e., SR_x). In this chapter, we consider the channel gain for the interference state between the PT_x and SR_x as g_p. Likewise, for effectual states of HRC and MRC devices, channel gain between ST_H and SR_x is g_h, and between ST_M and SR_x channel gain is g_m. In terms of our model, we derived the channel gain characteristics for all communicative devices from the pathloss model established in [45, 46]. As proposed, our model considers the mmWave band for its uplink communication. Thus, it is possible to gain over 10–2000 m of distance from the secondary transmitters (i.e., ST_H, ST_M) to the BS; over 2–6 GHz of frequency, we can express the line-of-sight (LOS) pathloss (Pl_{T2R}^{LOS}), and non-line-of-sight (NLOS) pathloss (Pl_{T2R}^{NLOS}) as follows:

$$Pl_{T2R}^{LOS}(d_{T2R})[dB] = 22 \log_{10}(d_{T2R}) + 28 + 20 \log_{10}(f_{c(mm)}) \tag{4.2}$$

$$Pl_{T2R}^{NLOS}(d_{T2R})[dB] = 36.7 \log_{10}(d_{T2R}) + 22.7 + 26 \log_{10}(f_{c(mm)}), \tag{4.3}$$

where T and R denote all the transmitters and receivers in Figure 4.1, and where d_{T2R} is their distance element. Additionally, $f_{c(mm)}$ exemplifies as the carrier frequency (2–6 GHz) over the mmWave band. Now, by utilizing the above expressions, we can develop the average pathloss over d_{T2R} as follows:

$$Pl\,(d_{T2R}) = \omega * Pl_{T2R}^{LOS} + (1 - \omega) * Pl_{T2R}^{NLOS}, \tag{4.4}$$

where ω is the coefficient probability of the LOS link, over d_{T2R} between all transmitters (ST_H, ST_M, PT_x) and BS.

4.3.4 Mathematical Model Evaluation

This section will examine the performance metrics of HRC and MRC devices, such as their corresponding throughputs and energy efficiency. This will ultimately lead to the optimum solution for our system model through optimization techniques. Moreover, in our proposed model, we have assumed that the HRC and MRC devices observe the channel activity to sense the presence of PT_x. Since the presence of PT_x results in two different types of throughputs at the receiver station (SR_x), we establish the effectual and interference throughput (for both the HRC and MRC device) with their corresponding energy efficiency aspect.

4.3.4.1 Effectual Throughput
As mentioned in Section 4.3.3, if the presence of PT_x is undiscovered, meaning only ST_H and ST_M is solely transmitting, then we denote it as an effectual state having effectual throughputs. Based on such states, we derive effectual throughputs for both HRC and MRC devices, by using $z = 0$ as an indicator for the absence of PT_x. Moreover, the effectual throughput will also imply a seamless spectrum sensing, since there is no interference from the PT_x [47]. Hereafter, for $z = 0$, we can derive the effectual throughput (S_H^0) for HRC device as follows:

$$S_H^0 = \left(\frac{t_t}{t_t + t_{se}}\right) P_x(z = 0)\,(1 - p_F) b \sum_{n=1}^{N_H} \log_2\left(1 + \frac{P_{H,n}\,|g_{h,n}|^2}{n_p\,b}\right), \tag{4.5}$$

where, t_t and t_{se} are transmission time and sensing time, respectively, P_H is the power transmitted from ST_H to SR_x for any n number of HRC devices, n_p is the noise power spectral density, b is the bandwidth, and g_h is the channel gain between ST_H to SR_x for any n number of HRC devices. Moreover, in the expression, p_F is a probability output that exceeds a certain threshold when there is no signal (noise only), which represents an alarm of false detection. Next, $p_x(z = 0)$ is the probability of the inactive state of PT_x. Henceforth, we represent the term, $p_x(z = 0)(1 - p_F)$ as perfect detection probability, which implies that there is no interference from PT_x between the channels ST_H to SR_x.

Equally, for $z = 0$, we can derive the effectual throughput (S_M^0) for any n number of MRC device as follows:

$$S_M^0 = \left(\frac{t_t}{t_t + t_{se}}\right) p_x(z = 0)(1 - p_F) b \sum_{n=1}^{N_M} \log_2 \left(1 + \frac{P_{M,n} |g_{m,n}|^2}{n_p b + P_{H,n} |g_{h,n}|^2}\right), \quad (4.6)$$

where P_M is the transmission power from ST_M to SR_x, and g_m is the channel gain between them. Uniquely, for our uplink system model, higher power signal ($P_H |g_h|^2$) in Equation (4.6) of any HRC device from ST_H to SR_x, will pose diverse levels of interference to the lower power signals of any MRC device (ST_M to SR_x).

4.3.4.2 Interference Throughput

In a similar manner, we can derive throughputs for interference state. However, on this occasion the HRC and MRC device senses the presence of PT_x, implying that PT_x is also transmitting along with ST_H and ST_M, causing interference to the system. Hereafter, we can derive interference throughput (S_H^1) for the HRC device, using $z = 1$ as an indicator for the presence of PT_x as follows:

$$S_H^1 = \left(\frac{t_t}{t_t + t_{se}}\right) p_x(z = 1)(1 - p_D) b \sum_{n=1}^{N_H} \log_2 \left(1 + \frac{P_{H,n} |g_{h,n}|^2}{n_p b + P_P |g_P|^2}\right), \quad (4.7)$$

where S_H^1 is associated with imperfect spectrum sensing, since PT_x is active on transmission. However, when compared to non-HRC or non-MRC devices, S_H^1 still provides the optimal throughput in the presence of interference.

Moreover, in Equation (4.7), P_P is the transmission power for PT_x, and g_P is the channel gain between PT_x and SR_x. Now, p_D symbolizes the probability of sensing the presence of a targeted signal, meaning the probability output is less than a certain threshold, and $p_x(z = 1)$ represents the probability of PT_x being active on transmission. Thus, $p_x(z = 1)(1 - p_D)$ is the imperfect detection probability for any HRC or MRC device, facing interference from PT_x.

Similarly, for $z = 1$, we can derive the interference throughput (S_M^1) for any n number of MRC device as follows:

$$S_M^1 = \left(\frac{t_t}{t_t + t_{se}}\right) p_x(z = 1)(1 - p_D) b \sum_{n=1}^{N_M} \log_2 \left(1 + \frac{P_{M,n} |g_{m,n}|^2}{n_p b + P_{H,n} |g_{h,n}|^2 + P_P |g_P|^2}\right)$$

$$(4.8)$$

As for uplink communication, MRC devices will encounter an additional interference from the high-powered HRC device (ST_H) and the active primary transmitter (PT_x). As per the decision made by the IEEE 802.22 committee, when SNR ≤ -20 dB, $p_D \geq 0.9$ and $p_F \leq 0.1$ [48].

4.3.4.3 Energy Efficiency

In this chapter, we define the energy efficiency as the ratio of throughputs (i.e., effectual and interference) to the total power usage by the devices (i.e., HRC, and MRC). Hence, we can develop energy efficiency expression (EE_H^z) for an HRC device, in both effectual $(z = 0)$, and interference states $(z = 1)$ as follows:

$$EE_H^0 = \frac{S_H^0}{P_H + P_{cp} + P_{sp}} \tag{4.9}$$

$$EE_H^1 = \frac{S_H^1}{P_H + P_{cp} + P_{sp}}, \tag{4.10}$$

where P_H is the power required for transmission from ST_H to SR_x, P_{cp} is the power consumed by the HRC device circuit, and P_{sp} is the power to sense the spectrum by the cognitive IoT-based HRC device.

Similarly, we can establish energy efficiency expression (EE_M^z) for the MRC device, in both effectual $(z = 0)$, and interference states $(z = 1)$ as follows:

$$EE_M^0 = \frac{S_M^0}{P_M + P_{cp} + P_{sp}} \tag{4.11}$$

$$EE_M^1 = \frac{S_M^1}{P_M + P_{cp} + P_{sp}}, \tag{4.12}$$

where P_M is the transmission power for the MRC device from ST_M to SR_x, and P_{cp} and P_{sp} is the circuit, spectrum sensing power for the MRC device.

4.3.4.4 Optimum Power

In this subsection, we derive the expression for optimum power transmission in both effectual and interference state. Now, from the energy efficiency equation for HRC and MRC device in Equations (4.9) and (4.11) at effectual state $(z = 0)$, we construct the following generalized single-objective optimization problems (SOP) as follows:

$$\max_{P_H} \; EE_H^0(P_H) \tag{4.13}$$

$$\max_{P_M} \; EE_M^0(P_M) \tag{4.14}$$

Likewise, for HRC and MRC devices in interference state $(z = 1)$, we can write Equations (4.10) and (4.12) as:

$$\max_{P_H} \; EE_H^1(P_H) \tag{4.15}$$

$$\max_{P_M} \; EE_M^1(P_M), \tag{4.16}$$

where in Equations (4.15) and (4.16), as interference from PT_x is present, the energy efficiency value that we determine is the optimum energy efficiency.

4.3.4.4.1 Optimum Power Derivation for HRC

Rewriting Equation (4.9), we obtain the optimum transmission power $(P_{H,n}^E)$ for the effectual state as follows:

$$EE_H^0 = \frac{\left(\frac{t_i}{t_i + t_{se}}\right) P_x(z = 0)\,(1 - p_F)b \sum_{n=1}^{N_H} \log_2\left(1 + \frac{P_{H,n}^E \, |g_{h,n}|^2}{n_p\, b}\right)}{P_H^E + P_{cp} + P_{sp}} \tag{4.17}$$

Thus, $P^E_{H.n}$ for any n number of HRC devices is given by

$$P^E_{H.n} = \frac{n_p b \left(\frac{P_{cp} g^2_{h.n} + P_{sp} g^2_{h.n} - n_p b}{W\left(0, \left(\frac{P_{cp} g^2_{h.n} + P_{sp} g^2_{h.n} - n_p b}{n_p b}\right) \exp(-1)\right)} - 1 \right)}{g^2_{h.n}},$$ (4.18)

Proof: Refer to Appendix 4.A.1

where $W(\cdot)$ is a Lambert function, which gives the solution to expression like $ae^a = b$. Here "b" acts as a function or a variable and "a" itself can be a constant or a complex variable, where a is acting both as an exponent and base. Moreover, we can solve such equations using W to substitute any complex variable (a) by real variable (b). Henceforth, we can express the solution for such cases as: $W(b) = a$ [49].

Now, from Equation (4.10), we rewrite EE^1_H for deriving optimum transmission power in interference state (P^I_H) as follows:

$$EE^1_H = \frac{\left(\frac{t_t}{t_t+t_{se}}\right) P_x(z=1)\,(1-p_D)\, b\sum_{n=1}^{N_H} \log_2\left(1 + \frac{P^I_{H.n}\,|g_{h.n}|^2}{n_p\,b + P_p\,|g_p|^2}\right)}{P^I_H + P_{cp} + P_{sp}}$$ (4.19)

Thus, the resulting $P^I_{H.n}$ for any n number of HRC device is given by

$$P^I_{H.n} = \frac{P_{cp} g^2_{h.n} + P_{sp} g^2_{h.n} - (n_p\,b + P_p\,g^2_p)}{W\left(0, \left(\frac{P_{cp} g^2_{h.n} + P_{sp} g^2_{h.n} - (n_p\,b + P_p\,g^2_p)}{n_p\,b + P_p\,g^2_p}\right) \exp(-1)\right) g^2_{h.n}} - \frac{(n_p\,b + P_p\,g^2_p)}{g^2_{h.n}}$$ (4.20)

Proof: Refer to Appendix 4.A.2

4.3.4.4.2 Optimum Power Derivation for MRC

Correspondingly, rewriting Equation (4.11), we obtain the optimum transmission power ($P^E_{M,n}$) for the effectual state as follows:

$$EE^0_M = \frac{\left(\frac{t_t}{t_t+t_{se}}\right) P_x(z=0)\,(1-p_F)\, b\sum_{n=1}^{N_M} \log_2\left(1 + \frac{P^E_{M.n}\,|g_{m.n}|^2}{n_p\,b + P_{H.n}\,|g_{h.n}|^2}\right)}{P^E_M + P_{cp} + P_{sp}}$$ (4.21)

The resulting $P^E_{M.n}$ for any n number of MRC device is

$$P^E_{M.n} = \frac{P_{cp} g^2_{m.n} + P_{sp} g^2_{m.n} - (n_p\,b + P_H\,g^2_{h.n})}{W\left(0, \left(\frac{P_{cp} g^2_{m.n} + P_{sp} g^2_{m.n} - (n_p\,b + P_H\,g^2_{h.n})}{n_p\,b + P_H\,g^2_{h.n}}\right) \exp(-1)\right) g^2_{m.n}} - \frac{(n_p\,b + P_H\,g^2_{h.n})}{g^2_{m.n}}$$

(4.22)

Proof: Refer to Appendix 4.A.3

Next, we rewrite EE^1_M of Equation (4.12) for deriving optimum transmission power in interference state ($P^I_{M,n}$) as follows:

$$EE_M^1 = \frac{\left(\frac{t_t}{t_t+t_{se}}\right) p_x(z=1)\,(1-p_D)b \sum_{n=1}^{N_M} \log_2\left(1 + \frac{P_{M.n}^I\,|g_{m,n}|^2}{n_p\,b+P_{H,n}\,|g_{h,n}|^2+P_P\,|g_p|^2}\right)}{P_M^I + P_{cp} + P_{sp}} \tag{4.23}$$

Hence, the resulting $P_{M.n}^I$ for any n number of the MRC device is

$$P_{M.n}^I = \frac{P_{cp}\,g_{m.n}^2 + P_{sp}\,g_{m.n}^2 - (n_p\,b + P_H\,g_{h.n}^2 + P_p\,g_p^2)}{W\left(0,\left(\frac{P_{cp}\,g_{m.n}^2+P_{sp}\,g_{m.n}^2-(n_p\,b+P_H\,g_{h.n}^2+P_p\,g_p^2)}{n_p\,b+P_H\,g_{h.n}^2+P_p\,g_p^2}\right)\exp(-1)\right)g_{m.n}^2}$$
$$-\frac{(n_p\,b+P_H\,g_{h.n}^2 + P_p\,g_p^2)}{g_{m.n}^2} \tag{4.24}$$

Proof: Refer to Appendix 4.A.4

As a result, to the optimum power expressions discussed above, throughput for HRC (S_H^0) and MRC (S_M^0) devices is at an effectual state and has been improved, and similarly the interference throughputs for HRC (S_H^1) and MRC (S_M^1) are also improved.

4.4 Simulation Results

To evaluate the performance of our system model, through Figures 4.2–4.5, we have applied Equations (4.18), (4.20), (4.22), and (4.24) to enhance both the throughput and energy efficiency of HRC and MRC devices in both effectual and interference states. There are five HRC and MRC devices at various distances. Specifically, in this chapter, we utilize the MATLAB software to configure appropriate codes, and to achieve numerical simulation data. Next, the parameters selected for simulation are: $f_{c(mm)} = 5$ GHz, $b = 1$ MHz, $c = 3 \times 10^8$ m/sec, $t_t = 0.125 \times 10^{-3}$ sec, $t_{se} = 0.125 \times 10^{-3}$ sec, $p_x(z=0) = 0:0.01:1$, $p_x(z=1) = 0:0.01:1$, $P_F = 0.1$, $p_D = 0.9$, $n_p = -174$ dBm, $P_{cp} = 99$ dBm, $P_{sp} = 1$ dBm, $P_p = 50$ dBm, $P_H = (0.7 \times 30)$ dBm, and $P_M = (0.3 \times 30)$dBm.

Figure 4.2 presents the average throughputs at effectual state ($z=0$) for both the HRC and MRC devices at several $p_x(z=0)$. Now, $p_x(z=0)$ represents that primary transmitter (PT_x) is not causing any interference to the system, and with more accurate $p_x(z=0)$ (1 being the maximum value), throughputs for both HRC and MRC devices will also increase. Additionally, we enhance the original throughputs of HRC and MRC devices in Equations (4.5) and (4.6) respectively, by applying the optimum powers derived in Equations (4.18) and (4.19) for HRC and MRC devices. Thus, for a low-latency and high-powered HRC device, original or conventional S_H^0 in Equation (4.5) at $p_x(z=0) = 0.5$ is 1.409×10^6 bps and gradually increasing as the probability gets closer to 1. Now, we enhance the throughput in Equation (4.5) by applying the optimum power (P_H^E) from Equation (4.18) to acquire optimized S_H^0 for HRC device, which is 8.835×10^6 bps, at $p_x(z=0) = 0.5$ and renders 83.13% optimized throughput over the original S_H^0. Similarly, for an MRC device requiring moderate power and latency, the original S_M^0 in Equation (4.6) is 1.864×10^5bps and increasing, at $p_x(z=0) = 0.5$, which is still considerably lower than the original S_H^0 in

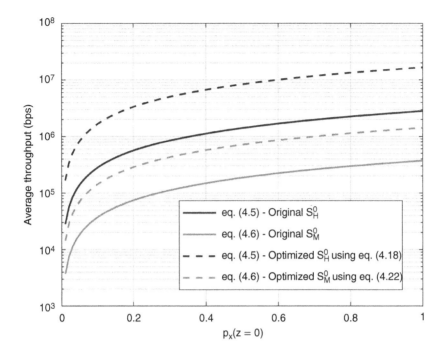

Figure 4.2 Average effectual throughput for both HRC and MRC devices at numerous $p_x(z = 0)$.

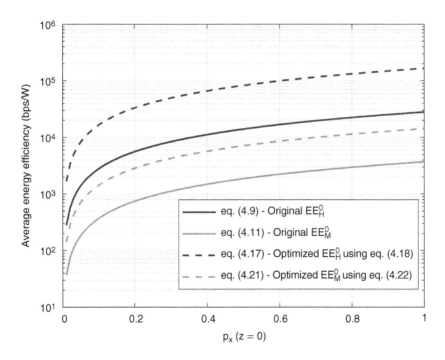

Figure 4.3 Average effectual energy efficiency for both HRC and MRC devices at various $p_x(z = 0)$.

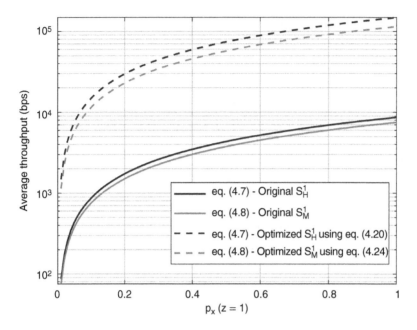

Figure 4.4 Average interference throughput for both HRC and MRC devices at various $p_x(z = 1.)$.

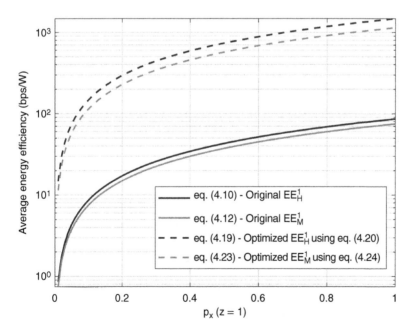

Figure 4.5 Average interference energy efficiency for both HRC and MRC devices at various $p_x(z = 1)$.

Equation (4.5). Then we apply P^E_M from Equation (4.22) to increase S^0_M in Equation (4.6) such as, at $p_x(z = 0) = 0.5$ optimized throughput for the MRC device is 7.157×10^5 bps, which is significantly better (73.96%) than the original S^0_M in Equation (4.6). Furthermore, both the original and optimized throughputs for an MRC device (S^0_M) will be lower than the average HRC throughputs (S^0_H), since an HRC device requires higher power to produce greater throughput with low latency.

Figure 4.3. displays the effectual energy efficiency for both the HRC and MRC devices at various $p_x(z = 0)$. As throughput for a device (HRC or MRC) increases with higher detection probability ($p_x(z = 0)$), the energy efficiency for that particular device will also rise if $p_x(z = 0)$ increases. Moreover, in Figure 4.3, original energy efficiency (EE^0_H) from Equation (4.9) for an HRC device at effectual state is 1.4×10^4 bps/W at $p_x(z = 0) = 0.5$ and increasing progressively. We then improve this energy efficiency by applying P^E_H from Equation (4.18) to obtain optimized energy efficiency (EE^0_H) in Equation (4.17), which is 8.3×10^4 bps/W at $p_x(z = 0) = 0.5$ and delivers 83.11% improvement over original EE^0_H. Likewise, in Figure 4.3., original energy efficiency for the MRC device in Equation (4.11) gives 1858 bps/W at $p_x(z = 0) = 0.5$ and after applying P^E_M from Equation (4.22), we obtain the improved energy efficiency (EE^0_M) in Equation (4.21). The optimized EE^0_M is 7136 bps/W at $p_x(z = 0) = 0.5$, which gives 73% improvement over the original energy efficiency for an MRC device in Equation (4.11). Here also, both the original and optimized energy efficiency (EE^0_H) graphs for an HRC device are greater than all the energy efficiency levels of an MRC device, since throughput for the HRC device is greater at all $p_x(z = 0)$ points.

Figure 4.4 depicts the original and optimized throughputs of HRC and MRC devices when interference is present (PT_x is active on transmission), with increasing $p_x(z = 1)$. Even with interference, higher detection probability ($p_x(z = 1)$) will provide greater throughputs. Now, for an HRC device facing interference, original throughput (S^1_H) in Equation (4.7) is 4330 bps and original throughput (S^1_M) for an MRC device in Equation (4.8) is 3753 bps, both at $p_x(z = 1) = 0.5$. Here, both the devices are moderately affected by interference, resulting in more familiar throughputs but still distinguishable at the receiving end, as an HRC device will have higher throughputs with larger power for critical response toward a patient. Moreover, after applying optimum powers (P^I_H and P^I_M) from Equations (4.20) and (4.24) in Equations (4.7) and (4.8), we acquire optimized throughputs for an HRC and MRC device, such as 7.419×10^4 bps and 5.72×10^4 bps, both at $p_x(z = 1) = 0.5$ respectively. Furthermore, in Figure 4.4, it is clear that our optimized throughputs for both the HRC and MRC devices are still superior when interference is present, and throughputs for the HRC and MRC devices have improved by 94.16 and 93.43%, respectively over the original S^1_H and S^1_M.

Figure 4.5 displays the original and optimized energy efficiency for both the HRC and MRC devices at various $p_x(z = 1)$. Similarly, in Figure 4.4, where interference throughputs for an HRC and MRC device are more similar to each other, their resultant energy efficiencies (Figure 4.5) will also be slightly familiar. Moreover, at $p_x(z = 1) = 0.5$, the original EE^1_H in Equation (4.10) and EE^1_M in Equation (4.12) are 42.99 and 37.41 bps/W, respectively and they are increasing with detection that is more accurate ($p_x(z = 1) \rightarrow 1$). Then we apply P^I_H from Equation (4.20) and P^I_M from Equation (4.24) to obtain optimized energy efficiencies for the HRC and MRC devices. Now from Figure 4.5, the optimized EE^1_H in Equation (4.19) is 736.8 bps/W by 94.16% improvement over the original and optimized EE^1_M is 570.3 bps/W by 93.44% improvement, for an HRC and MRC device respectively.

Overall, for both an HRC and MRC device, the effectual throughputs and energy efficiency graphs are significantly greater than all the throughputs and energy efficiencies in the interference state. This is because at interference state, PT_x, is actively in communication with SR_x, causing external interference to the system. However, even in the interference state, our optimized throughputs and energy efficiency still proves to be superior over the original. From Figures 4.2–4.5, all the optimized results (Throughput and Energy efficiency) contribute more than 70% improvement over their original outcomes in uplink communication during both effectual and interference states. Henceforth, for a patient needing critical response or surgery, HRC devices will render efficient communication, whereas MRC devices will deliver adequate connection necessary for monitoring regular patients over both the effectual and interference state.

4.5 Conclusion

In conclusion, in this chapter, we introduced a cognitive IoT-based system, which utilizes NOMA uplink communication to communicate successfully with the BS, to execute and process information depending on a patient's state of urgency. Here, we have generated the throughput expressions with their energy efficacy considering the presence and absent state of any interference. Then we have applied the derived optimum power expression to enhance both the throughput and energy efficiency for any HRC and MRC device. However, in the presence of any interference, our optimized throughput improves, by 94.16 and 93.43%, respectively for both HRC and MRC devices, with more than 93% improved energy efficiency collectively (both HRC and MRC). Similarly, for the effectual state, the throughput was enhanced vastly, with 78.53% average (both HRC and MRC) improved energy efficiency. Hence, concerning the appropriate response toward patients, our system structure serves the optimum energy efficiency with or without any presence of interference. Moreover, concerning the massive usage of communicative IoT devices in all types of institutions, our system will provide a solid wireless communication medium over any cooperative network structure, which will ensure proper spectrum utilization [50]. Finally, our system already confirms significant strides in the energy efficiency factor, and we will further boost this with energy harvesting methods from natural resources that we will apply alongside wireless sensors for optimal energy-efficient performance [44, 51].

4.A Appendix

4.A.1 Proof of Optimum Power Transmission for HRC Device at Effectual State ($z = 0$)

For the simplicity of the derivation, let us consider $P_{H.n} \rightarrow P_H$ & $g_{h,n} \rightarrow g_h$ for Equation (4.5). Now deriving EE_H^0 in Equation (4.9) with respect to P_H we obtain

$$\frac{dEE_H^0}{dP_H} = 0$$

$$\implies \frac{p_x(z=0)\, t_t\, (p_F-1)\, b \log\left(1 + \frac{P_H\, g_h^2}{n_p\, b}\right)}{\log(2)\, (t_t + t_{se})\, (P_H + P_{cp} + P_{sp})^2}$$

$$- \frac{p_x(z=0)\, t_t\, (p_F-1)\, g_h^2}{\log(2)\, (t_t + t_{se})\, n_p \left(1 + \frac{P_H\, g_h^2}{n_p\, b}\right)(P_H + P_{cp} + P_{sp})} = 0$$

$$\implies \frac{b \log\left(1 + \frac{P_H\, g_h^2}{n_p\, b}\right)}{P_H + P_{cp} + P_{sp}} = \frac{g_h^2}{n_p \left(1 + \frac{P_H\, g_h^2}{n_p\, b}\right)}$$

$$\implies \frac{\log\left(1 + \frac{P_H\, g_h^2}{n_p\, b}\right)}{P_H + P_{cp} + P_{sp}} = \frac{g_h^2}{b \left(n_p + \frac{P_H\, g_h^2}{b}\right)}$$

$$\implies \log\left(1 + \frac{P_H\, g_h^2}{n_p\, b}\right) = \frac{P_H\, g_h^2 + P_{cp}\, g_h^2 + P_{sp}\, g_h^2}{P_H\, g_h^2 + n_p\, b}$$

$$\implies \log\left(1 + \frac{P_H\, g_h^2}{n_p\, b}\right) - 1 = \frac{P_H\, g_h^2 + P_{cp}\, g_h^2 + P_{sp}\, g_h^2}{P_H\, g_h^2 + n_p\, b} - 1$$

$$\implies \log\left(1 + \frac{P_H\, g_h^2}{n_p\, b}\right) + \ln e^{-1} = \frac{P_{cp}\, g_h^2 + P_{sp}\, g_h^2 - n_p\, b}{P_H\, g_h^2 + n_p\, b}$$

$$\implies \log\left(\frac{P_H\, g_h^2 + n_p\, b}{n_p\, b}\right) exp(-1) = \frac{P_{cp}\, g_h^2 + P_{sp}\, g_h^2 - n_p\, b}{P_H\, g_h^2 + n_p\, b}$$

$$\implies \left(\frac{P_H\, g_h^2 + n_p\, b}{n_p\, b}\right) exp(-1) = exp\left(\frac{P_{cp}\, g_h^2 + P_{sp}\, g_h^2 - n_p\, b}{P_H\, g_h^2 + n_p\, b}\right)$$

$$\implies \left(\frac{P_{cp}\, g_h^2 + P_{sp}\, g_h^2 - n_p\, b}{n_p\, b}\right) * exp(-1) = \left(\frac{P_{cp}\, g_h^2 + P_{sp}\, g_h^2 - n_p\, b}{P_H\, g_h^2 + n_p\, b}\right)$$

$$* exp\left(\frac{P_{cp}\, g_h^2 + P_{sp}\, g_h^2 - n_p\, b}{P_H\, g_h^2 + n_p\, b}\right) \tag{A.1}$$

Applying the Lambert method to rewrite Equation A.1 as

$$[a = be^b \implies W(a) = b]$$

$$\implies W\left(\left(\frac{P_{cp}\, g_h^2 + P_{sp}\, g_h^2 - n_p\, b}{n_p\, b}\right) exp(-1)\right) = \left(\frac{P_{cp}\, g_h^2 + P_{sp}\, g_h^2 - n_p\, b}{P_H\, g_h^2 + n_p\, b}\right)$$

$$\implies P_H\, g_h^2 + n_p\, b = \frac{P_{cp}\, g_h^2 + P_{sp}\, g_h^2 - n_p\, b}{W\left(\left(\frac{P_{cp}\, g_h^2 + P_{sp}\, g_h^2 - n_p\, b}{n_p\, b}\right) exp(-1)\right)}$$

$$\implies P_H = \frac{P_{cp}\, g_h^2 + P_{sp}\, g_h^2 - n_p\, b}{W\left(\left(\frac{P_{cp}\, g_h^2 + P_{sp}\, g_h^2 - n_p\, b}{n_p\, b}\right) exp(-1)\right) g_h^2} - \frac{n_p\, b}{g_h^2}$$

Thus, resulting $P_{H.n}^E$ for any n number of HRC devices is

$$\therefore P_{H.n}^E = \frac{n_p\, b\left(\dfrac{P_{cp}\, g_{h.n}^2 + P_{sp}\, g_{h.n}^2 - n_p\, b}{W\left(0,\left(-\dfrac{P_{cp}\, g_{h.n}^2 + P_{sp}\, g_{h.n}^2 - n_p\, b}{n_p\, b}\right)\exp(-1)\right)} - 1\right)}{g_{h.n}^2}\qquad(A.2)$$

4.A.2 Proof of Optimum Power Transmission for HRC Device in Interference State ($z = 1$)

For the simplicity of the derivation, let us consider $P_{H.n} \rightarrow P_H$ & $g_{h,n} \rightarrow g_h$ for Equation (4.7). Now deriving EE_H^1 in Equation (4.10) with respect to P_H we obtain

$$\frac{dEE_H^1}{dP_H} = 0$$

$$\Rightarrow \frac{p_x(z=1)\, t_t\, (p_D - 1)\, b\log\left(1 + \dfrac{P_H\, g_h^2}{n_p\, b + P_p\, g_p^2}\right)}{\log(2)\, (t_t + t_{se})\, (P_H + P_{cp} + P_{sp})^2}$$
$$- \frac{p_x(z=1)\, t_t\, (p_D - 1)\, b\, g_h^2}{\log(2)\, (t_t + t_{se})\, (n_p\, b + P_p\, g_p^2)\left(1 + \dfrac{P_H\, g_h^2}{n_p\, b + P_p\, g_p^2}\right)(P_H + P_{cp} + P_{sp})} = 0$$

$$\Rightarrow \frac{\log\left(1 + \dfrac{P_H\, g_h^2}{n_p\, b + P_p\, g_p^2}\right)}{P_H + P_{cp} + P_{sp}} = \frac{g_h^2}{(n_p\, b + P_p\, g_p^2)\left(1 + \dfrac{P_H\, g_h^2}{n_p\, b + P_p\, g_p^2}\right)}$$

$$\Rightarrow \log\left(1 + \frac{P_H\, g_h^2}{n_p\, b + P_p\, g_p^2}\right) - 1 = \frac{P_H\, g_h^2 + P_{cp}\, g_h^2 + P_{sp}\, g_h^2}{P_H\, g_h^2 + P_p\, g_p^2 + n_p\, b} - 1$$

$$\Rightarrow \log\left(1 + \frac{P_H\, g_h^2}{n_p\, b + P_p\, g_p^2}\right) + \ln e^{-1} = \frac{P_{cp}\, g_h^2 + P_{sp}\, g_h^2 - (n_p\, b + P_p\, g_p^2)}{P_H\, g_h^2 + P_p\, g_p^2 + n_p\, b}$$

$$\Rightarrow \log\left(\frac{P_H\, g_h^2 + P_p\, g_p^2 + n_p\, b}{n_p\, b + P_p\, g_p^2}\right)\exp(-1) = \frac{P_{cp}\, g_h^2 + P_{sp}\, g_h^2 - (n_p\, b + P_p\, g_p^2)}{P_H\, g_h^2 + P_p\, g_p^2 + n_p\, b}$$

$$\Rightarrow \left(\frac{P_H\, g_h^2 + P_p\, g_p^2 + n_p\, b}{n_p\, b + P_p\, g_p^2}\right)\exp(-1) = \exp\left(\frac{P_{cp}\, g_h^2 + P_{sp}\, g_h^2 - (n_p\, b + P_p\, g_p^2)}{P_H\, g_h^2 + P_p\, g_p^2 + n_p\, b}\right)$$

$$\Rightarrow \left(\frac{P_{cp}\, g_h^2 + P_{sp}\, g_h^2 - (n_p\, b + P_p\, g_p^2)}{n_p\, b + P_p\, g_p^2}\right)$$

$$* \exp(-1) = \left(\frac{P_{cp}\, g_h^2 + P_{sp}\, g_h^2 - (n_p\, b + P_p\, g_p^2)}{P_H\, g_h^2 + P_p\, g_p^2 + n_p\, b}\right)$$

$$* \exp\left(\frac{P_{cp}\, g_h^2 + P_{sp}\, g_h^2 - (n_p\, b + P_p\, g_p^2)}{P_H\, g_h^2 + P_p\, g_p^2 + n_p\, b}\right)\qquad(A.3)$$

Utilizing the Lambert method to rewrite Equation A.3 as

$$[a = be^b \implies W(a) = b]$$

$$\therefore W\left(\left(\frac{P_{cp}\, g_h^2 + P_{sp}\, g_h^2 - (n_p\, b + P_p\, g_p^2)}{n_p\, b + P_p\, g_p^2}\right) exp(-1)\right)$$

$$= \left(\frac{P_{cp}\, g_h^2 + P_{sp}\, g_h^2 - (n_p\, b + P_p\, g_p^2)}{P_H\, g_h^2 + P_p\, g_p^2 + n_p\, b}\right)$$

$$\Rightarrow P_H\, g_h^2 + P_p\, g_p^2 + n_p\, b = \frac{P_{cp}\, g_h^2 + P_{sp}\, g_h^2 - (n_p\, b + P_p\, g_p^2)}{W\left(\left(\frac{P_{cp}\, g_h^2 + P_{sp}\, g_h^2 - (n_p\, b + P_p\, g_p^2)}{n_p\, b + P_p\, g_p^2}\right) exp(-1)\right)}$$

Thus, resulting $P_{H.n}^I$ for any n number of HRC devices is

$$\therefore P_{H.n}^I = \frac{P_{cp}\, g_{h.n}^2 + P_{sp}\, g_{h.n}^2 - (n_p\, b + P_p\, g_p^2)}{W\left(0, \left(\frac{P_{cp}\, g_{h.n}^2 + P_{sp}\, g_{h.n}^2 - (n_p\, b + P_p\, g_p^2)}{n_p\, b + P_p\, g_p^2}\right) exp(-1)\right) g_{h.n}^2} - \frac{(n_p\, b + P_p\, g_p^2)}{g_{h.n}^2} \quad (A.4)$$

4.A.3 Proof of Optimum Power Transmission for MRC Device at Effectual State ($z = 0$)

For the simplicity of the derivation, let us consider $P_{M.n} \rightarrow P_M$ & $g_{m,n} \rightarrow g_m$ for Equation (4.6). Now deriving EE_M^0 in Equation (4.11) with respect to P_M we obtain

$$\frac{dEE_M^0}{dP_M} = 0$$

$$\Rightarrow \frac{p_x(z = 0)\, t_t\, (p_F - 1)\, b \log\left(1 + \frac{P_M\, g_m^2}{n_p\, b + P_H\, g_h^2}\right)}{\log(2)\, (t_t + t_{se})\, (P_M + P_{cp} + P_{sp})^2}$$

$$- \frac{p_x(z = 0)\, t_t\, (p_F - 1)\, b\, g_m^2}{\log(2)\, (t_t + t_{se})\, (n_p\, b + P_H\, g_h^2)\left(1 + \frac{P_M\, g_m^2}{n_p\, b + P_H\, g_h^2}\right)(P_M + P_{cp} + P_{sp})} = 0$$

$$\Rightarrow \frac{\log\left(1 + \frac{P_M\, g_m^2}{n_p\, b + P_H\, g_h^2}\right)}{P_M + P_{cp} + P_{sp}} = \frac{g_m^2}{(n_p\, b + P_H\, g_h^2)\left(1 + \frac{P_M\, g_m^2}{n_p\, b + P_H\, g_h^2}\right)}$$

$$\Rightarrow \log\left(1 + \frac{P_M\, g_m^2}{n_p\, b + P_H\, g_h^2}\right) = \frac{P_M\, g_m^2 + P_{cp}\, g_m^2 + P_{sp}\, g_m^2}{P_M\, g_m^2 + P_H\, g_h^2 + n_p\, b}$$

$$\Rightarrow \log\left(1 + \frac{P_M\, g_m^2}{n_p\, b + P_H\, g_h^2}\right) - 1 = \frac{P_M\, g_m^2 + P_{cp}\, g_m^2 + P_{sp}\, g_m^2}{P_M\, g_m^2 + P_H\, g_h^2 + n_p\, b} - 1$$

$$\Rightarrow \log\left(1 + \frac{P_M\, g_m^2}{n_p\, b + P_H\, g_h^2}\right) + \ln e^{-1} = \frac{P_{cp}\, g_m^2 + P_{sp}\, g_m^2 - (n_p\, b + P_H\, g_h^2)}{P_M\, g_m^2 + P_H\, g_h^2 + n_p\, b}$$

$$\Rightarrow \left(\frac{P_M g_m^2 + P_H g_h^2 + n_p b}{n_p b + P_H g_h^2} \right) exp(-1) = exp\left(\frac{P_{cp} g_m^2 + P_{sp} g_m^2 - (n_p b + P_H g_h^2)}{P_M g_m^2 + P_H g_h^2 + n_p b} \right)$$

$$\Rightarrow \left(\frac{P_{cp} g_m^2 + P_{sp} g_m^2 - (n_p b + P_H g_h^2)}{n_p b + P_H g_h^2} \right)$$

$$* exp(-1) = \left(\frac{P_{cp} g_m^2 + P_{sp} g_m^2 - (n_p b + P_H g_h^2)}{P_M g_m^2 + P_H g_h^2 + n_p b} \right)$$

$$* exp\left(\frac{P_{cp} g_m^2 + P_{sp} g_m^2 - (n_p b + P_H g_h^2)}{P_M g_m^2 + P_H g_h^2 + n_p b} \right) \tag{A.5}$$

Applying the Lambert method to rewrite Equation A.5 as

$$[a = be^b \Longrightarrow W(a) = b]$$

$$\therefore W\left(\left(\frac{P_{cp} g_m^2 + P_{sp} g_m^2 - (n_p b + P_H g_h^2)}{n_p b + P_H g_h^2} \right) exp(-1) \right)$$

$$= \left(\frac{P_{cp} g_m^2 + P_{sp} g_m^2 - (n_p b + P_H g_h^2)}{P_M g_m^2 + P_H g_h^2 + n_p b} \right)$$

$$\Rightarrow P_M g_m^2 + P_H g_h^2 + n_p b = \frac{P_{cp} g_m^2 + P_{sp} g_m^2 - (n_p b + P_H g_h^2)}{W\left(\left(\frac{P_{cp} g_m^2 + P_{sp} g_m^2 - (n_p b + P_H g_h^2)}{n_p b + P_H g_h^2} \right) exp(-1) \right)}$$

Thus, resulting $P_{M.n}^E$ for any n number of MRC devices is

$$\therefore P_{M.n}^E = \frac{P_{cp} g_{m.n}^2 + P_{sp} g_{m.n}^2 - (n_p b + P_H g_{h.n}^2)}{W\left(0, \left(\frac{P_{cp} g_{m.n}^2 + P_{sp} g_{m.n}^2 - (n_p b + P_H g_{h.n}^2)}{n_p b + P_H g_{h.n}^2} \right) exp(-1) \right) g_{m.n}^2} - \frac{(n_p b + P_H g_{h.n}^2)}{g_{m.n}^2} \tag{A.6}$$

4.A.4 Proof of Optimum Power Transmission for MRC Device in Interference State (z = 1)

For the simplicity of the derivation, let us consider $P_{M.n} \rightarrow P_M$ & $g_{m,n} \rightarrow g_m$ for Equation (4.8)). Now deriving EE_M^0 in Equation (4.12) with respect to P_M we obtain

$$\frac{dEE_M^1}{dP_M}$$

$$\Rightarrow \frac{p_x(z = 1) t_t (p_D - 1) b \log \left(1 + \frac{P_M g_m^2}{n_p b + P_H g_h^2 + P_p g_p^2} \right)}{\log(2) (t_t + t_{se}) (P_M + P_{cp} + P_{sp})^2}$$

$$- \frac{p_x(z = 1) t_t (p_D - 1) b g_m^2}{\log(2) (t_t + t_{se}) (n_p b + P_H g_h^2 + P_p g_p^2) \left(1 + \frac{P_M g_m^2}{n_p b + P_H g_h^2 + P_p g_p^2} \right) (P_M + P_{cp} + P_{sp})} = 0$$

$$\Rightarrow \frac{\log\left(1 + \frac{P_M\, g_m^2}{n_p\, b + P_H\, g_h^2 + P_p\, g_p^2}\right)}{P_M + P_{cp} + P_{sp}} = \frac{g_m^2}{(n_p\, b + P_H\, g_h^2 + P_p\, g_p^2)\left(1 + \frac{P_M\, g_m^2}{n_p\, b + P_H\, g_h^2 + P_p\, g_p^2}\right)}$$

$$\Rightarrow \log\left(1 + \frac{P_M\, g_m^2}{n_p\, b + P_H\, g_h^2 + P_p\, g_p^2}\right) - 1 = \frac{P_M\, g_m^2 + P_{cp}\, g_m^2 + P_{sp}\, g_m^2}{P_M\, g_m^2 + P_H\, g_h^2 + P_p\, g_p^2 + n_p\, b} - 1$$

$$\Rightarrow \log\left(1 + \frac{P_M\, g_m^2}{n_p\, b + P_H\, g_h^2 + P_p\, g_p^2}\right) + \ln e^{-1} = \frac{P_{cp}\, g_m^2 + P_{sp}\, g_m^2 - (n_p\, b + P_H\, g_h^2 + P_p\, g_p^2)}{P_M\, g_m^2 + P_H\, g_h^2 + P_p\, g_p^2 + n_p\, b}$$

$$\Rightarrow \left(\frac{P_M\, g_m^2 + P_H\, g_h^2 + P_p\, g_p^2 + n_p\, b}{n_p\, b + P_H\, g_h^2 + P_p\, g_p^2}\right) exp(-1)$$
$$= exp\left(\frac{P_{cp}\, g_m^2 + P_{sp}\, g_m^2 - (n_p\, b + P_H\, g_h^2 + P_p\, g_p^2)}{P_M\, g_m^2 + P_H\, g_h^2 + P_p\, g_p^2 + n_p\, b}\right)$$

$$\Rightarrow \left(\frac{P_{cp}\, g_m^2 + P_{sp}\, g_m^2 - (n_p\, b + P_H\, g_h^2 + P_p\, g_p^2)}{n_p\, b + P_H\, g_h^2 + P_p\, g_p^2}\right) * exp(-1)$$
$$= \left(\frac{P_{cp}\, g_m^2 + P_{sp}\, g_m^2 - (n_p\, b + P_H\, g_h^2 + P_p\, g_p^2)}{P_M\, g_m^2 + P_H\, g_h^2 + P_p\, g_p^2 + n_p\, b}\right)$$
$$* exp\left(\frac{P_{cp}\, g_m^2 + P_{sp}\, g_m^2 - (n_p\, b + P_H\, g_h^2 + P_p\, g_p^2)}{P_M\, g_m^2 + P_H\, g_h^2 + P_p\, g_p^2 + n_p\, b}\right) \tag{A.7}$$

Utilizing the Lambert method to rewrite Equation A.7 as

$$[a = be^b \implies W(a) = b]$$

$$\Rightarrow W\left(\left(\frac{P_{cp}\, g_m^2 + P_{sp}\, g_m^2 - (n_p\, b + P_H\, g_h^2 + P_p\, g_p^2)}{n_p\, b + P_H\, g_h^2 + P_p\, g_p^2}\right) exp(-1)\right)$$
$$= \left(\frac{P_{cp}\, g_m^2 + P_{sp}\, g_m^2 - (n_p\, b + P_H\, g_h^2 + P_p\, g_p^2)}{P_M\, g_m^2 + P_H\, g_h^2 + P_p\, g_p^2 + n_p\, b}\right)$$

$$\Rightarrow P_M\, g_m^2 + P_H\, g_h^2 + P_p\, g_p^2 + n_p\, b = \frac{P_{cp}\, g_m^2 + P_{sp}\, g_m^2 - (n_p\, b + P_H\, g_h^2 + P_p\, g_p^2)}{W\left(\left(\frac{P_{cp}\, g_m^2 + P_{sp}\, g_m^2 - (n_p\, b + P_H\, g_h^2 + P_p\, g_p^2)}{n_p\, b + P_H\, g_h^2 + P_p\, g_p^2}\right) exp(-1)\right)}$$

Finally, the resulting $P_{M.n}^I$ for any n number of MRC devices is

$$P_{M.n}^I = \frac{P_{cp}\, g_{m.n}^2 + P_{sp}\, g_{m.n}^2 - (n_p\, b + P_H\, g_{h.n}^2 + P_p\, g_p^2)}{W\left(0, \left(\frac{P_{cp}\, g_{m.n}^2 + P_{sp}\, g_{m.n}^2 - (n_p\, b + P_H\, g_{h.n}^2 + P_p\, g_p^2)}{n_p\, b + P_H\, g_{h.n}^2 + P_p\, g_p^2}\right) exp(-1)\right) g_{m.n}^2}$$
$$- \frac{(n_p\, b + P_H\, g_{h.n}^2 + P_p\, g_p^2)}{g_{m.n}^2} \tag{A.8}$$

References

1 Laxminarayan, S. and Istepanian, R.S.H. (2000). Unwired E-MED: the next generation of wireless and internet telemedicine systems. *IEEE Transactions on Information Technology in Biomedicine* 4 (3): 189–193.

2 Llamas, R.T. Worldwide Wearable Computing Device Forecast, 2020–2024: CY 1Q20 [Online] Available at: https://www.idc.com/getdoc.jsp?containerId=US45811920

3 Istepanian, R.S.H., Jovanov, E., and Zhang Y.T. (2004). Guest editorial introduction to the special section on M-health: Beyond seamless mobility and global wireless health-care connectivity. *IEEE Transactions on Information Technology in Biomedicine* 8 (4): 405–414.

4 Bonato, P. (2003). Wearable sensors/systems and their impact on biomedical engineering. *IEEE Engineering in Medicine and Biology Magazine,* 22 (3): 18–20.

5 Park, P. and Jayaraman, S. (2003). Enhancing the quality of life through wearable technology. *IEEE Engineering in Medicine and Biology Magazine,* 22 (3): 41–48.

6 IEEE 802.15 Working Group for WPAN / Bluetooth [Online]. Available: http://www.ieee802.org/15/pub/TG1.html

7 IEEE Standard for Local and metropolitan area networks – Part 15.6: Wireless Body Area Networks, IEEE Std 802.15.6–2012, 1–271, 29 February 2012.

8 Foteinos, V., Kelaidonis, D., Poulios, G. et al. (2013). Cognitive management for the internet of things: a framework for enabling autonomous application. *IEEE Vehicular Technology Magazine* 8 (4): 90–99.

9 Mezghani E., Exposito, E., and Drira, K. (2017). A model-driven methodology for the Design of Autonomic and Cognitive IoT-based systems: application to healthcare. *IEEE Transactions on Emerging Topics in Computational Intelligence* 1 (3): 224–234.

10 Amin, S.U., Hossain, M.S., Muhammad, G. et al. (2019). Cognitive smart healthcare for pathology detection and monitoring. *IEEE Access* 7: 10745–10753.

11 Alhussein, M., Muhammad, G., Hossain, M.S. et al. (2018). Cognitive IoT-cloud integration for smart healthcare: case study for epileptic seizure detection and monitoring. *Mobile Networks and Applications* 23 (6): 1624–1635.

12 Hossain, M.S. and Muhammad, G. (2019). Emotion recognition using deep learning approach from audio–visual emotional big data. *Information Fusion* 49: 69–78.

13 Hossain, M.S. and Muhammad, G. (2016). Cloud-assisted industrial internet of things (IIoT)-enabled framework for health monitoring. *Computer Networks* 101: 192–202.

14 Du, D., Hu, F., Wang, F. et al. (2015). A game theoretic approach for inter-network interference mitigation in wireless body area networks. *China Communications* 12 (9), 150–161.

15 Yoo, H.-J. (2013). Wireless body area network and its healthcare applications. Asia-Pacific Microwave Conference Proceedings (APMC), 89–91, Seoul.

16 Qi, X., Wang, K., Huang, A. et al. (2015). A harvesting-rate oriented self-adaptive algorithm in energy-harvesting Wireless Body Area Networks. Proceedings of the 2015 IEEE 13th International Conference on Industrial Informatics (INDIN), 966–971, Cambridge.

17 Qi, X., Wang, K., Yue, D. et al. (2015). Adaptive TDMA-based MAC protocol in energy harvesting Wireless Body Area Network for mobile health. Proceedings of the

IECON 2015 - 41st Annual Conference of the IEEE Industrial Electronics Society, 004175–004180, Yokohama.

18 Chen, D. (2017). A QoS bandwidth allocation method for coexistence of Wireless Body Area Networks. In Proceedings of the 2017 25th Euromicro International Conference on Parallel, Distributed and Network-based Processing (PDP), 251–254, St. Petersburg.

19 Ambigavathi. M. and Sridharan, D. (2017). A review of channel access techniques in Wireless Body Area Network. Proceedings of the 2017 Second International Conference on Recent Trends and Challenges in Computational Models (ICRTCCM), 106–110, Tindivanam.

20 Pande. H. and Kharat, M.U. (2016). Adaptive energy efficient MAC protocol for increasing life of sensor nodes in Wireless Body Area Network. Proceedings of the 2016 International Conference on Internet of Things and Applications (IOTA), 349–352, Pune.

21 Wang, J., Xie, Y., and Yi, Q. (2015). An all dynamic MAC protocol for Wireless Body Area Networks. Proceedings of the 11th International Conference on Wireless Communications, Networking and Mobile Computing (WiCOM 2015), 1–6, Shanghai.

22 Prameela, S. and Ponmuthuramalingam, P. (2016). A robust energy efficient and secure data dissemination protocol for Wireless Body Area Networks. Proceedings of the 2016 IEEE International Conference on Advances in Computer Applications (ICACA), 131–134, Coimbatore.

23 ul Huque, M.T.I., Munasinghe, K.S., and Jamalipour, A. (2015). Body node coordinator placement algorithms for Wireless Body Area Networks. *IEEE Internet of Things Journal* 2 (1): 94–102.

24 Sipal, V., Gaetano, D.,McEvoy, P. et al. (2015). Impact of hub location on the performance of wireless body area networks for fitness applications. *IEEE Antennas and Wireless Propagation Letters* 14: 1522–1525.

25 Zhang, R., Moungla, H., and Mehaoua, A. (2015). Delay Analysis of IEEE 802.15.6 CSMA/CA Mechanism in Duty-Cycling WBANs. Proceedings of the 2015 IEEE Global Communications Conference (GLOBECOM), 1–6, San Diego, CA.

26 Zhang, R., Moungla, H., Yu, J. et al. (2017). Medium access for concurrent traffic in wireless body area networks: Protocol design and analysis. *IEEE Transactions on Vehicular Technology*, 66 (3): 2586–2599.

27 Akbar, M.S., Yu, H., and Cang, S. (2017). TMP: tele-medicine protocol for slotted 802.15.4 with duty-cycle optimization in wireless body area sensor networks. *IEEE Sensors Journal* 17 (6): 1925–1936.

28 Yang, X., Wang, L., and Zhang, Z. (2018). Wireless Body Area Networks MAC protocol for energy efficiency and extending lifetime. *IEEE Sensors Letters* 2 (1): 1–4, 7500404.

29 Marinkovic, S.J., Popovici, E.M., Spagnol, C. et al. (2009). Energy-efficient low duty cycle MAC protocol for wireless body area networks. *IEEE Transactions on Information Technology in Biomedicine* 13 (6): 915–925.

30 Alshaheen, H. and Rizk, H.T. (2017). Improving the energy efficiency for a WBSN based on a coordinate duty cycle and network coding. Proceedings of the 2017 13th International Wireless Communications and Mobile Computing Conference (IWCMC), 1215–1220, Valencia.

31 Hussain, Z., Karvonen, H., and Iinatti, J. (2017). Energy efficiency evaluation of wake-up radio based MAC protocol for Wireless Body Area Networks. Proceedings of the 2017 IEEE 17th International Conference on Ubiquitous Wireless Broadband (ICUWB), 1–5, Salamanca

32 Samouni, N., Jilbab, A., and Aboutajdine, D. (2015). Energy efficient of ARQ and FEC in wireless body area networks. Proceedings of the 2015 Third World Conference on Complex Systems (WCCS), 1–5, Marrakech.

33 Hossain, M.U., Dilruba, M., Kalyan, M. et al. (2014). Multi-dimensional traffic adaptive energy-efficient MAC protocol for Wireless Body Area Networks. Proceedings of the 2014 9th International Forum on Strategic Technology (IFOST), 161–165, Cox's Bazar.

34 Rahman, A.F.A., Ahmad, R., and Ramli, S.N. (2014). Forensics readiness for Wireless Body Area Network (WBAN) system. Proceedings of the 16th International Conference on Advanced Communication Technology, 177–180, Pyeongchang.

35 Al Rasyid, M.U.H., Prasetyo, D. ,Nadhori, I.U. et al. (2015). Mobile monitoring of muscular strain sensor based on Wireless Body Area Networks. Proceedings of the 2015 International Electronics Symposium (IES), 284–287, Surabaya.

36 Golestani, N. and Moghaddam, M. (2018). Improving the Efficiency of Magnetic Induction-Based Wireless Body Area Network (WBAN). Proceedings of the 2018 IEEE International Microwave Biomedical Conference (IMBioC), 166–168, Philadelphia, PA.

37 Zhang, R., Moungla, H., Yu, J. et al. (2017). Multichannel broadcast in duty-cycling WBANs via channel hopping. *IEEE Internet of Things Journal* 4 (6): 2351–2361.

38 Zhang, R., Moungla, H., Yu, J. et al. (2017). Multi-channel broadcast in asymmetric duty cycling wireless body area networks. Proceedings of the 2017 IEEE International Conference on Communications (ICC), 1–6, Paris.

39 Ahmed, T., Ahmed, F., and Le Moullec, Y. (2016). Optimization of channel allocation in wireless body area networks by means of reinforcement learning. Proceedings of the 2016 IEEE Asia Pacific Conference on Wireless and Mobile (APWiMob), 120–123, Bandung.

40 Abbaspour-Asadollah, M. and Soleimani-Nasab, E. (2016). Performance analysis of cooperative wireless body area networks over Gamma fading channels. Proceedings of the 2016 8th International Symposium on Telecommunications (IST), 380–385, Tehran.

41 Han, S. and Park, S.K. (2011). Performance analysis of wireless body area network in indoor off-body communication. *IEEE Transactions on Consumer Electronics* 57 (2): 335–338.

42 Di Franco, F., Tachtatzis, C., Graham, B. et al. (2010). The effect of body shape and gender on wireless Body Area Network on-body channels. Proceedings of the IEEE Middle East Conference on Antennas and Propagation (MECAP 2010), 1–3, Cairo.

43 Ali, M.S., Tabassum, H., and Hossain, E. (2016). Dynamic user clustering and power allocation for uplink and downlink Non-Orthogonal Multiple Access (NOMA) systems. *IEEE Access* 4: 6325–6343.

44 Sabuj, S.R. and Hamamura, M. (2018) Two-slope path-loss design of energy harvesting in random cognitive radio networks. *Computer Networks* 142: 128–141.

45 Yildirim, I., Uyrus, A., Basar, E. et al. (2019). Propagation Modeling and Analysis of Reconfigurable Intelligent Surfaces for Indoor and Outdoor Applications in 6G Wireless Systems. [online] https://arxiv.org/abs/1912.07350

46 Evolved Universal Terrestrial Radio Access (E-UTR) (2010). Further Advancements for E-UTRA Physical Layer Aspects (Release 9), 3GPPTR- 36.814.

47 Rahul, A.R., Sabuj, S.R., Akbar, M.S. et al. (2020). An optimization based approach to enhance the throughput and energy efficiency for cognitive unmanned aerial vehicle networks. *Wireless Networks* 27 (1): 1–19.

48 Sabuj, S.R. and Hamamura, M. (2017). Outage and energy-efficiency analysis of cognitive radio networks: A stochastic approach to transmit antenna selection. *Pervasive and Mobile Computing* 42: 444–469.

49 Lambert function [Online] Available: https://en.wikipedia.org/wiki/Lambert_W_function

50 Sabuj, S.R and Islam, M.S. (2012). Performance analysis of SFBC and data conjugate in MIMO-OFDM system over Nakagami Fading Channel. *Journal of Communication* 7 (11): 790–794.

51 Nazneen, S., Chowdhury, M.M.J., and Sabuj, S.R. (2019). Analysis of delay-sensitive performance in cognitive wireless sensor networks. *Internet Technology Letters* 2 (3): 1–6.

5

Overview of Resource Management for Wireless Ad Hoc Network

Mehajabeen Fatima[1] and Afreen Khursheed[2]

[1] *SIRT, Bhopal, Madhya Pradesh, India*
[2] *IIIT, Bhopal, Madhya Pradesh, India*

5.1 Introduction

The popularity of mobile computing and communication devices has increased significantly, facilitated by the developments in software and processing power of devices. The technological innovation is so fast that it is expected that the majority of the world's population will be online in the next few years. Most wireless devices like laptops, cell phones or smart phones, handheld digital devices, or wearable computers like the tablet, interact with humans and connect to the global Internet. In the future, a considerable number of wireless devices will work with little or no human intervention. Laptops, cell phones, smart phones, and tablets are considered to generate temporary networks or an ad hoc network, entirely for instantaneous communication without any external involvement. An ad hoc network may be understood as the cooperative engagement of collection of mobile nodes without any centralized access or control point in which each node acts as a router. It is supposed to interact with mobile nodes using a dynamic network in which nodes can join and leave arbitrarily. Currently, complications are faced because of the structure of the Internet. Ad hoc networks can assist to address these difficulties. The two devices that are in instant range of each other still have to use routers and switches to forward packets between them. Ad hoc networks may be able to change this by directly connecting multiple wireless devices. Ad hoc networks may be employed in companies, university campuses, hospitals, navy, civilian, and the military services, and may become the first choice for connecting devices that are located nearby. Also, unlike regular Internet networks, ad hoc networks have several additional constraints such as battery life, mobility, and relatively lower processing power. To solve such problems, several different protocols have been designed for ad hoc networks. All these protocols have their own advantages and disadvantages. It is important to note that even though the intention is to generate an ad hoc network for mobile nodes, these nodes should be compatible with the regular Internet too. For this reason, the standard network TCP/IP architecture is implemented on all of these nodes in WANET. The standard TCP/IP architecture poses new design challenges, mainly due to mobility and limited resources of nodes.

Smart and Sustainable Approaches for Optimizing Performance of Wireless Networks: Real-time Applications, First Edition.
Edited by Sherin Zafar, Mohd Abdul Ahad, Syed Imran Ali, Deepa Mehta, and M. Afshar Alam.
© 2022 John Wiley & Sons Ltd. Published 2022 by John Wiley & Sons Ltd.

The whole life-cycle of ad hoc networks can be categorized into the first-, second-, and third-generation ad hoc network systems. Present ad hoc networks systems are considered to be the third generation. The first generation goes back to 1972. At that time, they were called PRNET (Packet Radio Networks). In conjunction with ALOHA (Areal Locations of Hazardous Atmospheres) and CSMA (Carrier Sense Multiple Access), they were approaches for medium access control and a kind of distance-vector routing. A PRNET was used on a trial basis to provide different networking capabilities in a combat environment. The IEEE 802.11 subcommittee adopted the term "ad hoc networks" and the research community had started to look into the possibility of deploying ad hoc networks in other areas of application. Meanwhile, the work is ongoing to advance the previously built ad hoc networks. GloMo (Global Mobile Information Systems) and the Near-term Digital Radio are examples of the results of these efforts. GloMo was designed to provide an office environment to connect at any hour from any place with Ethernet-type multimedia in handheld devices. In 2020, WANET is looking like the demand of the time and it is required to develop it.

In the first part of the chapter, Communication System Design, and Wired and Wireless Network Design Approach architectures are discussed. This part will discusses about the layered approach which was applied in the Open System Interconnect (OSI) reference model. TCP/IP and cross-layer architectures are also discussed here.

The second part of the chapter gives an overview of History, Spectrum of Wireless Ad Hoc Network, General Concepts of Ad Hoc Network, and Enabling Technologies and Taxonomy. WANET's various technologies are Zigbee, Bluetooth, IEEE 802.11, and IEEE802.16. A comparison based on data rate, range, configuration, and applications of these technologies is presented in this section.

The third part of this chapter investigates various emerging models of ad hoc networks, discussing their distinctive properties and highlighting various research issues arising due to these properties. We specifically provide discussions on Mobile Ad Hoc Network (MANET), Vehicular Ad Hoc Network (VANET), Wireless Mesh Network (WMN), and Wireless Sensor Network (WSN). MANET provides communication among mobile nodes, but lacks any infrastructure. There are no base stations, no fixed routers, and no centralized administration. All nodes may move randomly and connect dynamically to each other. Therefore, all nodes are operating as routers, capable to discover and maintain routes and propagate packets accordingly in the network. Due to the limited propagation range of the wireless environment, routes in ad hoc networks are multihop and mobile nodes establish routes dynamically to form their own network.

An ad hoc network can be used to make vehicles smart and intelligent and this can be done with the help of VANET. VANET turns every participating vehicle into a wireless router or node, allowing vehicles approximately 300 m to 1 km apart from each other to connect and, in turn, create a network with a wider range. VANET enables communication between the vehicles and roadside infrastructures. It makes transportation systems more intelligent. Since the movement of vehicles are restricted by roads and traffic regulations, we can deploy fixed infrastructure at critical locations. For this, a mesh type of network is required which can use infrastructure and also can function without it. Thus, WMNs come into the picture.

WMNs are dynamically self-organized and self-configured, with the nodes in the network automatically establishing an ad hoc network and maintaining the mesh connection. Mesh networks are built on a mix of fixed and mobile nodes interconnected via wireless links to form a multihop ad hoc network. Even though mesh networks are quite recent, they have already shown great potential in the wireless market. A wireless mesh network is a fully wireless network that employs multihop communications to forward traffic or route to and from wired Internet entry points. A mesh network introduces a hierarchy into the network architecture with the implementation of dedicated nodes (called wireless routers) communicating with each other and providing wireless transport services to data traveling from users to either other users or access points (access points are special wireless routers with a high-bandwidth wired connection to the Internet backbone).

Nowadays, temperature, humidity, etc. information is available in smart phones. Heart rate and blood pressure readings are made available through smart bands. All this happens due to the availability of cheap cost sensors. This can be accomplished with the help of WSN. Therefore, WSN is described in this chapter and a brief note is provided on this. A WSN is a wireless network consisting of spatially distributed autonomous devices using sensors to cooperatively monitor physical or environmental conditions, such as temperature, sound, vibration, pressure, motion, or pollutants, at different locations. It is a collection of sensing devices that can communicate wirelessly. Even so, wireless sensors have limited resources in memory, computation power, bandwidth, and energy. With small physical size, such a device can be embedded anywhere in the physical environment.

For communication, issues like coverage limitation, mobility of nodes, channel access, dynamic routing, etc., should be considered. Thus, Network layer and MAC layer challenges of wireless ad-hoc networks are discussed. An overview on the communication management requirements of WANET is also given.

The chapter ends with a summary of current research on ad hoc networks, neglected research areas, and directions for further research, improvement in data transmission, addressing the issues arising in multi-channel MAC, routing, transport protocols, and security. The research is at initial stages and extensive efforts are required in development of techniques that can address spectrum management, mobility management, signaling, routing, transport protocol, and security issues. Finally, new application models, implementation tools, analytical models, and benchmarks have to be developed.

5.1.1 Wired and Wireless Network Design Approach

The layered approach was applied in the OSI reference model. It defines seven layers of network stack (physical, data link, network, transport, session, presentation, and application layer). The seven layers are reduced to either five or four layers. This model is called TCP/IP. Now, the wired and wireless network is based on this layered architecture. The worldwide success of the Internet has led to the domination of the layered architecture [1]. TCP/IP is a hierarchical model in which each layer of the protocol stack operates independently and exchanges information with adjacent layers only, as shown in Figure 5.1. Communication between nonadjacent layers is not allowed, whereas adjacent layers communicate through static interfaces, independent of the individual network constraints and applications. The lower layer presents only a service interface to an adjacent upper

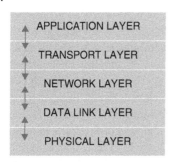

APPLICATION LAYER

TRANSPORT LAYER

NETWORK LAYER

DATA LINK LAYER

PHYSICAL LAYER

Figure 5.1 Information exchange for traditional approach.

layer and hides any other information, which can lead to poor performance. To avoid this, the protocol can reveal implementation information that would normally be hidden behind a layer boundary.

The layered network was originally designed for wired links, characterized by high bandwidth, low delay, low probability of packet loss or high reliability, static routing, and no mobility [2]. But in the wireless links, performance and resources are limited by the scarce availability of transmission spectrum, the employed modulation type, and the limited transmission power.

In a wireless network, the physical layer, MAC layer, and routing layer together contend for the network resource [3]. At the physical layer, transmission power and data rate that is decided upon affects MAC and routing decisions. The MAC layer is responsible for scheduling and allocating the wireless channel, and it will determine the available transmitter bandwidth and the packet delay [4]. The routing layer also depends on bandwidth and delay to select the link [5]. The routing layer chooses the route by which to send the data packets to the destination. The routing decision will change the contention level at the MAC layer and accordingly the physical layer parameters [6]. Because of the direct dependencies among lower layer and the upper layers, the traditional and strict layered design is not flexible enough to cope with the dynamics of next-generation communications which will be dominated by wireless ad hoc networks. The cross-layer design has emerged as a flexible architecture. In the cross-layer design, all the layers communicate with each other and can make decisions accordingly [7]. Although cross-layer design sometimes can lead to better performance of protocols, the vast majority of communication algorithms for wired networks conform to a strictly layered architecture scheme. In contrast, the cross-layer approach has gained significant popularity in wireless networks, because the wireless medium allows modalities of communication that are not possible in wired networks.

The cross-layer approach has been applied in the development of communication protocols for ad hoc networks. We can distinguish numerous factors motivating cross-layer design. The necessity of an infrastructureless, hop-by-hop communication, a lightweight protocol in ad hoc network because of constrained resources available at each node, motivates the use of the cross-layer approach, thus the prevalent trend in building a more flexible communication architecture.

Since inflexibility and sub-optimality of the layered architecture design usually result in poor performance of a network, a cross-layer protocol design that supports adaptivity and optimization across multiple layers of the protocol stack is needed [8]. Cross-layer

Figure 5.2 Information exchange in cross layer approach.

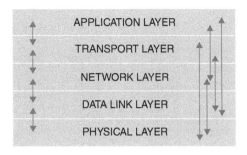

design breaks away from traditional network design. The dependent design between the layers blurs the boundary between the two adjacent layers [9]. It allows exchange of information of a layer with any other possibly non-adjacent layer in the protocol stack. In the cross-layer approach, the information is exchanged among different layers of the protocol stack, as shown in Figure 5.2. The end-to-end performance is optimized by adapting each layer against extracted information.

So instead of using TCP/IP, cross-layer architecture can be a more promising architecture for ad hoc networks. Cross-layer architectures may provide a more flexible solution, breaking the traditional structure by allowing interactions between two or more non-adjacent layers [10]. The most current approaches depend purely on local information and provide only poor and inaccurate information gathered at the global scale. The possible use of cross-layering architectures on the local view can have potential importance on the global view for autonomic communications. It is clear from recent initiatives that there is a need to make future networks in an optimal way with endogenous management and control with minimum human interference. To attain such a self-optimization system with existing strictly layered approaches may be possible, but will not leverage all the possible optimizations. Cross-layer architectures to be better suited to achieve such self-optimization [11].

Any design changes in the protocol stack when adding interaction between different layers may have an effect on the whole system, which may lead to a spaghetti design [12]. So, it is required to use cross-layer design with caution. Careful exploitation of cross-layer protocol interactions can lead to more efficient performance of the protocol stack and hence better application layer performances. There are many cross-layer design proposals in the literature. A survey of several cross-layer design proposals is presented. The layered architecture can be violated in the following basic ways. The cross-layer design proposals in the literature may fit into one of these basic categories:

- **Creation of new interfaces** (Figures 5.3a–c): Several cross-layer designs create new interfaces for information sharing between the layers at runtime. The new interface not available in the layered architecture is a violation of its basic design. It is divided into three subcategories, depending on the direction of information flow along the new interfaces [13]:
  - Upward: From lower layer(s) to a higher layer
  - Downward: From higher layer(s) to a lower layer
  - Back and forth: Iterative flow between two layers

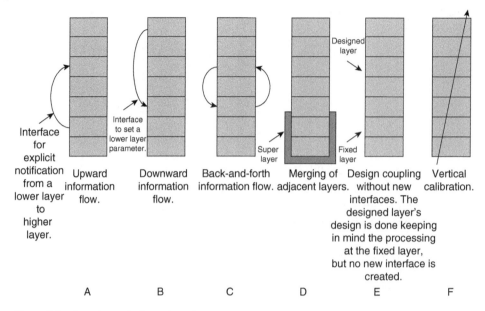

Figure 5.3 Cross layer design categories.

- **Merging of adjacent layers** (Figure 5.3d): In this category, two or more adjacent layers are together such that the service provided by the new superlayer is in unison with the services provided by the constituent layers. The super layer can be interfaced with the rest of the stack by using the interfaces that already exist in the original architecture.
- **Design coupling without new interfaces** (Figure 5.3e): Another category of cross-layer design involves coupling two or more layers at the design stage without creating any new interfaces for information sharing at runtime, as illustrated in Figure 5.3e. Without new interfaces, the architectural cost may be such that it may not be possible to replace one layer without making corresponding changes to another layer.
- **Vertical calibration across layers** (Figure 5.3f): Here, the parameters are adjusted that span across layers, as illustrated in Figure 5.3f. Basically, the performance seen at the level of the application is a function of the parameters at all the layers below it.

5.1.2 History

The concept of ad hoc networks is not new. Its history can be dated back to the Department of Defense (DoD)-sponsored Packet Radio networks (PRNET) research for military purposes in 1970s, which evolved into the Survivable Adaptive Radio networks (SURAN) program in the early 1980s [14]. The Defense Advanced Research Projects Agency's Strategic Technology Office DARPA [15, 16] wants proposals that will help it to develop tactical wireless networks that have little centralized control or infrastructure and limited or no reliance on aerial relay nodes, with a throughput of greater than 300 kilobits/sec for a network with 200 nodes. The PRNET used a combination of ALOHA and CSMA approaches for medium access, and a kind of distance-vector routing. SURAN employed routing protocols which were based on hierarchical link-state and were highly scalable. In the early 1990s, ad hoc networks entered a new phase of development due to the popularity of notebook computers as communication equipment based on RF and infrared grew. The idea of an infrastructureless collection of mobile hosts was proposed, and the IEEE 802.11 subcommittee adopted the term "ad hoc networks." Novel non-military applications were suggested. At around the same time, the DoD continued to found programs such as the Global Mobile Information Systems (GloMo), and the Near-term Digital Radio (NTDR) [15]. The goal of GloMo was to provide office-environment Ethernet-type multimedia connectivity in handheld devices. Channel access approaches were CSMA/CA and TDMA, and several novel routing and topology control schemes were deployed. The NTDR used clustering and linkstate routing, and self-organized into a two-tier ad hoc network. Presently, NTDR is the only "real" (non-prototype) ad hoc network used by the US Army. Since the mid-1990s, a lot of work has been done on the ad hoc standards [16]. Within the IETF, the mobile ad hoc networking (MANET) working group was born, and made efforts to standardize routing protocols for ad hoc networks. The IEEE 802.11 subcommittee standardized a medium access protocol that was based on collision avoidance.

5.1.3 Spectrum of Wireless Ad Hoc Network

Wireless Spectrum is divided into different communication system. Figure 5.4 demonstrates the spectrum used for broadcast TV, Digital TV, and FM Radio. Figure 5.5 shows the allotment of bandwidth to 3G broadband wireless, cellular phone, and personal communication services [17].

Figure 5.6 demonstrates the allocation of bandwidth to wireless LAN, Bluetooth, and local multipoint distribution services [18].

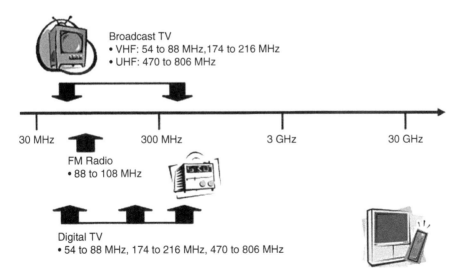

Figure 5.4 Broadcast TV spectrum.

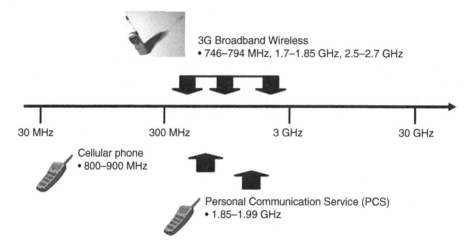

Figure 5.5 3G Broadband spectrum. Source: Based on D.B. Johnson and D.A. Maltz [17].

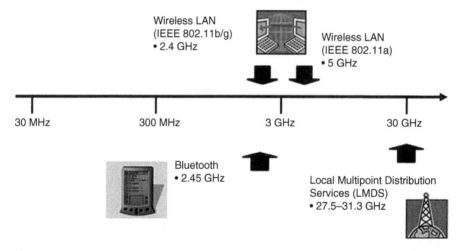

Figure 5.6 Wireless LAN spectrum. Source: Based on Lei Chen and Wendi B. Heinzelman [18].

5.1.4 Enabling and Networking Technologies

The single hop communication between devices is in the direct communication range of each other [18]. Wireless ad hoc communication technologies are:

- Bluetooth (IEEE 802.15.1) – used for personal area networks
- ZigBee (IEEE.15.4) – Short range (~ 100 m), low data rate (< 250 kb/s)
- Ad hoc Network (IEEE 802.11) – High-speed medium range MANETs
- High-speed wide range (IEEE 802.16)

IEEE 802.11 is the most widely-used standard. There are several specifications in the 802.11 family:

802.11 – applies to wireless LANs and provides 1 or 2 Mbps transmission in the 2.4 GHz band using either frequency hopping spread spectrum (FHSS) or direct sequence spread spectrum (DSSS).

802.11a – an extension to 802.11 that applies to wireless LANs and provides up to 54-Mbps in the 5 GHz band. 802.11a uses an orthogonal frequency division multiplexing encoding scheme rather than FHSS or DSSS.

802.11b – an extension to 802.11 that applies to wireless LANS and provides 11 Mbps transmission in the 2.4 GHz band. 802.11b uses only DSSS. 802.11b was a 1999 ratification to the original 802.11 standard, allowing wireless functionality comparable to Ethernet.

802.11e – a wireless draft standard that defines the Quality of Services support for LANs, and is an enhancement to the 802.11a and 802.11b wireless LAN (WLAN) specifications. 802.11e adds QoS features and multimedia support to the existing IEEE 802.11b and IEEE 802.11a wireless standards, while maintaining full backward compatibility with these standards.

802.11g – applies to wireless LANs and is used for transmission over short distances of up to 54-Mbps in the 2.4 GHz bands.

802.11n – 802.11n builds upon previous 802.11 standards by adding multiple input multiple output (MIMO). The additional transmitter and receiver antennas allow for increased data throughput through spatial multiplexing. A comparison of 802.11 standards is given Table 5.1.

Network technologies use the one-hop direct communication capabilities provided by enabling technologies to give reliable communication from sender to receiver, who may not be in direct range. The functions are routing, channel sharing, channel handshaking, and end-to-end reliable transmission of data.

Table 5.1 IEEE standard 802.11 comparison.

802.11 protocols-> Parameters	802.11 a	802.11b	802.11g	802.11 n
Frequency (GHz)	5	2.4	2.4	2.4/5
Data rate (Mbps)	54	11	54	700
Range (feet)	50	100	100	50
Bandwidth (MHz)	20	20	20	40
Modulation	OFDM	DSSS	OFDM, DSSS	OFDM
No. of overlapping channels	4 (indoor/outdoor)	3 (indoor/outdoor)	3 (indoor/outdoor)	4 (indoor/outdoor)
Compatibility	Wi-Fi	Wi-Fi	Wi-Fi at 11Mbps	Wireless Ethernet compatibility alliance
MIMO streams	1	1	1	4

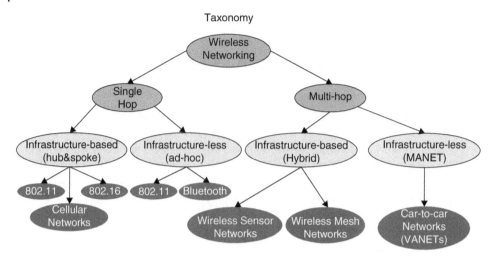

Figure 5.7 WANET taxonomy.

5.1.5 Taxonomy of Wireless Ad Hoc Network (WANET)

WANET consists of single- and multi-hop infrastructure-based and infrastructureless networks. Infrastructure-based relies on fixed infrastructure such as base station. Base stations or access points are fixed and centralized, e.g., Cellular networks, WLANS. Infrastructureless networks, also called ad hoc networks, have no centralized access points [19]. The nodes can move at any time from the area covered. The taxonomy of different types of wireless networks is shown in Figure 5.7. Only ad hoc networks are considered in this chapter. Different ad hoc networks are:

1. MANET
2. VANET
3. Wireless sensor Nodes
4. Wireless Mesh Network

5.2 Mobile Ad Hoc Network (MANET)

The requisite of human talk and communication has led to the development of wireless networks, which opens up a new vista of research in mobile ad hoc networks. Ad hoc networks with no communication infrastructure may be used in areas for emergency searches, rescue operations, or places where people wish to quickly share information [20]. An ad hoc network lacks any infrastructure. There are no base stations, no fixed routers, and no centralized administration. All nodes may move randomly and are connecting dynamically to each other. Therefore, all nodes are operating as routers, capable to discover and to maintain routes, and to propagate packets accordingly in the network. Due to the limited propagation range of the wireless environment, routes in ad hoc networks are multi-hop, and mobile nodes in this network dynamically establish routing to form a network.

5.2.1 Introduction to MANET

Wireless communication is currently one of the fastest growing technologies worldwide, due to recent advances in mobile computing devices and wireless technology. Mobile devices such as laptops, personal digital assistants (PDAs), mobile phones, or smart phones, etc., have become lightweight and portable enough to be carried everywhere. Wireless communication networks have a number of benefits over their traditional wired counterparts. In principle, wireless networks allow connectivity whenever required. They can be deployed in areas without a preexisting wired-communication infrastructure or where it is difficult to lay cables.

Figure 5.8 shows the difference between Ad Hoc Mode and Infrastructure Mode. In ad hoc mode, all mobile nodes can communicate with each other. Nodes communicate through a base station in infrastructure mode.

Mobile Ad hoc network is shown in Figure 5.9. MANET has lack of communication infrastructure may be used in areas for emergency searches and rescue operations, or places where people wish to quickly share information [5]. MANETs are distinguished from other types of networks by their physical characteristics, organizational format, and dynamic topology.

There are no base stations, no fixed routers, and no centralized administration. All nodes may move randomly and are connecting dynamically to each other, as shown in Figure 5.9. Therefore, all nodes are operating as routers, capable to discover and maintain routes, and to propagate packets accordingly in the network. Due to the limited propagation range of the wireless environment, routes in ad hoc networks are multi-hop and mobile nodes in this network dynamically establish routing among themselves to form their own network. Thus, connectivity is a challenging task of MANET.

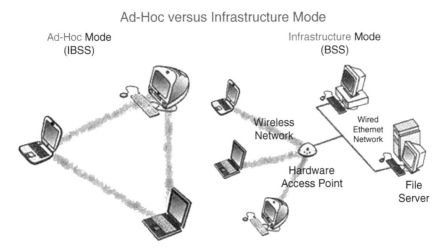

Figure 5.8   Ad hoc mode Vs infrastructure mode.

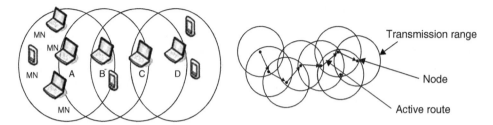

Figure 5.9 Mobile ad hoc network.

5.2.2 Common Characteristics of MANET

- Autonomous and infrastructureless: MANET does not depend on any established infrastructure or centralized administration. Each node operates in distributed peer-to-peer mode, acts as an independent router, and generates independent data. MANET's ability to self-form and self-manage eliminates the need for intensive central management of network links. MANET technologies allow a force of mobile nodes to more easily share data and attain greater situational awareness than a non-networked force [20].
- Multihop routing: No dedicated routers are necessary; every node acts as a router and forwards each other' packets to enable information sharing between mobile hosts.
- Mobility: Each node is free to move while communicating with other. Links are created and destroyed in an unpredictable way. Therefore, the network status can change quickly, causing hosts to have imprecise knowledge of the current network state.
- Physical characteristics: Wireless channels are inherently error-prone, due to effects such as multipath fading, interference, and shadowing, causing unpredictable link bandwidth and packet delay.
- Organizational format: The distributed nature of MANET means that channel resources cannot be assigned in a predetermined way.
- Decentralized and robust: If one of the base stations is not working in an infrastructure network, all users of that base station will lose connectivity to other networks. The ad hoc networks can avoid such problems. If one node leaves the network or is not working, it can still have connectivity to other nodes and maybe can use these nodes to multi-hop the message to the destination nodes, as long as there is at least one route to the desired node [21].

5.2.3 Disadvantages

The shortcomings of the wireless link impacts the ad hoc network, such as fading and bandwidth availability [22]. The following are the disadvantages of ad hoc networks:

- The routing table entries in an ad hoc network change much faster than they do in regular networks.
- Devices in MANET use batteries as their power supply. The route may fail frequently due to excessive battery power consumption. Thus, power limitation in mobile devices is a serious factor.

- Lower data rate is one of biggest problem of ad hoc networks. The speed of data will be less if data rate is low. This is inconvenient for non-real time communication.
- Ad hoc networks have the problem of dynamic topology and scalability. Since an ad hoc network does not allow the same kind of aggregation techniques that are available to standard Internet routing protocols, they are vulnerable to scalability problem.
- Large networks can have excessive latency (high delay), which affects real-time applications.

5.2.4 Applications of MANET

Some of the application areas [23, 24] are listed in Table 5.2.

Table 5.2 Applications of MANET.

Tactical networks	Military communication and operations, Automated Battlefields
Sensor networks	Collection of embedded sensor devices used to collect real-time data to automate everyday functions. Data highly correlated in time and space, e.g., remote sensors for weather, earth activities, sensors for manufacturing equipment. Can have between 1000 and 100 000 nodes, each node collecting sample data, then forwarding data to centralized host for processing using low homogeneous rates.
Emergency services	Search-and-rescue operations, as well as disaster recovery, e.g., early retrieval and transmission of patient data (record, status, diagnosis) from/to the hospital, replacement of a fixed infrastructure in case of earthquakes, hurricanes, fire, etc.
Home and enterprise	Home/office wireless networking (WLAN), e.g., shared white-board networking
	application, using PDA to print anywhere, trade shows, Personal Area Network (PAN), Body Area Network (BAN).
Commercial	E-Commerce, e.g., electronic payments from anywhere (i.e., in a taxi). Dynamic Business environment – access to customer files stored.
	In a central location on the fly provides consistent databases for all agents' mobile office, e.g., transmission of news, road conditions, weather, music, and local ad hoc network with nearby vehicles for road/accident guidance.
Educational	Set up virtual classrooms or conference rooms applications. Set up ad hoc communication during conferences, meetings, or lectures.
Entertainment	Multiuse games, Robotic pets, Outdoor Internet access.
Location-aware services	Follow-on services, e.g., automatic call forwarding, transmission of the actual workspace to the current location.
	Information services: Push, advertise location-specific service, like gas stations, Pull, e.g., location-dependent travel guide, Services (printer, fax, phone), availability of information; etc.

5.2.5 Major Issues of MANET

A. Autonomous – No centralized administration entity is available to manage the operation of the different mobile nodes [23].
B. Dynamic topology – Nodes are mobile and can be connected dynamically in an arbitrary manner. Links of the network vary with time and are based on the proximity of one node to another node.
C. Device discovery – Identifying relevant newly moved nodes and informing about their existence need dynamic updates to facilitate automatic optimal route selection.
D. Bandwidth optimization – Wireless links have significantly lower capacity than wired links.
E. Limited resources – Mobile nodes rely on battery power, which is a scarce resource. Also, storage capacity and power are severely limited.
F. Scalability – Scalability can be broadly defined as whether the network is able to provide an acceptable level of service, even in the presence of a large number of nodes.
G. Limited physical security – Mobility implies higher security risks such as peer-to-peer network architecture or a shared wireless medium accessible to both legitimate network users and malicious attackers. Eavesdropping, spoofing, and denial-of-service attacks should be considered.
H. Infrastructureless and self-operated, self-healing feature demands MANET should realign itself to blanket any node moving out of its range.
I. Poor Transmission Quality – This is an inherent problem of wireless communication caused by several error sources that result in degradation of the received signal.
J. Ad hoc addressing – Challenges in standard addressing scheme to be implemented.
K. Network configuration – The whole MANET infrastructure is dynamic and is the reason for dynamic connection and disconnection of the variable links.
L. Topology maintenance – Updating information of dynamic links among nodes in MANETs is a major challenge.

5.3 Vehicular Ad Hoc Network (VANET)

With the increasing number of vehicles equipped with computing technologies and wireless communication devices, inter-vehicle communication is becoming a promising field of research, standardization, and development. VANET is an application of an ad hoc and self-configuring network of vehicles.

5.3.1 Introduction to VANET

In VANET, nodes can be connected by wireless links. Vehicles known as nodes free to move in all directions. Every node sends, receives, and retransmits data that includes speed, location, and direction. The mobile nodes or vehicles attached with sophisticated equipment traveling on roads [24]. These nodes communicate with each other for information exchange through vehicle to infrastructure (V2I), vehicle to vehicle (V2V), or inter-vehicle

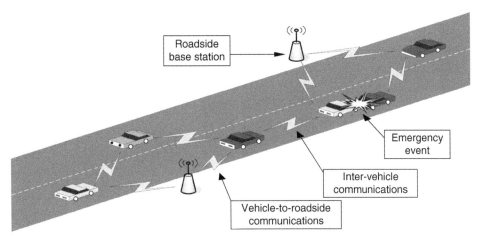

Figure 5.10 VANET.

communication, and vehicle to fixed roadside base station, as shown in Figure 5.10. Communications requires standard wireless to handle the fast dynamic network changes and to provide real-time results with high accuracy, such as IEEE 802.11P.

VANETs are characterized by high mobility, rapidly changing topology, and are temporary. VANETs are also featured by the movement and self-organization of the nodes (i.e., vehicles). The growing number of vehicles and users on the roads has led to a range of different problems in society [25]. These problems include road congestion, loss of human lives due to road accidents, and increased air pollution. Furthermore, the continuous use of vehicles results in a shortage of fossil fuels. To solve such issues, researchers and industries have worked together to present the idea of a VANET. VANET turns every participating vehicle into a wireless router or node, allowing vehicles approximately 300 m to 1 km apart from each other to connect and, in turn, create a network with a wide range. VANET enables communication among the vehicles and roadside infrastructures. It makes transportation systems more intelligent. Since the movement of vehicles are restricted by roads and traffic regulations, we can deploy fixed infrastructure at critical locations. The major components of a VANET are On-Board Unit (OBU), Road Side Unit (RSU), and Trusted Authority (TA) [18].

5.3.2 Common Features of VANET

VANET is characterized by high mobility, interactions of vehicles, dynamic topology, and self-organization of nodes. Characteristics are:

- Wireless communication: VANET is designed for the wireless environment. Nodes are connected and exchange their information via wireless.
- Network topology: Due to high node mobility and random speed of vehicles, the position of a node changes frequently.
- High mobility: The nodes in VANETs usually are moving at high speed. This makes it harder to predict a node's position and to protect the node's privacy.

- Unbounded network size: VANET can be implemented for one city, several cities, or for countries.
- Frequent exchange of information: The ad hoc nature of VANET motivates the nodes to gather information from the other vehicles and roadside units [26].

5.3.3 Pros, Cons, Applications

V2V communications have the following pros:

 (i) minimizes latency in transmission of data
 (ii) allows short- and medium-range communications
 (iii) presents lower deployment costs
 (iv) supports short messages delivery

V2I presents following advantages:

 (i) VANET/cellular interoperability
 (ii) information dissemination for VANETs using advanced antennas
 (iii) the integration of WiMAX (Worldwide Interoperability for Microwave Access) and Wi-Fi technologies seems to be a better option for better and cheaper wireless coverage extension in vehicular networks

V2V communications present the following shortcomings that can be solved by integrating with V2I, such as:

 (i) frequent topology partitioning due to high mobility
 (ii) problems in long-range communications
 (iii) problems using traditional routing protocols
 (iv) broadcast storm problems in high density scenarios [27]

Different applications of VANET are:

1. Driving experience and the safety of transportation:
   - Safety Application (to enhance driving safety)
   - EEBL – Emergency Electronic Brake Light (sudden braking)
   - PCN – Post Crash Notification
   - RFN – Road Feature Notification (ex. downhill curve)
   - LCA – Lane Change Assistance
2. Commercial/Infotainment Application:
   - Free flow Tolling
   - Social Networking
   - Multimedia Content Exchange
3. Convenience Application (for better driving experience):
   - Road Congestion Notification
   - Dynamic Route/Travel Time Planning
   - Finding Parking Spots
   - Passengers oriented applications – for offering new services, onboard Internet access, distributed games, chats, tourist information, city leisure information, movies announcement downloads

4. Infrastructure oriented applications – optimizing their management: transit management, freeway management, emergency organization
5. Vehicle oriented applications – for increasing road safety incident management, crash prevention, collision avoidance, driver assistance, automatic/adaptive settings
6. Driver oriented services – for improving road usage, traffic jams, road works information, traveler payment, ride duration estimate

The bottlenecks in deployment are unavailability of basic roadside infrastructure, privacy of the vehicles, lack of coordination among manufacturing giants, and security.

5.4 Wireless Mesh Network (WMN)

A wireless mesh network (WMN) is a fully wireless network that employs multihop communications to forward traffic on route to and from wired Internet entry points, as shown in Figure 5.11.

5.4.1 Preface of WMN

A mesh network introduces a hierarchy into the network architecture with the implementation of dedicated nodes (called wireless routers) communicating with each other and providing wireless transport services to data traveling from users to either other users or access points (access points are special wireless routers with a high-bandwidth wired connection to the Internet backbone) [28].

The network of wireless routers forms a wireless backbone (tightly integrated into the mesh network), which provides multihop connectivity between nomadic users and

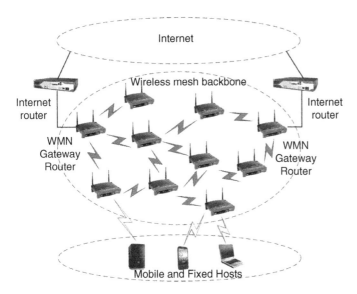

Figure 5.11 Mesh network.

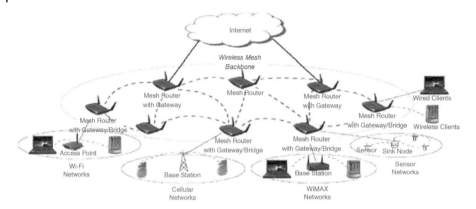

Figure 5.12 Wireless mesh network architecture. Source: Based on A.A. Pirzada, M. Portmann, and J. Indulska [29].

wired gateways. The meshing among wireless routers and access points creates a wireless backhaul communication system, which provides each mobile user with a low-cost, high bandwidth, and a seamless multihop interconnection service with a limited number of Internet entry points and with other wireless mobile users. Specifically, in the mesh case, the traffic is originated in the users' devices, traverses the wireless backbone, and is distributed over the Internet network. Figure 5.12 illustrates the mesh network architecture, highlighting the different components and system layers [29].

5.4.2   Common Traits of WMN

WMN do not require centralized access points to mediate the wireless connection. WMNs are dynamically self-organized and self-configured systems (one node can automatically establish and maintain the mesh connectivity). WMN increases the coverage area and link robustness of existing Wi-Fi's; if the correspondent nodes are not in the wireless transmission range of each other WMNs wireless nodes can be mobile or fixed. Basic types of nodes are mesh routers (MR) and mesh clients (MC), where MR establishes an infrastructure backbone for clients [30]. WMN may have Wireless Mesh routers, Gateways, Printers, Servers, Mobile or Stationary clients (terminals). They can form a client mesh network among themselves and/or together with MRs. Traffic is user-to-gateway or user-to-user.

The WMN may be connected to:

- the Internet through gateway/routers
- other networks through Gateways/Bridges
- end hosts and routing nodes might be distinct
- routers

The Wireless Mesh Networks Technologies is an extension of IEEE 802.11 WLANs. The hybrid technologies can cooperate in mesh topology. The APs (access points) are considered as the nodes of a mesh, being heterogeneous and interconnected in a hierarchical fashion. The integration and interconnection of different technologies in/with WMNs (Internet, cellular, IEEE 802.11, IEEE 802.15, IEEE 802.16, sensor networks, etc.) can be achieved through the gateway and bridging functions in the MRs [31].

5.4.3 WMN Open Issues and Research Challenges

All wireless networks present different challenges to make a sensor network stronger. Different challenges are:

- Architectures and Protocol layers
- PHY and MAC specific issues
- Scheduling, Multi-channel, multi-radio
- Cognitive radio
- MIMO, directional and smart antennas systems
- Capacity and coverage
- Scalability (major point): To improve the scalability of WMNs, designing a hybrid MAC scheme with CSMA/CA and TDMA or CDMA is an interesting and challenging research issue
- Routing
- Resource management and QoS capabilities
- Cross-layer optimization
- Security, etc.
- **Inter-operability and integration in architectures**

WMNs usually support network access for both conventional and mesh clients. WMNs need to be backward compatible with conventional client nodes. Integration of WMNs with other wireless/wireline networks requires appropriate routers/gateways/bridges.

- **Control and Management issues**

Protocols that are needed to make WMN autonomous include power management, self-organization, dynamic topology control, robust to temporary link failure, fast network subscription, and User authentication procedures to maintain the operation, monitor the performance, and configure the WMNs parameters [31].

5.4.4 Performance Metrics

Hop Count: is the most used metric in wireless multihop networks. To avoid looping in the Wireless network, a hop count is assigned to each node. Every time packets pass from one node to another, the hop count is decremented until it becomes zero [31].

 Expected Transmission Count: ETX is the transmission count for delivering a packet over a wireless link successfully. ETX of a path is the sum of ETXs of all links of this path:

$$ETX = \sum_{k=1}^{\infty} k * s(K) = I/1 - P$$

Expected Transmission Time (ETT): We define the ETT of a link as "bandwidth-adjusted ETX." In other words, we start with ETX and multiply by link bandwidth to obtain the time spent in transmitting the packet. Let S denote the size of packet and B bandwidth of link:

$$ETT = ETX * S/B$$

Weighted Cumulative Expected Transmission Time: This time depends on the loss rate and bandwidth. In WCETT, is the sum of expected times to successfully transmit a packet along the route:

$$WCETT = \sum_{k=1}^{n} ETT$$

5.4.5 Advantages and Disadvantages

Reduction of installation costs
Currently, one of the major efforts to provide wireless Internet beyond the boundaries of indoor WLANs is through the deployment of Wi-Fi hotspots. Basically, a hotspot is an area that is served by a single WLAN or a network of WLANs, where wireless clients access the Internet through an 802.11-based access point.

Large-scale deployment
In recently standardized WLAN technologies (i.e., 802.11a and 802.11g), increased data rates have been achieved by using more spectrally efficient modulation schemes. However, for a specific transmitting power, shifting toward more efficient modulation techniques reduces coverage.

5.4.6 Prominent Areas and Challenges of WMN

Broadband home networking
Currently broadband home networking is realized through IEEE 802.11 WLANs. An obvious problem is the location of the access points. Without a site survey, a home (even a small one) usually has many dead zones without service coverage. Solutions based on site survey are expensive and not practical for home networking, while installation of multiple access points is also expensive and not convenient because of Ethernet wiring from access points to backhaul network access a modem or hub. Moreover, communications between end nodes under two different access points have to go all the way back to the access hub. This is obviously not an efficient solution, especially for broadband networking. Mesh networking, as shown in Figure 5.13, can resolve all these issues in the home.

The access points must be replaced by wireless mesh routers with mesh connectivity established among them. Therefore, the communication between these nodes becomes much more flexible and more robust to network faults and link failures. Dead zones can be eliminated by adding mesh routers, changing locations of mesh routers, or automatically adjusting power levels of mesh routers. Communication within home networks can be realized through mesh networking without continuously going back to the access hub. Thus, network congestion due to backhaul access can be avoided. In this application, wireless mesh routers have no constraints on power consumptions and mobility. Thus,

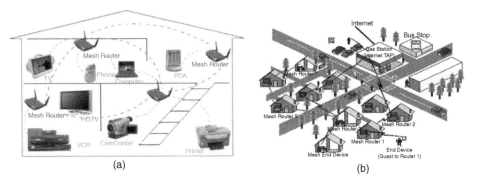

Figure 5.13   (a) Broadband WMN. (b) Community networking.

protocols proposed for mobile ad hoc networks and wireless sensor networks are too cumbersome to achieve satisfactory performance in this application. On the other hand, Wi-Fi's are not capable of supporting ad hoc multi-hop networking. As a consequence, WMNs are well suited for broadband home networking.

Community and neighborhood networking

In a community, the common architecture for network access is based on cable or DSL connected to the Internet, and the last-hop is wireless by connecting a wireless router to a cable or DSL modem. This type of network access has several drawbacks. Even if the information must be shared within a community or neighborhood, all traffic must flow through the Internet. This significantly reduces network resource utilization. A large percentage of areas in between houses is not covered by wireless services. An expensive but high bandwidth gateway between multiple homes or neighborhoods may not be shared and wireless services must be set up individually. As a result, network service costs may increase. Only a single path may be available for one home to access the Internet or communicate with neighbors. WMNs mitigate the above disadvantages through flexible mesh connectivities between homes, as shown in Figures 15.3a and b. WMNs can also enable many applications such as distributed file storage, distributed file access, and video streaming

5.5   Wireless Sensor Network (WSN)

A wireless sensor network (WSN) is a wireless network consisting of distributed autonomous devices using sensors to cooperatively monitor physical or environmental conditions, such as temperature, sound, vibration, pressure, motion, or pollutants, at different locations [32]. This is a collection of sensing devices that can communicate wirelessly, even though wireless sensors have limited resources in memory, computation power, bandwidth, and energy. With small physical size, it can be embedded in the physical environment and self-organizing multihop ad-doc networks [33].

5.5.1   Overview of WSN

As shown in Figure 5.14, WSN includes a large number of circulating, self-directed, minute, low-powered devices named sensor nodes or motes. These networks certainly

Figure 5.14 Wireless sensor network.

cover a huge number of spatially distributed, little battery-operated embedded devices that are networked to collect, process, and transfer data to the operators and with controlled capabilities of computing and processing [34]. Nodes are the tiny computers that work jointly to form the networks.

The most common WSN architecture follows the OSI architecture model. The architecture of the WSN includes five layers and three cross-layers. Five layers, namely, application, transport, network, data link, and physical layer [35]. The three cross-planes are named power management, mobility management, and task management. These layers of the WSN are used to accomplish the network and make the sensors work together in order to raise the complete efficiency of the network [36].

5.5.2 Common Properties of WSN

The properties of WSN [37]

- Mobility of nodes
- Power consumption constraints for nodes using batteries or energy harvesting
- Ability to cope with node failures (resilience)
- Scalability to large scale of deployment
- Heterogeneity of nodes
- Ability to withstand harsh environmental conditions
- Ease of use
- Cross-layer design

5.5.3 Benefits, Harms, and Usage of WSN

The benefits and harms of WSN (38) include the following:

- Network arrangements can be carried out without immovable infrastructure
- Execution pricing is inexpensive
- It avoids plenty of wiring
- It can be opened using a centralized monitoring

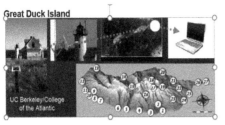

Great Duck Island

UC Berkeley/College
of the Atlantic

- 150 sensing nodes deployed throughout the island relay data
  temperature, pressure, and humidity to a central device.
- Data was made available on the Internet through a satellite link.

Zebranet: a WSN to study the behavior of zebras

- Special GPS-equipped collars were attached to zebras
- Data exchanged with peer-to-peer info swaps
- Coming across a few zebras gives access to the data

Figure 5.15 Environment monitoring through sensors.

- Flexible if there is a casual situation when an additional workstation is required
- App for the non-reachable places like mountains, over the sea, rural areas, and deep forests
- It might provide accommodations for new devices at any time

Some of the prominent applications of WSN include:

- Military applications
- Environmental applications
- Home and office applications
- Health applications
- Automotive applications
- Other commercial applications
- Area monitoring
- Forest fire detection
- Water quality monitoring
- Industrial monitoring
- Landslide detection
- Environmental/Earth sensing
- Air pollution monitoring

Some of the existing examples of Environment Monitoring through sensors are shown in Figure 5.15.

5.6 Intelligent Resource Management Concerns in WANET

In ad hoc networks, several interesting and difficult problems arise due to the shared nature of the wireless medium, limited transmission power, limited transmission range, node mobility, battery limitations and bandwidth limitation, etc. The limited transmission range of a wireless network coupled with the highly dynamic routing infrastructure requires extra care. Mobility also creates a lot of concerns. It is required to manage the mobility, coverage, and battery power. For communication, issues such as dynamic routing, efficient channel access, and quality-of-service (QoS) support, bandwidth, synchronization, distributed nature, and lack of central coordination should be considered. In this section, we will discuss the above-mentioned challenges at Network layer and MAC layer in wireless ad hoc networks. For this, the MAC layer, network layer protocols, and their issues are investigated

Different architectures like the cross-layer approach are proposed by many researchers. The cross-layer is in its early stages of research. This has not been implemented until now. The WANET is trying to manage through TCP/IP architecture. TCP/IP uses five layers for communication, as discussed in Section 5.1. The MAC, Network, and Transport layer design issues of the TCP/IP model is discussed in this section. Power management is also discussed here. In a wireless network, the physical layer, MAC layer, and routing layer together contend for the network resource. How the physical layer transmission power and data rate is decided affects MAC and routing decisions. The MAC layer is responsible for scheduling and allocating the wireless channel, and will determine the available transmitter bandwidth and the packet delay. The routing layer also depends on bandwidth and delay to select the link. The routing layer chooses the route to send the data packets to their destination. The routing decision will change the contention level at the MAC layer and accordingly the physical layer parameters [6]. The traditional protocol stack has direct dependencies between the physical layer and the upper layers. A strict layered design is trying to cope with the dynamics of next-generation communications, which will be dominated by wireless ad hoc network.

This section enlightens on the communication management requirements in WANET. Also, possible power management is provided.

5.6.1 Major Issues of WANET

The major issues that affect the design and performance of a WANET are as follows:

1) Hardware and Operating System
2) Wireless Radio Communication Characteristics
3) Medium Access Schemes
4) Deployment
5) Localization
6) Synchronization
7) Calibration
8) Network Layer
9) Transport Layer
10) Data Aggregation and Data Dissemination
11) Database Centric and Querying
12) Architecture
13) Programming Models for Sensor Networks
14) Middleware
15) Quality-of-Service
16) Security

5.6.2   Challenges of MAC Protocols

MAC protocol is used for coordination and scheduling of transmissions among competing neighbors and its goals are low latency, good channel utilization, best effort services, and real-time support. MAC protocol coordinates transmissions from different stations in order to minimize/avoid collisions. Various MAC layer protocols are available. Random Access is done in CSMA and MACA, and Channel Partitioning is done in TDMA, FDMA, and CDMA. The 802.11 standard specifies a common media access control (MAC) layer, which provides a variety of functions that support the operation of 802.11-based wireless LANs. Main functions of the 802.11 MAC layer are Scanning, Authentication, Association, WEP, RTS/CTS, Power Save Mode, and Fragmentation.

6.1 Challenges in MAC layer: The main issues to be considered while designing a MAC layer protocol for ad-hoc wireless network are given below.

6.1.1 Bandwidth efficiency and overhead: Since the radio spectrum is limited, the bandwidth available for communication is very limited. The MAC protocol must be designed in such a way that the scarce bandwidth is utilized in an efficient manner [5]. The control overhead involved must be kept to as much a minimum as possible.

6.1.2 Quality of Support (QoS): The nodes are mobile most of the time. Providing QoS support to data sessions in such networks is very difficult. Bandwidth reservation made at one time point may become invalid once the node moves out of the region where the reservation was made. QoS support is essential for supporting time critical traffic sessions.

6.1.3 Synchronization: A MAC protocol must take into consideration the synchronization between nodes in the network and bandwidth reservation by nodes. The exchange of control packets may be required for achieving time synchronization among nodes. 802.11 packets must not consume too much of the work bandwidth.

6.1.4 Hidden and exposed terminal problems: The hidden terminal problem refers to the collision of packets at a receiving node due to simultaneous transmission of those nodes that are not within the direct transmission range of the sender, but are within the transmission range of the receiver. Collision occurs when both nodes transmit packets at the same time [6]. The hidden and exposed terminal problems significantly reduce the throughput of a network when the traffic load is high. It is therefore desirable that the MAC protocol be free from hidden and exposed terminal problems.

6.1.5 Error-prone shared broadcast channel: Because of the broadcasting nature of transmission, collisions may occur. A MAC protocol should grant channel access to nodes in such a manner that collisions are minimized. Also, the protocol should ensure that all nodes are treated fairly with respect to bandwidth allocation.

6.1.6 Distributed nature/lack of central coordination: In MANET, nodes move continuously, therefore nodes must be scheduled in a distributed fashion for gaining access to the channel. This may require exchange of control information. The MAC protocol must make sure that the additional overhead, in terms of bandwidth consumption, incurred due to this control information exchange is not very high.

6.1.7 Mobility of nodes: Nodes are mobile most of the time in a wireless network. The bandwidth reservation made or control information exchange may end up being of no use if node mobility is very high. The MAC protocol has no role to play in influencing the mobility. The MAC protocol design must take this mobility factor into consideration, such that the performance of the system is not significantly affected due to node mobility.

5.6.3 Routing Protocols

Prominent studies are ongoing on routing protocols. An ad hoc routing protocol is a standard for controlling node decisions when routing packets traverse a MANET between devices. A node in the network, or one trying to join, does not know about the topology of the network. It discovers the topology by announcing its presence and listening to broadcasts from other nodes (neighbors) in the network. The process of route discovery is performed differently, depending on the routing protocol implemented in a network.

The conventional routing protocols are not well suited to ad hoc environments for several reasons. Firstly, they are designed for static topology, which is not the case in an ad hoc network. Secondly, they are highly dependent on periodic control messages, which is in contradiction with a resource limited ad hoc environment. Moreover, classical protocols try to maintain routes to all reachable destinations, which waste resources. Another limitation comes from the use of bidirectional links, which is not always the case in an ad hoc environment. There is a need for new routing protocols, adapted to the dynamic topology and the wireless links limitations. Routing protocols in such networks should provide a set of features including: distributed operation, loop freedom, on-demand-based operation, unidirectional link support, power conservation, multiple routes, efficiency, scalability, security, and QoS support. However, none of the proposed protocols have all the above desired properties. Thus, protocols are still under development and are probably being enhanced and extended with more functionality. Until now, no standard has been adopted and many critical issues remain to be solved. Routing protocols used in ad hoc networks must automatically adjust to environments that can vary. Such protocols must operate in an on-demand fashion and must carefully limit the number of nodes required to react to a given topological change in the network.

5.6.3.1 Challenges of Routing Protocols

Being one of the most popular fields of study during the last few years, almost every aspect of ad hoc networks has been explored to some level of detail. Yet, no ultimate resolution to any of the problems has been found or, at least, agreed upon. On the contrary, more questions have arisen than have been answered [27]. The major challenges and their description are given:

1. Mobility
2. Scalability
3. Quality-of-Service
4. Security
5. Node Cooperation

5.6.3.1.1 Scalability

Scalability can be broadly defined as whether the network is able to provide an acceptable level of service, even in the presence of a large number of nodes in the network. It is one of the most important open issues of ad hoc networks. Firstly, ad hoc networks suffer, by nature, from scalability problems in capacity. In a non-cooperative network, where omni-directional antennas are used, the throughput decreases at a rate where N is the number of nodes [28]. That is, in a network with 100 nodes, a single device gets approximately one-tenth of the theoretical data rate of the network interface card at maximum. This problem, however, cannot be solved, except by physical layer improvements such as smart antennas. Routing protocols also set limits for the scalability of ad hoc networks. Route acquisition, service location, and encryption key exchange are examples of tasks that will require considerable overheads, which will grow rapidly with the network size. Proactive routing is not applicable in a dynamic environment, due to the huge amount of broadcast messages of topology changes. Re-active protocols allow deploying large networks at the expense of increased route acquisition latency. The minimum route acquisition latency is the product of maximum network diameter and minimum node traversal time for route request.

Correspondingly, demands for short latencies for route acquisition drastically limit the network size. There is still much work to be done to optimize the trade-off between capacity and scalability in different scenarios and applications separately for a general solution.

5.6.3.1.2 Quality of Service

Quality of Service (QoS) is being developed to meet the emerging requirements of heterogeneous applications in the Internet, which is able to provide only best effort service. QoS is a guarantee by the network to provide a certain performance for a flow in terms of the quantities of bandwidth, delay, jitter, packet loss probability, etc. QoS of fixed wireless networks is still an open problem [8]. Moreover, ad hoc networks make the QoS appear an even more challenging problem than ever before, despite that some reactive routing protocols can be configured to return to only paths that comply with certain desired QoS parameters. RF channel characteristics often change unpredictably, along with the difficulty of sharing the channel medium with many neighbors, each with its own potentially changing QoS requirements. Routes are using links with different quality and stability, which are often asymmetrical. There are numerous multi-layer attempts to improve the QoS problems, from the service contracts to the MAC layer. A promising method for satisfying QoS requirements is a more unified approach of cross-layer or vertical-layer integration. The idea is different from many of the traditional layering styles, to allow different parts of the stack to adapt to the environment in a way that takes into account the adaptation and available information at other layers. QoS routing policies, algorithms, and protocols with multiple, including preemptive, priorities to be researched in the future [29]. Due to the nature of ad hoc networks, QoS cannot be guaranteed for a long time because of the link quality variation. Methods to detect and report changes in the connection quality should be investigated in the future. For example, Perkins suggested the addition of a new ICMP message (QOS_LOST) to be defined to inform the end point at which a new route discovery should be initiated.

5.6.3.1.3 Security

Security is a critical issue of ad hoc networks that is still a largely unexplored area. Since nodes use the open, shared radio medium in a potentially insecure environment, they are particularly prone to malicious attacks, such as denial of service (DoS). Lack of any centralized network management or certification authority makes the dynamically changing wireless structure very vulnerable to infiltration, eavesdropping, interference, etc. Security is often considered to be the major "roadblock" in commercial applications of ad hoc network technology. Traditional methods of protecting the data with cryptographic methods face a challenging task of key distribution and refresh. Accordingly, the research efforts on security have mostly concentrated on secure data forwarding. However, many security risks are related to peculiar features of ad hoc networks. The most serious problem is probably the risk of a node being captured and compromised. This node would then have access to structural information on the network, relayed data, but it can also send false routing information which would paralyze the entire network very quickly [31]. One of the current approaches to the security problems is building a self-organized public-key infrastructure for ad hoc networks cryptography. Key exchange, however, raises again the scalability issue. Another common approach is secure routing, which has the appealing idea of dividing the data into N pieces, which are sent along separate routes and, at the destination, the original message is reconstructed out of any (M-out-of-N) pieces of the message. Nevertheless, security is indeed one of the most difficult problems to be solved. Having received only modest attention so far, its "golden age" of research can be expected after the functional problems on the underlying layers have been agreed upon. Other challenging issues on ad hoc networks are node cooperation, interoperation with the Internet, aggregation, multicast, as well as the theoretical limitation, of ad hoc networks. Technologies such as smart antennas and software radios also bring new research problems along with impetus to ad hoc networks.

5.6.4 Energy and Battery Management

Energy efficiency has become an important issue in all layers of communication protocols, as well as on a wide range of technological applications. Until recently, research and development in the field of communication networks was mainly targeted at their functionality and performance. Now energy efficiency has drawn significant attention among the researchers due to the introduction of battery-operated devices. Energy efficiency is one of the most crucial design criteria as nodes are battery operated. If a node runs out of battery power, its ability to route traffic becomes affected. This adversely affects the network lifetime as well as degrading the performance. The network lifetime can be enhanced by minimizing the power consumption and/or maximizing the battery power of a node. Though considerable progress has been made in battery technology in recent years, it is incomparable with the progress made in semiconductor technology. Battery life has not kept pace with advances in mobile devices. To enhance the lifetime of mobile devices, it necessitates the requirement of power conservation techniques to enhance the lifetime of the network. Such techniques can be applied to different layers of the protocol stack.

Nodes in wireless ad hoc networks are battery operated, which must be judiciously used to increase the lifetime of the network. To increase the network's lifetime, attempts should be made to minimize the power consumption at nodes. Nodes in ad hoc network act as a source as well as a router to relay traffic from other nodes in the network. These relay nodes play a vital role in determining the network's lifetime. If the proportions of relay nodes are large in a network, then it may lead to faster depletion of the node's battery power. For a lesser number of relay nodes in the network, nodes have to transmit with maximum power. Transmitting with maximum power not only depletes a node's energy but also increases interference. Interference not only degrades the network throughput and node's lifetime but also the main cause of bad carrier sensing. Selection of optimal transmission power can reduce the interference to a greater extent.

Carrier sensing is a challenging task in wireless network. A node expends maximum energy in carrier sensing, which is also affected by hidden and exposed terminals. Re-transmission occurs due to collision, and this increases per packet energy consumption. Re-transmission also affects other network parameters such as end-to-end delay, jitter, throughput, etc. Continuous monitoring of an idle channel for possible traffic consumes energy without doing any useful work. Experimental results have shown that idle power consumption is nearly equal to received power consumptions. Therefore, it cannot be ignored from a power consumption point of view. Reducing idle power consumption is a major task of all energy efficient protocols in ad hoc networks. When a node transmits a packet, it is overheard by neighbors within its transmission range. Thus, a node expends its energy by receiving a packet, even though it is not destined to the node. Higher bit rate and larger packet size consume more power as compared to lower bit rate and smaller packet size. Larger packet size also increases the probability of re-transmission due to collision. Message flooding has disadvantages: (i) due to an overlapping region, a node may receive many redundant messages from its neighbors; (ii) when a broadcast message is transmitted by a set of nodes in an overlapping region, there may be severe contention for the medium in that region; and (iii) the flooding mechanism does not consider the available battery power at the time of transmission. Due to node mobility, frequent path break may take place. This necessitates route discovery and route maintenance, which consumes energy.

In recent years, various techniques have been proposed to achieve energy efficiency at the protocol level. These techniques adopt different approaches to achieve energy efficiency. A few of them are energy-efficient path selection, adjusting transmission power dynamically, reducing maximum transmission power at node level, adaptive sleeping, and use of directional antenna, etc.

5.7 Future Research Directions

We believe that the development of ad hoc networks and related technologies will eventually lead toward an ubiquitous computing environment. Ubiquitous computing enables the assimilation of information processing in a human's life in seamless ways. The emergence of 4G wireless networks and corresponding technological developments are already paving the way. Ad hoc networks will play a key role in realization this vision by enabling spontaneous erection of networks. These networks have the essential ingredients of self-organization, self-management, and self-healing required for operations in ubiquitous environments. However, a lot of research efforts are still required before the true realization of WANET. The following paragraphs outline opportunities for further research in WANET.

MAC layer protocols require several modifications. Future research should focus on energy efficient, multi-channel, and cooperative MAC protocols and medium access control in the presence of directional antennas. Since ad hoc networks are zero-configuration systems, various research efforts need to be done to devise secure and scalable approaches for addressing assignments. The various routing protocols must consider the power constraints, the characteristics of channels along the path, and other routing metrics to improve the routing operation. At the transport layer, not much work has been done. Protocols are required that utilize network events information from the lower layers for congestion and flow control. Security is also a challenging research area and research is required on addressing various associated challenges. Existing security solutions cannot counter all types of security attacks, as almost all of them are developed for a particular type of security threat. Generalized solutions are therefore required that can cope with any type of security threat. In addition, research should also be done on designing realistic mobility models that mimic real-world situations. Future proposals should also exploit cross-layer operations for energy-efficient and QoS-based MAC, routing, and security solutions for ad hoc networks. WANET require cooperation enforcement mechanisms to enable distributed operation of various protocols. In this direction, various efforts are being put into design error-free incentive-based mechanisms to ensure the node's cooperation. Simulators and testbeds are vital tools for validation of any research proposal specifically for ad hoc networks. Different proposals have evolved in recent years. The problem with current simulator is that results obtained from a particular simulator are usually not similar when the same algorithm is executed on a different simulator. Research efforts are therefore required on accuracy and reliability of simulators. Data management has recently emerged as a hot topic of research. Several frameworks have recently emerged. However, most of the frameworks are still immature. Future work is required on designing a complete framework encompassing solutions for its various issues, i.e., discovery, knowledge management, semantic data representation, and consistency management, etc. Finally, the standardization of ad hoc networks is an area that also requires attention. Various models of ad hoc networks are emerging. These new models provide various innovative applications; however, their potential is still to be explored. At the same time, several additional challenges require attention from the research community. For example, the ad hoc grid requires novel resource discovery and scheduling algorithms to keep up with the dynamically changing QoS requirements of the ad hoc environment. In addition, new models for security and economics are required to ensure cooperation among nodes. Similarly, ad hoc cloud requires new approaches for ensuring service provisioning, market-oriented management of resources, data interoperability, and privacy, etc. Research efforts are required to improve data transmission, and address the issues arising in multi-channel MAC, routing, transport protocols, and security. For WANET, the research is at initial stages and extensive efforts are required in development of techniques that can address spectrum management, mobility management, signaling, routing, transport protocol, and security issues. Finally, new application models, implementation tools, analytical models, and benchmarks have to be developed to assess the promise of all of these emerging models of ad hoc networks.

5.8 Conclusion

Wireless network is an interconnection of many systems capable of providing service to mobile users. In wireless network, data are carried by electrical wave (e.g., radio wave) from one node to another. There is no physical cable/wire connecting one computer to another. An ad hoc wireless network is a collection of two or more devices equipped with wireless communications and networking capabilities. Such devices can communicate with another node that is immediately within their radio range or even one that is outside their radio range. A wireless ad-hoc network is a decentralized type of wireless network. The network is ad hoc because it does not rely on a preexisting infrastructure, such as routers in wired networks or access points in managed (infrastructure) wireless networks. Each node participates in routing by forwarding data for other nodes, and so the determination of which nodes forward data is made dynamically based on the network connectivity.

MANET, VANET, WSN, and WMN comes under WANET. This chapter provides a survey of these different types of networks. Basic concepts, features, applications, and research trends of these networks are discussed in this chapter.

Communication system design, wired and wireless network design approach, architectures, overview of history, spectrum of wireless ad hoc network, general concepts of ad hoc network, enabling technologies, and taxonomy are discussed. Several problems arise due to the shared nature of the wireless medium, limited transmission power, limited transmission range, node mobility, battery limitations, and bandwidth limitation, etc. These problems can be solved through proper designing of MAC, Network, and transport layer protocols. Thus, design issues of protocols are considered here. The chapter ends by presenting a summary of current research on ad hoc networks, neglected research areas, and leads to further investigations.

References

1 Belding-Royer, E.M. and Perkins, C.E. (2003). Evolution and future directions of he ad hoc on-demand distance-vector routing protocol. *Journal of Ad Hoc Networks* 1: 125–150.

2 Chlamtac, I., Conti, M., and Liu, J.J.-N. (2003). Mobile ad hoc networking: imperatives and challenges. *Journal of Ad Hoc Networks* 1: 13–64.

3 Dressler, F. (2008). A study of self-organization mechanisms in ad hoc and sensor networks. *Journal of Computer Communications* 31: 3018–3029.

4 Dollas, A., Ermis, I., Koidis, I. et al. (2005). An open TCP/IP core for reconfigurable logic. *Proceedings of the 13th Annual IEEE Symposium on Field-Programmable Custom Computing Machines (FCCM'05)*. 2: 140–144.

5 Srivastava, V. and Motani, M. (2005). Cross-layer design: a survey and the road ahead. *IEEE Communications Magazine* 43: 112–119.

6 Qin, L. and Kunz, T. (2004). Survey on mobile ad hoc network routing protocols and cross-layer design. Carleton University, Systems and Computer Engineering, Technical Report SCE-04-14, 01–26.

7 Carneiro, G., Ruela, J., and Ricardo, M. (2004). Cross-layer design in 4G wireless terminals, wireless communications. *IEEE Wireless Communications* 11: 7–13.

8 Foukalas, F., Gazis, V., and Alonistioti, N. (2008). Cross-layer design proposals for wireless mobile networks: a survey and taxonomy. *IEEE Communication Surveys and Tutorials* 10: 70–85.

9 Shakkottai, S., Rappaport, T.S., and Karlsson, P.C. (2003). Cross-layer design for wireless networks. *IEEE Communication Magzine* 41: 74–80

10 Kliazovich, D., Redana, S., and Granelli, F. (2011). Cross-layer error recovery in wireless access networks: the ARQ proxy approach. *International Journal of Communication Systems* 25 (4) : 415–426.

11 Razzaque, M.A., Dobson, S., and Nixon, P. (2007). Cross-layer architectures for autonomic communications. *Journal of Network and Systems Management* 15 (1) : 13–27.

12 Kawadia, V. and Kumar, P.R. (2005). A cautionary perspective on cross-layer design. *IEEE Wireless Communications* 12 (1): 3–11.

13 Raisinghani, V.T. and Iyer, S. (2004). Cross-layer design optimizations. *Journal of Computer Network and Communication* 27: 720–24.

14 Ramanathan, R. and Redi, J.(2002). A brief overview of ad hoc networks: challenges and directions. *IEEE Communications Magzine* 40: 611–617.

15 James, A. and Macker Freebersyser, J.P. (2007). Overview of CBMANET, ITMANET for WAND proposer day meeting. *Advanced Technology Office Defense Advanced Research Projects Agency* 06–12.

16 Jubin, J. and Tornow, J.D. (1987). The DARPA packet radio network protocols. *Proceedings of the IEEE* 75: 21–32.

17 Johnson, D.B. and Maltz, D.A. (1997). Dynamic source routing in ad hoc wireless networks. *Journal of Mobile Computing* 353 : 153–181.

18 Chen. L. and Heinzelman, W.B. (2005). QoS aware routing based on bandwidth estimation for mobile ad hoc networks. *IEEE Journal on Selected Areas in Communications* 23 (3): 561–572.

19 Toh, C.K. (1996). A novel distributed routing protocol to support ad hoc mobile computing. *Proceedings of the IEEE Fifteenth Annual International Phoenix Conference on Computers and Communications: California* 5: 480–486.

20 Toh, C.K. (2002). *Book on Ad Hoc Mobile Wireless Networks: Protocols and Systems*. Cliff, NJ: Prentice Hall Englewood. 07632.

21 Cordeiro, C. and Agrawal. D.P. (2006). *Book on Ad Hoc and Sensor Networks: Theory and Applications*. World Scientific Publishing.

22 Goyal, N. and Gaba, A. (2013). A review over MANET – issues and challenges. *International Journal of Enhanced Research in Management & Computer Applications* 25: 24–36.

23 Car 2 Car Communication consortium (2007). C2C-CC Manifesto Version 1.1.

24 Jiang, D. and Delgrossi, L. (2008). IEEE 802.11p: towards an international standard for wireless access in vehicular environments. *IEEE Vehicular Technology Conference* 2036–2040.

25 Torrent-Moreno, M., Mittag, J., Santi, P. et al. (2009). Vehicle-to-vehicle communication: fair transmit power control for Safety Critical information. *IEEE Transactions on Vehicular Technology* 58 (7): 3684–3703.

26 Sadakale, R., Ramesh, N.V.K., and Patil, R. (2020). TAD-HOC routing protocol for efficient VANET and infrastructure-oriented communication network. *Journal of Engineering* 20: 1–12.

27 Akyildiz, I.F., Wang, X., and Wang, W. (2005). Wireless mesh networks: A survey. *Computer Networks* 47: 445–487.

28 Akyildiz. I.F. and Wang. X. (2008). Cross-layer design in wireless mesh networks. *IEEE Transactions on Vehicular Technology* 5 (2): 1061–1076.

29 Pirzada, A.A., Portmann, M., and Indulska, J. (2006). Evaluation of multi-radio extensions to AODV for wireless mesh networks. *Proceedings of the 4th ACM International Workshop on Mobility Management and Wireless Access (MobiWAC '06), Torremolinos, Spain.* 45–51.

30 Hamid, Z. and Khan, S.A. (2006). An augmented security protocol for wireless MAN mesh networks, communications and information technologies, ISCIT '06. *International Symposium.* 30: 861–865.

31 Dandekar, D.R. and Deshmukh, P.R. (2013). Energy balancing multiple sink optimal deployment in multi-hop wireless sensor networks. *Proceedings of the 3rd International Advance Computing Conference (IACC '13), Ghaziabad, India.* 408–412.

32 Wang, Z.M., Basagni, S., Melachrinoudis, E. et al. (2005). Exploiting sink mobility for maximizing sensor networks lifetime. *Proceedings of the 38th Annual Hawaii International Conference on System Sciences.* pp. 1–9.

33 Gatzianas, M. and Georgiadis, L. (2008). A distributed algorithm for maximum lifetime routing in sensor networks with mobile sink. *IEEE Transactions on Wireless Communications* 7 (3): 984–994.

34 Yu, F., Lee, E., and Kim, S.-Ha. (2010). Elastic routing: a novel geographic routing for mobile sinks in wireless sensor networks. *IET Communications* 4 (6): 716–727.

35 Akkaya, K . and Younis, M. (2005). A survey on routing protocols for wireless sensor network. *Elsevier Ad Hoc Network Journal* 3 (3): 325–349.

36 Kumar, S.P. (2003). Sensor networks: evolution, opportunities, and challenges. *Proceedings of the IEEE* 91 (8): 1247–1256.

37 Al-Karaki, J.N . and Kamal, A.E. (2004). Wireless sensor networks routing techniques in wireless sensor networks: A survey. *IEEE Wireless Communications* 4: 6–28.

38 Tomić, I. and McCann, J.A. (2017). A survey of potential security issues in existing wireless sensor network protocols. *IEEE Internet of Things Journal* 4 (6): 1910–1923. doi: 10.1109/JIOT.2017.27498832017)

6

A Survey: Brain Tumor Detection Using MRI Image with Deep Learning Techniques

Chalapathiraju Kanumuri[1,2] and CH. Renu Madhavi[1]

[1]*RV College of Engineering, Bengaluru, Karnataka, India*
[2]*S.R.K.R Engineering College, Bhimavaram, Andhra Pradesh, India*

6.1 Introduction

Brain tumors [1] form due to abnormal cell development [2]. They are tumors that occur in the skull or central portion of the spinal cord [3]. A chance of a tumor forming depends on a combination of components, including form, area, scale, and how it grows and propagates. Glioma is the primary cause of a brain tumor in adults. The tumors derive from glial cells and are classified into five grades. Grades I and II are considered low grade (LG), and grades III to V are high-grade gliomas (HG) [4]. According to the American Association of Brain Tumors, in 2016, about 80 000 cases of brain tumor were reported [5]. Because of the complexity and size of clinical images, experts and radiologists are increasingly challenged in the time-efficiency and precise analysis of these images. For the volume of interest (VOI) and region of interest (ROI) identification, an automatic method is therefore required to help explain medical images [6, 7]. Over the last decade, the tumor has become the deadliest perpetrator in the world [8]. Invasive methods for the identification, analysis, treatment, and simulation of tumors, have been replaced by better options such as radiographs, which include Magnetic Resonance Imaging (MRIs), Computed Tomography (CTs), and Positron Emission Tomography (PET) [9].

Glioblastoma multiforme (GBM) is a lethal type of brain tumor in adults, with poor prognosis. In general, GBM diagnosis and treatment is driven by immune histochemistry and histopathology [10]. Non-invasive techniques also include multiple pictures to determine the condition of the tumor. Mainly, since MRI produces no ionization radiation [11], it is more robust. However, a single MRI is insufficient to give accurate results in segmentation. The MRI series is used to achieve more precise and reliable segmentation data. T1 of T1c MRI tumor borders appear sharper due to the contrast agent's aggregation in the brain's blood vessels. An MRI contrast agent is used to improve illumination of the body's internal structure and to provide more useful information on the active and necrotic tumor region. Similarly, the edema area appears light in T2 pictures. Flair is a human IRM modality, with an adequate differentiation between cerebrospinal fluid (CSF) and edema areas [12].

Early brain tumor diagnosis is the key problem that requires initial surveillance and screening. Proper brain imaging assists in early diagnosis and recovery and improves

Smart and Sustainable Approaches for Optimizing Performance of Wireless Networks: Real-time Applications, First Edition.
Edited by Sherin Zafar, Mohd Abdul Ahad, Syed Imran Ali, Deepa Mehta, and M. Afshar Alam.

patients' recovery rate – many CAD programs aid in the automatic diagnosis of brain pathology [13, 14]. The brain abnormality identification CAD system uses MRIs, which play an essential role in documenting and diagnosing specific abnormalities. The automated systems can distinguish regular and irregular MR images and reduce the ophthalmologist's workload by examining MR images found with suspected system-wide abnormalities [2, 15]. The CAD method provides radiologists and physicians with the aid of a clinical image to identify anomalies [16, 17].

The N.N. approach is used to differentiate between regular and irregular images, with 83% success [18]. Deep Networks have proven their usefulness in enhancing predictive analysis performance. The implementation of DNN allows a better prediction of the brain tumor using MR Imaging as the method. This chapter reflects on the possibilities of MRI in tandem with deep neural networks as the principal modality.

This chapter is organized into six sections. After the Introduction, the context in Section 6.2 is explored in conjunction with the use of MRI and machine learning in Section 6.3; the discovery and testing discrepancies are later addressed in Section 6.4. Recommendations for the study are given in Section 6.5. Finally, Section 6.6 addresses these findings.

6.2 Background

In an ideal world, we can diagnose and treat patients without harmful side effects. Medical simulation is one of the best approaches to accomplish this and see what is happening without surgery or other invasive treatment to the body. In reality, it is something that tends to be taken for granted. Diagnostic and therapeutic visualization can be used, making it one of the most powerful means to treat patients successfully. Typical imaging types include CT (Computed Tomography), MRI, ultrasound, and X-rays for diagnostic purposes (Table 6.1).

6.2.1 Types of Medical Imaging

Ultrasound: The best medical imaging method is ultrasound, which has a wide range of applications [19]. Ultrasound has no adverse consequences and is one of our best medical imaging techniques, regardless of specialties or circumstances. Ultrasound neither uses ionizing radiation nor requires waves of sound. High-frequency sound waves are transmitted from the electrode into the body via the conveyor pad, which rebound when they hit different body diagnostic components. Doppler is another common type of ultrasound, using very different sound waves to show blood flow through the arteries and the veins. Due to the low likelihood of ultrasound, it is the first choice for conception. Still, for treatments to be completed, such as physiological treatments to spinal or internal organs, for many people, it seems to be one of the first options.

X-Ray Imaging: the earliest but one of the most frequently used imagery styles. Most of us will have had at least one X-ray in our lives. Radiation was considered to be nuclear radiation in 1895 [20]. X-rays are a range and wavelength that cannot be observed by the human eye, but they can reach through the skin to produce an image of actual events occurring inside the human body. Besides, X-rays can also detect cancer by mammography,

Table 6.1 Comparison of CT and MRI in the assessment of brain tumors.

Modality	Indications	Limitations
Computed Tomography (CT)	Shorter time for imaging	Poor edema description
	Low testing costs	Just one plane and mainly non-isotropic acquisition
	Adequate spatial resolution	Chance of X-ray radiation risk
	Suitable for extra-axial brain tumor treatment	Poor characterization of tissue
	Superior in calcification identification, erosion of the skull, penetration, degradation	Bone objects limit the picture of the posterior fossa
Magnetic Resonance Imaging (MRI)	Useful for displaying parenchyma edema (early indication for diagnosis of tumor)	Insufficient calcification and bone degradation detection
	The same deleting extent of edema and compression	Intraoperative testing cannot be carried out.
	Effective diagnosis and atrophy of toxic effects	Better spatial faithfulness
	A good description of neuroanatomical (tissue differentiation)	Often series takes a lot of time
	Precise identification of tumor vascularity (acquisition of different planes)	

barium swallowing, and enemas, and are also typically used to identify skeletal system issues. X-rays are widely used for low-cost, easy, and relatively safe patient care. The use of radiation in X-ray imaging, however, raises threats. A dose of radiation is received by the patient when they have an X-ray. This can cause cancer and cataracts later in life or fatally injure a pregnant patient's embryo. These risks can only be mitigated if X-rays are urgently required, and the body is protected as much as possible.

Computed Tomography (CT): CT or CAT scans are an X-ray procedure that provides a 3D visual image. The X-rays for cross-sectional body scans are used in CT or in CAT [20]. The CT scanner has a large circular space on a motorized bench for the patient. The X-ray source and a scanner then move around the patient to produce a brief "fan-like" light ray that provides a snapshot of the patient's body. These snapshots are then merged into one or more images of the inner tissues such as the heart. CT scans give a better picture of the inner tissues, muscles, soft tissue, and blood vessels of the body than regular X-rays. The drawbacks of CT scans greatly outweigh the advantages, including damage to the body tissues, or even cancer.

Using a CT scan often removes the need for experimentation. The radiation dose for scanning of children must be smaller than that for adults to protect the child from excessive radiation exposure. For this function, in many hospitals, you will see a pediatric CT scanner.

MRI: MRI [19] uses a substantial magnetic field and radio waves to generate images of the body that cannot be seen by the more often used X-rays or CT scans; i.e., it makes it possible for an image of a joint or ligament to appear. MRIs for internal tissue tests, such as stroke diagnosis, tumor, spinal cord injury, aneurysms, and brain functions have been used. The human body is made up almost entirely of water, and each water molecule consists of a nucleus of hydrogen (proton). An MRI scanner uses a powerful magnetic field to control proton rotations. A radiofrequency is used to "react" with the protons before initialization.

Protons return to their normal spins in different tissues of the body at differing rates, such that the MRI can differentiate between tissue types and so detect abnormalities. The molecular flip and return is registered and transformed into an image. MRI does not use ionizing radiation and so is widely used during breastfeeding without side-effects to the infant. However, there are dangers involved with MRI scanning and no intervention at the first level is recommended. Solid magnets may be used to move or heat a metal implant or an artificial joint, etc. into the magnetic field. Several cases have occurred in which patients with pacemakers have died due to MRI scanning. The noise generated by the scanner requires ear protection. In periods of rising rates of treatment and demand, we must be respectful of ourselves as medical practitioners when we use the best services available to fulfill patient needs. Thus, patients are deliberately considered for the right diagnostic terminology and potential prediction.

6.2.2 MR Imaging as a Modality

MRI is the preferred approach in which patients with signs of brain tumor and related symptoms are examined. Its multi-lane ability, superior contrast, and various techniques render it an invaluable instrument for evaluating tumor size and magnitude, performing biopsies, planning for optimal care, and interpreting therapeutic findings. The key protocol used by institutions consists of spin-echo T1 (T1WI), a proton-density-weighted image (PDWI), t2 (T2WI), and t1wI (T1WI) following administration of paramagnetic agents.

Contrast content in MRI increases the resolution of certain brain tumors and can help separate the tumor from the adjacent normal brain parenchyma. Chelates of gadolinium (Gd) are the most commonly-used contrast MRI agents for tumor imaging in the central nervous system (CNS). Although Gd cannot pass from intravascular to interstitial in a normal brain, Gd is transferred into the tumor's extracellular region in brain tumors, where normal BBB may be influenced. In T1WI, however, the tumor becomes more resilient in post-contrast than the normal brain tissue surrounding it, by shortening T1's relaxation time. However, histological analysis of tissues from brain tumor patients showed regions beyond the Gd-enhancing region's tumor cells.

6.2.3 Types of Brain Tumor MR Imaging Modalities

Brain tumors' neuroimaging requires three physiological MR imaging methods: DWI, proton MR spectroscopy, and infusion weighted imaging [19, 20]. Those approaches played a central role in the transition from a strictly morphological approach to incorporating the clinical MR imagery's structure and function.

Diffusion-weighted imaging is a unique tissue contrast technique based on the diffusion of water molecules that travel through random pathways (Brownian motion). A spin-echo EPI series is commonly used for DWI, where the same gradient pulses are inserted before and after the refocusing pulse has been applied. DWI's physical principle is similar to that of MRA phase contrast, and while PCA is considered as a macroscopic movement, DWI is known as a microscopic movement. More precisely, the first gradient pulse introduces phases of water molecules, while the second gradient cancels the process by replacing their spins entirely. Moving molecules become a phase shift because of their displacement during the time of diffusion gradients. In other words, stationary spins will be refocused, which means that no phase shifting is possible, and traveling spins will be partly refocused.

Proton magnetic resonance spectroscopy (1HMRS) is a non-invasive method for studying various chemical compounds found in human brain tissue. This method effectively discriminates between the brain's normal and dysfunctional tissues and presents information that is likely to be essential for differential diagnosis. There have been two approaches: Single-voxel spectroscopy (SVS) and Chemical Change Imaging (CSI). Like the first one, the 3D region or tissue volume is excited, and signals are converted into a continuum from this range. In the second method, multiple voxels are used in either a plane (2D CSI) or a volume (3D CSI); thus, a single experiment may research more detailed regions. Based on the details obtained from each voxel, metabolic maps can be determined.

Perfusion weighted imagery provides information on the microcirculation perfusion status. This technique involves the effective intravenous treatment of an MR contrast agent. During the paramagnetic contrast agent's intravascular pockets, magnetic susceptibility effects are produced by local field inhomogeneity, which can be determined by a decrease in the signals in pictures T2*. The reduction in the signal depends on the contrast agent's vascular concentration and the concentration of the small voxel vessels [37, 38]. Shifts of signal amplitude can be used to calculate the values. Echo planar MR imaging systems using incredibly rapidly shifting magnetic field gradients enable the simultaneous acquisition of multiple T2-weighted slices in contrast materials management.

6.2.4 Suitable Technologies Before Machine Learning

The basic principle of machine learning is to accurately predict diseases with better precision relative to humans' same mechanism. However, techniques for identifying brain tumors were available. As stated earlier, tools like X-ray, ultrasound, CT, MRI, and human observations, have been used. While computing machines were used, they were restricted to gathering the human skull's brain tumor status. The method takes time to print the film's picture, which often loses small pieces.

The specialist has to be qualified to examine all of the X-rays, ultrasound, CT, and MRI to determine the tumor status. Any research arising from a misinterpretation during the study could lead to incorrect care, loss of life, and legal problems. Therefore, the medical industry has welcomed machine learning to be involved in the detection and prediction processes that, once learned, produce better accuracy.

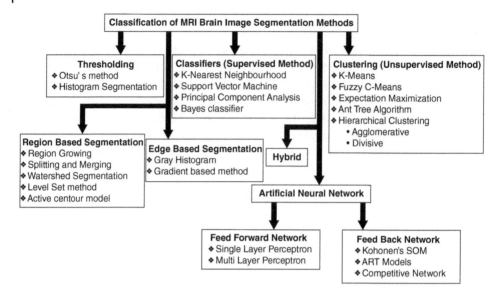

Figure 6.1   Classification of MRI brain image segmentation methods.

6.2.5   MRI Brain Image Segmentation

An overview of existing approaches, including most common image segmentation methods, is given in [42]. The study presents a detailed description of the MRI segmentation approach Figure 6.1). The methods of MRI brain image segregation are represented in a hierarchical structure, i.e., thresholding, region-based segmentation, edge-based segmentation; classifications such as KNN (K-Nearest Neighbor); SMV (Support vector machine); PCA (Main Component Analysis); Naive bay classifier; hybrid models; and artificial neural network models that include transmission feed Clustering approaches for MRI image segmentation, i.e., K-means, Fuzzy C-means, Ant Tree Algorithms, and Hierarchical clustering are also implemented.

Segmentation of MRI Medical Image Challenges:

1. Considerable sensing modality heterogeneity. This is compounded further by human anatomy.
2. Normal tissues overlap in density in the brain image, creating a partial volume effect (PVE) and the tumor tissues also have similar characteristics as PVE.
3. A noise signal from the sensor and related circuitry.
4. For different brain regions, the detected gray levels are very similar to one another.

6.3   Related Work

The dissertation of Lundervold et al. [21] had three objectives:

1. Offer a short introduction to meaningful learning with references to key sources.
2. Provide the application of deeper learning in the entire MRI processing chain, from acquisition to image recovery, from the segmentation of disease prediction.

3. Provide a starting point for people interested in working with and maybe contributing to deep learning in medical imaging by showcasing relevant educational materials, state-of-the-art open-source coding, and unique data sources and associated imagery issues.

A new fully automatic segmentation technique from MRI data containing *in vivo* brain gliomas was developed using a Deep Cascaded Neural Network (Authors). This method cannot only identify the whole tumor region but also accurately isolate the structure of the intratumor.

The incidence of brain tumors in multiple cancers is significant, so prompt diagnosis and correct care are essential to save a life. The identification of these cells is a challenging problem due to tumor cell development. A comparison of brain tumor from MRI therapy is essential. The pathological configurations of the human brain using basic MRI methods are impossible to conceive. In the last decade, ensemble methods have been pointed to as the most critical advancement in data mining and machine learning. They integrate many models into one, generally more detailed than the sum of their parts. The combined approaches incorporate the neural network method, the extreme learning machine (ELM), and the vector machine classifier.

Parasuraman Kumar et al. [22] suggested a different stage method: pre-processing, segmentation, analysis, and description of characteristics. Pre-processing is initially done using a filtering algorithm. Secondly, segmentation using the clustering algorithm is carried out. Thirdly, the Gray Degree Co-Occurrence Matrix (GLCM) removes functions. An ensemble grouping achieves the automated stage of the brain tumor. This process classifies brain images into tumors and non-tumors using Feed Forwarded Neural Artificial Network Classification. Experiments have demonstrated that the process is more comfortable and more reliable to update.

Havaei et al. [23] implemented modern CNN architecture, which differs from computer vision. CNN uses both geographic and more common investigative methods effectively. The networks often have a final layer, a fusion of a dynamically connected layer that requires 40 times more latency than the majority of traditional CNNs. We also describe a 2-phase preparation approach to handle tumor-related problems. Finally, a cascade paradigm is explored in which the raw CNN output is treated as a secondary source of information for a subsequent CNN. Results from the 2013 BRATS study data collection suggest that the software is improving with state-of-the-art launches, while it is more than 30 times quicker.

The 2-way, 11-layer wide, 3D Convolutional Neural Network has been proposed for a challenging brain injury separation task by Kamnitsas et al. [24]. A detailed review and the limitations of existing networks designed for related implementations result in the design methodology. We also developed an effective and efficient complicated training method to address the technical strain of performing medical 3D scans, integrating the performing of neighboring image patches into one move through the network, which automatically adapts them to intrinsic class imbalances. Besides, we study the creation in more depth, thus more biased, 3D CNNs. To combine local and more critical contextual details, we use a dual-path architecture that simultaneously processes input images on multiple scales. We use a 3D fully linked Conditional Random Field to post-process the network's soft segmentation, which virtually eliminates wrong positive elements. The pipeline has been thoroughly tested in three complex lesion segmentation tasks of traumatic brain injury, brain tumors, and ischemic stroke inpatient MRI multichannel results. With success at the

BRATS 2015 and ISLES 2015 public benchmarks, we boost the cutting-edge technologies for all three implementations. This approach is computationally practical and can be used in several scientific and clinical settings.

A prime concern, but a repetitive task carried out by radiologists or clinical specialists, is the segmentation, identification, and retrieval of affected tumor areas from magnetic resonance imaging (MRI). Your performance is based on your knowledge. Thus it is essential to use computer-aided technologies to solve these restrictions. In this study, Nilesh Bahadure et al. [25] analyzed the segmentation of the brain tumor based on the Berkeley wavelet transformation (BWT) to increase accuracy and minimize uncertainty in medical image segmentation. Also, to increase the classifier's accuracy and consistency based on the SVM, the related features are derived from each segmented tissue. The proposed experimental findings for efficiency and quality analysis of brain resonance magnetic images were tested and validated based on accuracy, sensitivity, specificity, and the Dice similarity index coefficient. The tests' findings showed the efficacy of the experimental methodology for detecting normal and pathological tissues in brain MR imagery, 96.51% accuracy, 94.2% precision, and 97.72% sensitivity. The experimental findings have obtained a coefficient averaging 0.82 times the similarity index, which suggests that a more significant comparison of automatic (machines) radiologists have extracted a tumor area with a manually extracted tumor area. The findings of the simulation show the value of consistency parameters and precision compared with state-of-the-art techniques.

Zacharaki et al. [26] have suggested an Adult Brain Tumors Classification Scheme using traditional MRI and rCBV maps. Form features, image intensity statistics, and rotational-invariant Gabor texture characteristics have been derived from the tumor, edematous, and necrotic regions in central and marginal areas. The system is fully automated, and no professional support is required except to track the ROIs. In general, we find that an objective and quantitative examination of brain tumors is a promising approach in the SVM classification of texture patterns. However, more massive datasets must be evaluated, checked, and enhanced by such classification systems, which perform better if more thoroughly qualified.

Soltaninejad et al. [27] suggests an educationally supervised method for tumor segmentation in MRIs' multimodal brain images. Supervoxels are measured using multimodal MRI image information fusion. A new histogram of the text description is collected on each super voxel of different MRI imaging modalities based on a set of 3D Gabor filters of different sizes and orientations. A random forest classifier is then used to classify into a tumor (including tumor core and edema) or normal brain tissue per super voxel. The approach shows promising results in brain tumor segmentation (core and edema). Texton demonstrates the benefits in providing essential information in 3D spaces to differentiate different patterns. Adding multimodal MRI images functionality will primarily improve super voxels' classification precision with a manually specified gold standard. The test results using the clinical dataset demonstrate that the combination of the p-map can further enhance the segmentation and classification efficiency – and q-map DTI protocols and C-MRI, which can be used to further segment tissue subtypes. Simultaneously, the outstanding success obtained using the BRATS 2013 dataset shows the strength of the process. The approach is closely associated with the specialist delineation of all glioma classes, resulting in a more comfortable and more efficient brain tumor diagnosis and delineation process to improve case care.

Any brain region may contain a tumor of any scale, form, and contrast. Multiple tumors of various types may occur concurrently in a human brain. Accurate segmentation of the tumor region is considered the primary stage of brain tumor care. Deep Learning is a collection of innovative strategies for segmenting a timed portion of a brain that may have better outcomes than non-deep learning techniques. A deep convolution neural network (CNN) for the MRI section of brain tumors is provided in the article by Iqbal et al. [28].

The network suggested using the BRATS challenge segmentation dataset comprising of photographs collected in four different modalities. We thus propose an expanded version of the current network to address the issue of segmentation. The network architecture consists of several neural network layers attached to Convolutional feature maps on a sequential order on a peer level. Experimental findings of BRATS 2015 assessments suggest the usefulness and effectiveness of the method presented in this field of study over the other approaches.

Naceur et al. [29] found three end-to-end Incremental Deep Convolutional Networks models for fully automated brain tumor segmentation. The models suggested are distinct from other CNN models using the test and error method technique, which does not use a guided approach to achieve the required hyperparameters. We also follow the Ensemble Learning methodology to build a more effective model. To address the CNNs model training issue, we suggest a new training approach that takes the essential hyperparameters into account by linking and building a roof to improve teaching. The findings of the experiment were published on BRATS-2017. The deep learning models proposed to achieve cutting-edge efficiency without any post-processing operations. In reality, their models hit an average of 0.88 Dice over the entire area.

The GPU's robust architecture features three profound learning models for average brain segmentation outcomes of 20 to 87 seconds. The proposed profound learning models are ideal for the segmentation of brain tumors and have highly reliable outcomes. Also, the models suggested could allow doctors to reduce the time of diagnosis.

6.4 Gaps and Observations

Several detection techniques, including semi-supervised and controlled approaches, for brain tumors are advised [30–32]. However, supervised techniques involve a distinct preparation phase for simple skills, while semi-supervised techniques include segmentation of client observation and interaction [33]. Principle component analysis is an unsupervised technique used to compress the data, which in turn helps to reduce the computational time.

Kadkhodaei et al. [34] developed a multichannel patch segmentation technique with many patches extracted from training results. In the test process, identical patches are described using a similarity measure from training results. The labeling method is achieved with a single pattern. The MAP method is used to segment the ROI [35]. First, generate the super voxels set and then apply it to the label using the Markov Random Field graph cutting process [36]. A semi-automatic segmentation technique is used for brain tumor diagnosis. This method is relatively stable in contrast to manual segmentation approaches [37].

The semi-automatic active contour model is used for tumor diagnosis in the brain in which the patient has a tumor ROI [38]. Grow Cut is a semi-supervised segmentation

Table 6.2 Analysis of work proposed for human brain data for detection of abnormalities.

Methodology	Datasets used	Prediction accuracy	Research gaps
HMRF and threshold [43]	T2-weighted MRI	94% accuracy	MRI has a variety of modalities, but the authors only tested the T2 MRI form.
Unsupervised auto-encoder [44]	T2-weighted MRI	0.92 ROC	Although the work is fine, only T2-weighted MRIs have validated their system.
Gaussian mixture model (GMM) with k-means clustering [31]	T1, T2, DWI, Flair	89.15% DSC, 89.88% SE, 98.10% SP, 91.97% PPV, 96.36% NPV	The findings suggest that the obtained values are smaller, so performance can be changed.
Otsu method [33]	T1c	0.84 Jaccard index, 0.91 DSC	The index of Jaccard and DSC has low scores, which can be changed. The process is not tested in all respects.
Saliency detection algorithm [34]	Flair, T2, T1c	68% DSC	The value of DSC is low, and a development deficit remains. The system may also be measured for other performance measuring parameters.
Markov random field model [36]	Flair, T2, T1c	83.44% DSC, 86.32% SP, 82.33%	The DSC, SP, and SE values are low, indicating that outcomes can be improved.
Edge-based active contours and level set [45]	Flair, T2, T1c	0.86 DSC, 0.76 Jaccard, 0.86 PPV, 0.86 SE	The DSC, Jaccard, and PPV index have poor performance and potential for change.

approach integrated into object regions [39]. This strategy improves the ROI by using a user-based cellular automation algorithm. For the diagnosis of brain tumors, Semantic Preserving Distance Metric Learning (SP-DML) is used. It codes the similarity of low-level characteristics in new function space, determined using the patches' learning distance metric. Relevant resemblance tests are not successful in vital (semantic) regions and low (visual) parallels. The hyper graphic learning system detects a brain tumor with better accuracy than expected similarities [40, 41].

Table 6.2 suggests an overview of human brain evidence to identify anomalies in their cognitive processes based on multiple experiments in related fields. Also, a supervised method for classifying brain image marks is used.

6.5 Suggestions

Several researchers have suggested a comprehensive tumor assessment based on texture characteristics and machine learning based on BRATS challenge datasets. Several works can be applied to enhance the deep neural network (DNN) model, which is already being used for segmentation specialization and intensity modification to diagnose sub-tumoral areas such as the entire, central, and tumor-enhancing. Enhancing lesions is the most critical and challenging task. It offers assistance in the segmentation phase. A new filter called PDDF has been added to improve the ROI.

However, enhancing the Otsu clustering algorithm will have several more useful results in a more reliable section of the abnormal area. The LBP and GLCM concatenation called C2LBPGLCM can be used to discriminate among MR pictures. Furthermore, findings on individual GLCM, LBP, and C2LBPGLCM functionality must be analyzed. The efficiency of the derived characteristics must be measured using multiple classifiers.

6.6 Conclusion

MRI brain tumor image segmentation is a challenging task. Several researches have been done on MRI image segmentation and localization. Brain tumors can appear in the brain anywhere and can be of varying proportions and morphology. Also, these tumors tend to be small and ambiguous. Therefore, it remains difficult to classify the brain tumor and intratumor with limited human involvement using MRI data. Our survey is conducted on several literature pieces published on a sample of deep learning approaches to identify brain tumors using MRI images [52]. These studies cannot only find the whole area of the tumor but also precisely segment the intratumor structure. Our survey discloses many facts based on the literature on deep learning neural convolution network models, consisting of two subnetworks: (i) the localization of tumors; and (ii) the classification of tumors. Wavelet decomposition will highly reduce the dimensionality of the data which simplifies the work of the Neural Network Classifier.

References

1 Amin, J., Sharif, M., Yasmin, M. et al. (2018). Big data analysis for brain tumor detection: deep convolutional neural networks. *Future Generation Computer Systems* 87: 290–297.

2 Amin, J., Sharif, M., Yasmin, M. et al. (2017). A distinctive approach in brain tumor detection and classification using MRI. *Pattern Recognition Letters* 139: 1–10.

3 Louis, D.N., Ohgaki, H., Wiestler, O.D. et al. (2007). 2007 WHO classification of tumours of the central nervous system. *Acta Neuropathologica* 114 (2): 97–109.

4 Bahadure, N.B., Ray, A.K., and Thethi, H.P. (2017). Image analysis for MRI based brain tumor detection and feature extraction using biologically inspired BWT and SVM. *International Journal of Biomedical Imaging* 2017: 1–12,

5 American Cancer Society (2012) *Cancer Facts & Figures for Hispanic*. Atlanta: American Cancer Society.

6 Desai, U., Nayak, C.G., Seshikala, G. et al. (2018). Automated diagnosis of tachycardia beats. *Smart Computing and Informatics* 77: 421–429.

7 Fernandes, S.L., Chakraborty, B., Gurupur, V.P. et al. (2016). Early skin cancer detection using computer aided diagnosis techniques. *Journal of Integrated Design and Process Science* 20 (1): 33–43.

8 Mitra, S. and Shankar, B.U. (2015). Medical image analysis for cancer management in natural computing framework. *Information Sciences* 306: 111–131. https://doi.org/10.1016/j.ins.2015.02.015

9 Gatenby, R.A., Grove, O., and Gillies, R.J. (2013). Quantitative imaging in cancer evolution and ecology. *Radiology* 269 (1): 8–14.

10 Holland, E.C. (2000). Glioblastoma multiform: the terminator. *Proceedings of the National Academy of Sciences* 97 (12): 6242–6244.

11 Kamnitsas, K., Ledig, C., Newcombe, V.F. et al. (2017). Efficient multi-scale 3D CNN with fully connected CRF for accurate brain lesion segmentation. *Medical Image Analysis* 36: 61–78.

12 Khan, M.W., Sharif, M., Yasmin, M. et al. (2016). A new approach of cup to disk ratio based glaucoma detection using fundus images. *Journal of Integrated Design and Process Science* 20 (1): 77–94.

13 Benson, C.C. and Lajish, V.L. (2014). Morphology based enhancement and skull stripping of MRI brain images. In Proceedings of the 2014 International Conference on Intelligent Computing Applications (ICICA) 254–257. IEEE, Coimbatore.

14 Demirhan, A., Törü, M., and Güler, I. (2015). Segmentation of tumor and edema along with healthy tissues of brain using wavelets and neural networks. *IEEE Journal of Biomedical and Health Informatics* 19 (4): 1451–1458.

15 Shil, S.K., Polly, F.P., Hossain, M.A. et al. (2017). An improved brain tumor detection and classification mechanism. In Proceedings of the 2017 International Conference on Information and Communication Technology Convergence (ICTC) 54–57. https://doi.org/10.1109/ictc.2017.8190941

16 Amin, J., Sharif, M., Yasmin, M. et al., Ali H (2017). A method for the detection and classification of diabetic retinopathy using structural predictors of bright lesions. *Journal of Computational Science* 19: 153–164.

17 Banerjee, S., Mitra, S., and Shankar, B.U. (2016). Single seed delineation of brain tumor using multi-thresholding. *Information Sciences* 330: 88–103.

18 Damodharan, S. and Raghavan, D. (2015). Combining tissue segmentation and neural network for brain tumor detection. *International Arab Journal of Information Technology* 12: 42–52.

19 Drevelegas, A. (1970). *Imaging Modalities in Brain Tumors*. Berlin, Heidelberg: Springer.

20 Cha, S. (2006). Update on brain tumor imaging: from anatomy to physiology. *American Journal of Neuroradiology* 27 (3): 475–487.

21 Lundervold, A.S. and Lundervold, A. (2018). An overview of deep learning in medical imaging focusing on MRI. *Zeitschrift Für Medizinische Physik, Urban & Fischer*, 29 (2): 102–127.

22 Parasuraman Kumar, B. and Kumar, V. (2019). Brain tumor MRI segmentation and classification using ensemble classifier, *International Journal of Recent Technology and Engineering (IJRTE)*, 8 (1S4): 244–252.

23 Havaei, M., Davy, A., Farley, A.B. et al. (2016). Brain tumor segmentation with deep neural networks. *Medical Image Analysis*, Elsevier.

24 Kamnitsas, K., Ledig, C., Newcombe, V.F.J. et al. (2016). Efficient multi-scale 3D CNN with fully connected CRF for accurate brain lesion segmentation. *Medical Image Analysis*, Elsevier.

25 Bahadure, N.B., Ray, A.K., and Thethi, H.P. (2017). Image analysis for MRI based brain tumor detection and feature extraction using biologically inspired BWT and SVM. *International Journal of Biomedical Imaging, Hindawi*, 6: 1–12.

26 Zacharaki, E.I., Wang, S., Chawla, S. et al. (2009). Classification of brain tumor type and grade using MRI texture and shape in a machine learning scheme. *Magnetic Resonance in Medicine, US National Library of Medicine* 62 (6): 1609–1618.

27 Soltaninejad, M., Yang, G., Lambrou, T. et al. (2018). Supervised learning based multimodal MRI brain tumour segmentation using texture features from Supervoxels. *Computer Methods and Programs in Biomedicine*, Elsevier, 11.

28 Iqbal, S., Usman Ghani, M., Saba, T. et al. (2018).Brain tumor segmentation in multi-spectral MRI using convolutional neural networks (CNN). *Microscopy Research and Technique. US National Library of Medicine* 81 (4): 419–427.

29 Naceur, M.B., Saouli, R., Akil, M. et al. (2018). Fully automatic brain tumor segmentation using end-to-end incremental deep neural networks in MRI images. *Computer Methods and Programs in Biomedicine*, Elsevier.

30 Banerjee, S., Mitra, S., and Shankar, B.U. (2018). Automated 3D segmentation of brain tumor using visual saliency. *Information Sciences* 424: 337–353.

31 Binczyk, F., Stjelties, B., Weber, C. et al. (2017). MiMSeg – an algorithm for automated detection of tumor tissue on NMR apparent diffusion coefficient maps. *Information Sciences* 384: 235–248.

32 Huang, Q., Yang, F., Liu, L. et al. (2015). Automatic segmentation of breast lesions for interaction in ultrasonic computer-aided diagnosis. *Information Sciences* 314: 293–310.

33 Fernandes, S.L., Chakraborty, B., Gurupur, V.P. et al. (2016). Early skin cancer detection using computer aided diagnosis techniques. *Journal of Integrated Design and Process Science* 20 (1): 33–43.

34 Kadkhodaei, M., Samavi, S., Karimi, N. et al. (2016). Automatic segmentation of multimodal brain tumor images based on classification of super-voxels. In IEEE 38th Annual International Conference of the Engineering in Medicine and Biology Society (EMBC) 2016. 5945–5948. IEEE.

35 Meier, R., Bauer, S., Slotboom, J. et al. (2013). A hybrid model for multimodal brain tumor segmentation. *Multimodal Brain Tumor Segmentation* 31: 31–37.

36 Zhao, L., Sarikaya, D., and Corso, J.J. (2013). Automatic brain tumor segmentation with MRF on supervoxels. *Multimodal Brain Tumor Segmentation* 51: 51–54.

37 Parmar, C., Velazquez, E.R., Leijenaar, R. et al. (2014). Robust radiomics feature quantification using semiautomatic volumetric segmentation. *PLoS One* 9 (7): e102107.

38 Guo, X., Schwartz, L., and Zhao, B. (2013). Semi-automatic segmentation of multimodal brain tumor using active contours. *Multimodal Brain Tumor Segmentation*, 27: 27–30.

39 Vezhnevets V, Konouchine V (2005). GrowCut: interactive multi-label N.D. image segmentation by cellular automata. Proceedings of Graphicon 1 (4): 150–156.

40 Yu, J., Tao, D., and Wang, M. (2012). Adaptive hypergraph learning and its application in image classification. *IEEE Transactions on Image Processing* 21 (7): 3262–3272. https://doi.org/10.1109/tip.2012.2190

41 Yu, J., Rui, Y., and Chen, B. (2014) Exploiting click constraints and multi-view features for image re-ranking. *IEEE Transactions on Multimedia* 16 (1): 159–168. https://doi.org/10.1109/tmm.2013.2284755

42 Menze, B.H., Jakab, A., Bauer, S. et al. (2015). The multimodal brain tumor image segmentation Benchmark (BRATS). *IEEE Transactions on Medical Imaging* 34 (10): 1993–2024. https://doi.org/10.1109/tmi.2014.2377694

43 Abdulbaqi, H.S., Jafri, M.Z.M., Omar, A.F. et al. (2015). Segmentation and estimation of brain tumor volume in computed tomography scan images using hidden Markov random Field Expected Maximization algorithm. *2015 IEEE Student Conference on Research and Development (SCOReD)* https://doi.org/10.1109/scored.2015.7449396

44 Xiaoran, C. and Ender, K. (2018). Unsupervised detection of lesions in brain MRI using constrained adversarial auto-encoders. arXiv preprint arXiv:1806.04972. https://doi.org/https://doi.org/10.3929/ethz-b-000321650

45 Khan, M.W., Sharif, M., Yasmin, M. et al. (2016). A new approach of cup to disk ratio based glaucoma detection under fundus images. *Journal of Integrated Design and Process Science* 20 (2): 77–94. https://doi.org/10.3233/jid-2016-0004

46 Amin, J., Sharif, M., Yarmin, M. et al. (2019). Use of machine intelligence to conduct analysis of human brain data for detection of abnormalities in its cognitive functions. *Multimedia Tools and Applications* 79 (15–16): 10955–10973. https://doi.org/10.1007/s11042-019-7324-y

47 Sarhan, A.M. (2020). Brain tumor classification in magnetic resonance images using deep learning and wavelet transform. *Journal of Biomedical Science and Engineering, Scientific Research Publishing* 13 (6): 102–112. https://doi.org/10.4236/jbise.2020.136010

48 Sultan, H.H., Salem, N.M., and Al-Atabany, W. (2019). Multi-classification of brain tumor images using deep neural network. *IEEE Access* 7: 69215–69225. https://doi.org/10.1109/access.2019.2919122

49 Jia, Z. and Chen, D. (2020). Brain tumor identification and classification of MRI images using deep learning techniques. *IEEE Access*, 1–1. https://doi.org/10.1109/access.2020.3016319

50 Sharma, M. and Miglani, N. (2019). Automated brain tumor segmentation in MRI images using deep learning: overview, challenges and future. 68: 347–383. https://doi.org/10.1007/978-3-030-33966-1_16

51 Chanu, M.M. and Thongam, K. (2020). Computer-aided detection of brain tumor from magnetic resonance images using deep learning network. *Journal of Ambient Intelligence and Humanized Computing* 12 (7): 6911–6922. https://doi.org/10.1007/s12652-020-02336-w

52 Das, S. (2020). Brain tumor segmentation from MRI images using deep learning framework. *Advances in Intelligent Systems and Computing* 1119: 105–114. https://doi.org/10.1007/978-981-15-2414-1_11.

7

Challenges, Standards, and Solutions for Secure and Intelligent 5G Internet of Things (IoT) Scenarios

Ayasha Malik[1] and Bharat Bhushan[2]

[1] *Noida Institute of Engineering Technology (NIET) Greater Noida, Noida, Uttar Pradesh, India*
[2] *School of Engineering and Technology (SET), Sharda University, Noida, Uttar Pradesh, India*

7.1   Introduction

The emergence of movable grids is designed to meet the fresh requirements to improve the enactment, load, durability, and power efficacy of the network facilities. 5G mobile networks are embracing fresh communication concepts to additional advancement of these characteristics [1]. Telecommunication verification themes work to incorporate novel communication concepts such as Software Defined Networking (SDN), Network Function Virtualization (NFV), fog computing, cloud computing, Multi-access Edge Computing (MEC), Machine Type Communication (MTC), and Network Slicing (NS) ideas on telecommunications webs [2]. The purpose of these exertions is to plan a new-fangled movable system that will assist in the development of innovative network services to meet the need for emerging movable systems [3]. SDN perception recommends removing controller planes and records for communication strategies. Network management and SDN-based network intelligence are integrated into the central controller. In addition, it can provide ambiguity under the network of control services and the business application framework [4]. NFV recommends a novel way to generate, organize, and accomplish communication amenities. This thought intends to remove system roles as of computer hardware, with instructions to execute them by means of software contexts [5]. Cloud computing, Fog computing, Edge computing, and MEC will deliver the need for network failures. Network slice-cuts enhance maintenance for different phases of traffic in 5G Network [6]. Defending security and privacy has become a major concern in these fresh communication networks, as risks can have far-reaching consequences. Figure 7.1 shows the four stages of network softwarization that pave the way to 5G. It shows how the overhead technology has allowed the disposition of a soft 5G network, ranging from flexible mobile design to energetic and fast network architecture [7]. These variations in construction of 5G are predictable to accelerate the digital alteration that is seen by the entire industry [8]. This will affect the creation of new facility models and new significant chains that will lead to greater social and financial impression. The safety connected through 5G technology is measured by important desires associated with both 5G and the above systems.

Smart and Sustainable Approaches for Optimizing Performance of Wireless Networks: Real-time Applications, First Edition.
Edited by Sherin Zafar, Mohd Abdul Ahad, Syed Imran Ali, Deepa Mehta, and M. Afshar Alam.
© 2022 John Wiley & Sons Ltd. Published 2022 by John Wiley & Sons Ltd.

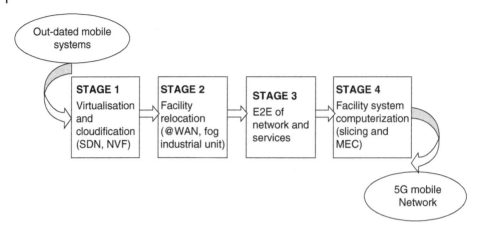

Figure 7.1 Network evaluation toward 5G.

In addition, many safety prototypes in pre-5G networks standards cannot be used in a straightforward manner in 5G because of its fresh design and new facilities [9]. However, certain security measures can be used with specific adjustment. In line with the preceding generation, the Open Air Interface (OAI) policy was deliberated in a broader 5G framework and a holistic view of the growth of the 5G safety protocol proposed in [10]. Earlier, the main desire for safety in telecommunication systems stood to safeguard the efficiency of payment organizations and safety of the receiver connector by encoding communication information. In the 3G network standard, two-mode certification is handed over to terminate the formation of a links with a bogus BS. Lastly, 4G grids apply progressive encrypting procedures to authenticate the consumer. Likewise, it provides security from corporeal attacks such as corporeal interference of sub-locations, which could be connected to open and consumer localities. In addition, certain confidentiality topics remain unresolved to some extent on the pre-5G network because user data was stored in the mobile company's data warehouse [11]. However, 5G security and privacy subjects overcome these trends due to variations in new facilities and structures. The safety of 5G standard and past 5G standard networks has three key features. Firstly, nearly all of the overhead safety threats and safety necessities associated with previous-5G groups are no longer relevant to 5G standards and beyond. Secondly, the 5G system will face fresh innovative safety experiments due to enlarged user count, diversity of allied strategies, novel system amenities, great consumer secrecy anxieties, and new-fangled investors, in addition to the need for critical IoT applications. Thirdly, network mitigation and the use of new innovative skills like SDN, MTC, NFV, and NS will announce a different trademark set of safety and confidentiality experiments [12].

In addition, there are fresh technologies or resolutions that will be used in 5G to encounter the needs of a wide variety of applications and linked devices. For instance, the ideas of cloud and fog computing, SDN, MTC, and NFV, are measured solutions to probable problems in terms of price tag and competence, [13] although some of these technologies have their personal safety tests. For example, key network organizations like Home-based Subscriber Hub (HSH) and Flexibility Administration Unit (FAU) that store consumer, private, and travel billing statistics, correspondingly embedded in the cloud, will disable all net activity in the event of security breaks [14]. Likewise, SDN focuses on the concept of system mechanism in SDN supervisors. These supervisors will be a much-loved option for enemies to affect an entire system with Double Denial of Service (DDoS) or a Source Destruction attack. The same goes for virtual machine monitoring on NFV. Consequently, it is very important and judicious to acknowledge the weaknesses that exists in this technology and to seek resolutions to those weaknesses [15]. The major contributions of this work are enumerated as follows:

- The work discusses the safety mechanisms required in various mobile networks.
- This work redefines the safety needs of 5G and beyond network in order to enhance security in mobile networks.
- This work explores the background, basic characteristics, requirements, and applications of various IoT ecosystems.
- The work highlights the expected benefits of IoT adoption in organization and summarizes non-5G technologies that address IoT environments.
- The work explains the 5G advanced security model needed to achieve security in 5G networks and the safety parameters that are required to secure three tiers of 5G network.

The remainder of this chapter is organized as follows. Section 7.2 discusses the security in wireless networks where security of Non-IP network, 3G networks, 4G networks, and 5G networks are discussed. Section 7.3 provides information about IoT background and requirements in the form of its basic characteristics. This section also discusses the expected benefits of IoT adoption in organization related to big data, openness of IoT, and linked aspects of IoT. Section 7.4 explores non-5G technologies that address the security concerns in IoT environments. Section 7.5 explores the advanced 5G security model to secure the 5G mobile network. Section 7.6 explains the security fundamentals of three-tiered 5G network highlighting the associated safety challenges and possible resolutions. Finally, the chapter concludes with Section 7.7 with suggestions for future research direction.

7.2 Safety in Wireless Networks: Since 1G to 4G

The safety of communicating systems has remained a challenge due to the complexity of the sub-network, security solutions designed to deal with areas that are difficult to handle, and weaknesses in administration of personnel administration. In addition, the construction of the Internet experiences difficulties from substructure that have matured with safety challenges and are grounded in invention of wireless system safety which have evolved steadily since the advancement of mobile networks. The largest variation occurs, approached with the outline of IP-based communications to wireless systems, where many net-based security experiments are streaming to wireless networks [16]. This section provides a summary of the various security conditions for wireless grids on or after 1G to 4G or from non-IP based wireless networks to IP-supported wireless standards, which are summarized in Table 7.1.

Table 7.1 Security evaluation from 1G to 4G.

Network	Security procedure	Security threat
1st Generation	Not any kind of privacy management nor explicit security	Snooping, call interference, and no confidentiality procedure
2nd Generation	Encryption based prevention, authentication and anonymity	Bogus base station, radio link security, one way authentication, and spamming
3rd Generation	Take on the 2G security parameters, safe admittance to network, presented Authenticated Key Arrangement (AKA), and two-way authentications	IP traffic security susceptibilities, encryption keys security, roaming security
4th Generation	Announced different encryption EPS-AKA and dependence process, security of cryptographic keys, non-3G Partnership Project (3GPP) admittance safety, and reliability prevention	Improved IP traffic flow attempted safety, such as DoS attacks, file reliability, Base Transceiver Stations (BTS) safety, and spying on extensive term solutions. Not appropriate for safety of new amenities and strategies, such as enormous IoT, foreseen in 5G

7.2.1 Safety in Non-IP Networks

First-generation (IG) mobile application has been used for analog signal handling and is mainly intended for speech facilities. A very effective 1G program named Progressive Phone Facility (PPF) was distributed by AT&T and Bell labs in 1983. Because of the working of analog communications, it was difficult to afford effective 1G safety facilities. This mobile facility had not used cryptographic techniques before, so there was not any information on safety of phone chats. Thus, almost the entire organization and consumers remained exposed to safety challenges, such as audio listening, illegitimate admission, composition, and consumer secrecy. Digital transportable organizations were projected to raise the efficacy of rate of recurrence range [17].

Global Scheme for Mobile (GSM) communication becomes effectively and extensively used in mobile transportations as a part of 2G phone systems. The GSM has recognized four aspects (Secrecy, Certification, signal wave protection, and user data protection) of security facilities that will be delivered through the GSM system. Anonymity is achieved by using momentary identifiers to make it easier to identify the user of the program. When the device is turned on then real identifiers are used by assigning a temporary identifier. The. network operator uses authentication to recognize the genuine user. It is done in a way that responds to a challenge. Signaling and protection of the user's data is done by encryption where the Subscriber Identity Module (SIM) has a key job in cryptographic keys. Conversely, 2G had many limitations or safety vulnerabilities. The machinists only validated UE in one way, and the UE had not any other option than to validate the machinist. Hence, it would be probable for an incorrect machinist to imitate an actual operator and carry out an attack on the middle person, called a third-party attack. In addition, the encoding set of rules is too restricted and correction algorithms are subject to numerous attacks. GSM could not afford data integrity with respect to hijacking channels in non-appearance of encryption, and they were at risk of a DoS attack. In addition, 2G schemes were not capable of improving their safety [18].

7.2.2   Safety in 3G

Third-generation (3G) mobile networks have been industrialized specifically to offer an advanced data range as compared to 2G networks. 3G applications also permit new facilities such as video phone calling and video streaming over mobile networks. The 3G network has projected improved security structure to reduce the risks to 2G network systems. There are three main security principles in a 3G network that are defined by Three-Generation Partnership Project (3GPP), namely: i) 3G security facilities will take over 2G safety facilities; ii) 3G safety will improve the 2G safety boundaries; and iii) 3G safety structures will enhanced other security features that were not accessible in 2G. 3GPP has established and retains a 3G mobile technology called the Universal Mobile Telecommunications System (UMTS). The UMTS security design has five sets of security structures described in TS33.102, better known as Release 99. These sets confirm that UE has safe admittance to 3G facilities and delivers defense from attack on the radio access link. The UMTS-AKA protocol is designed in a mode whereby compliance with GSM is enhanced. On the other hand, UMTS-AKA operates for supplementary protocol purposes, such as ensuring network compliance and as well as integrity key compliance, etc. Opposed to GSM's authenticity, UMTS carries two-country verification to remove the threat of a false channel. The contact of security services contains the user's uniqueness, which guarantees that the user cannot be detached from the radio access link. User privacy is also required to maintain user position privacy and user security. To accomplish these goals, the user is recognized by a momentary identifier or by a stable hidden identity. Likewise, the same user should not be recognized for an extensive time and some information that may disclose user uniqueness should be encoded [19].

Yu et al. [20] discussed the safety network's structure of 3G communication networks as a significant issue that is essential to be resolved quickly and attentively on the safety agenda of 3G communication networks. Pattekar et al. [21] proposed a summary that the 3G network has predictable enhanced safety arrangements to decrease the danger of 2G network systems. Li et al. [22] examine the safety admittance dominion in a 3G communication system that is susceptible to Man in the Middle (MITM) attack and Denial of Service (DoS), based on deep exploration of Authentication and Key Agreement (AKA) protocol.

7.2.3   Security in 4G

10th releases from 3GPP, better identified as LTE-Advanced (LTE-A), meet the necessities of 4G standard network quantified by the International Telecommunications Union – Telecommunications Division (ITU-R). The LTE-A system takes two components: Evolve Packet Core (EPC) and Evolve-Universal Terrestrial Radio Access Network (E-UTRAN). EPC is a whole network of IP backbone and package. The LTE-A system maintains non-3GPP admission systems. LTE-A organization networks also present new organizations and uses such as Machine-Type-Communication (MTC) and eNodeB for phones and transmission nodes. 3GPP defines the same groups of LTE-A safety structures. These are Admission Security, Network Dominion Security, User Dominion Security, Application Dominion Security, and Visibility or Security settings. On the other hand, every facility has been meaningfully improved to protect LTE-A systems. In addition, completely new security measures for MTC, domestic eNB, and transmission areas were described. The Evolve Packet System-AKA (EPS-AKA) had one main improvement over UMTS-AKA termed cryptographic web partition. This service minimizes several safety breaks on the network and reduces the chances of distribution attacks through the system. This is attained by requiring any EPS-related encryption public as well as private keys to the Serving Node (SN) ownership, to which the keys are transported. This functionality allows UE to verify SN. The UE does not give permission of authenticity of SN to UMTS, but only confirms that SN is approved by the UE home-based network [23].

3GPP has stated mobile security features inside E-UTRAN or in the middle of E-UTRAN and previous generations or non-3GPP organizations. Two non-3GPP access networks exist: Reliable (trusted) admission to non-3GPP admission network and unreliable (untrusted) non-3GPP access networks. With a non-3GPP unreliable access grid, UE has to transfer the data gateway to the trusted packets that are a portion of the EPC. Additionally, a novel significant system and significant administration provision have been presented to guarantee a safe navigation procedure in LTE. Because of the vulnerabilities of the first endpoint, eNB was more susceptible than 3G security builds. Alternatively, EPS configuration permits to place eNB exterior in the network operator's secure dominion, i.e., in physically unsafe areas. Consequently, eNB is at risk of physical attacks, flooding attacks, DoS attacks, or subtle attacks when searching for durable keys. To address this risk, 3GPP has announced a strong demand for eNB. 3GPP necessities contain secure configuration and formation of base station, safe key administration within BTS, and a safe environment for user data management and plane control, etc. [24].

Yarali et al. [25] demonstrated useful cost-effective methods to connect people all over the world, but they were inclined to several security holes. 4G safety needs to think through extra safety requirements (likened to earlier generations), such as SDN controller safety, hypervisor safety, orchestrator safety, cloud safety, and peripheral safety, etc. Bikos et al. [26] presented robust and secure safety precautions to improve any vulnerability found in 3G security, and the addition of original types to address the innovative security necessities of 4G services. Security still plays an important role in the effective and reliable implementation and innovation of new LTE/SAE technologies. Zeqiri et al. [27] revealed information about a security problem that had been solved by using many layers of encryption. There is a lack of power ingestion and a significant delay in broadcasting. In 4G, there is a sense of player security, where a single layer will be set up to perform data encryption.

7.2.4 Security in 5G

According to ITU-T, 5G recommendations-based present network design and the lack of security methods that may or may not have been upgraded or implemented. The reference high points warning notes that include such things as 5G standard with several doubts, lack of well-defined ideas, and end-to-end construction and unknown structures [28]. The approval emphasizes security experiments on access network systems, and cyber-attacks on customers and network substructure. Details of safety boundaries and approvals can be found and main ideas are explained in the subsections.

7.2.4.1 Flashy System Traffic and Radio Visual Security Keys

Fifth-generation (5G) systems should handle large traffic fluctuations and provide durability whenever this increase occurs, while still preserving satisfactory performance. Previous network structures include 4G, encryption keys radio made on the home-based system, and referred to the stayed web via doubtful associations that source a strong opinion of key disclosure. Hence, it is suggested that privacy must be protected before or not be referred via unprotected contacts like SS7/DIAMETER.

7.2.4.2 User Plane Integrity

Third-generation (3G) and fourth-generation (4G) organizations provide shelter for specific signature communications but do not offer cryptographic security for user data reliability. Therefore, it is recommended to pay for security in the transference layer or application layer that store over movable networks. On the other hand, end-to-end safety installation may involve excessive extraction of data transfer to packet headers. As a result, the exclusion from this might be the grid-level safety of IoT devices that are restricted to sensitive services or 5G latency services [29].

7.2.4.3 Authorized Network Security and Compliance with Subscriber Level Safety Policies

It is very likely that safety areas can be rearranged frequently for each user, as the user moves from one location to another or from a single operator system to an alternative operative system, as in the case of signal roaming. Network operatives are therefore important in defining safety policies and a certain level of subscription data provision. Safety is a versatile topic, and the sheer number of devices with 5G services makes it very difficult. The concept of securing 5G systems defined by NGMN [30] is built upon three ideologies. These are Elastic safety systems, High integral safety, and automatic safety equipment. The purpose of this dream is to provide the most security against cyber attacks by securing confidentiality and safety in the 5G network, and providing a complete security analysis of wireless networks with simple ideas and descriptions that encompass the entire representation, regardless of basic technology. This is delivered by ITU-T [31] in the custom of safety sanctions and provides a walking grit agenda for the safety of communiqué networks. For example, in 5G standards, two types of approvals are provided by ITU-T in the circumstance of system verification for facilities. The first one is Push manner, where the set-up admission mechanism grid is acquainted with application layer procedure, and therefore provides certification keys directly. Another is Pull manner, where the entire contact controller device does not recognize the default process, so the facility stage has to take certification outcomes from the 5G network [32].

7.2.5 Security in 5G and Beyond

Fifth-generation (5G) and beyond security will be paramount, with 5G and beyond services established to cover the furthermost regions of the UK and US in excess of the next few years. By way of the main network, operatives introduced new services everywhere in the world; a major controversy relating to the network was imminent for 5G and beyond independent security risks. Mobile security is high on the concerns outlined by both the UK and the US administrations. Besides, with 5G and beyond networks unlocking up additional openings in zones like as healthcare, industrialization, and conveyance, the fact is that it is attractive and progressively popular among cyber criminals, as it expands the existing risk zone and the significance of any harms [33]. The result is to set out the succeeding safety concerns found in the 5G and beyond release. These points are needed to be concerned and fulfilled to achieve security in 5G and beyond networks:

- More disclosure of attacks and potential intruders: With 5G and beyond networks in terms of software, the risks are associated with major security errors. Like errors found in improper software development procedures, inside providers are significant. And they can make it easier for threat players to contaminate the products and make it harder to spot them.
- Because of new 5G and beyond network design features and new functionality, definite portions of network tools or roles are converting to more complexes, such as basic channels or core network management functions.
- Increased coverage to threats associated with mobile network operatives depend on providers. This will lead to advanced attack methods that can be demoralizing by threatening characters and increase the likelihood of a potential impact of such an attack. Among the numerous probable performers, non-EU or government-sponsored countries are taken to be the most important and most possible to aim for in 5G and beyond networks.
- In the current framework of enlarged experience to provider-provided attacks, the threat outline of distinct dealers will be critical, considering the possibility that the dealer may be disrupted by a non-EU country.
- Increased risks arising from high reliance on providers: high dependence on a solitary provider raises the risk of supply chain disruption, subsequent in for example trading failure, and its penalties. It also magnifies the latent influence of weakness or susceptibility, as well as their exploitation by threatening characters, especially when dependence affects the provider who carries the greatest risk.
- Threats to access and integrity of networks will be major safety anxieties: in calculation of privacy and confidential threats, with 5G and beyond networks predicted to be the support of countless serious IT applications, integrity and access to individual networks will be major national safety issues and a main security experiment from an EU standpoint [34].

Anamalamudi et al. [35] describes the safety concepts of Wi-Fi and Li-Fi network connectivity via 5G systems. While planning 5G set-ups with Wi-Fi and Li-Fi connections, architectural contemplations are required to be conveyed by safety reflections, and these safety reflections will probably effect constructed decisions. Khan et al. [36] proposed the basic and dynamic technology that will be integrated into the 5G network; confidential data will go through all layers in upcoming wireless structures. A number of occurrences have exposed the risk posed by a wireless network infected with the virus, not only affects safety and privacy issues, but also interferes with the compound environmental conditions of communication. As a result, the severity of safety attacks that occurred in the previous creation need to be improved upon, so uncovering or hindrance of vandalism is a world-wide challenge. Dutta et al. [37] introduced the reliability of 5G reputation and the status of incorporating security capabilities from scratch, while 5G formats are well-defined and consistent. Safety necessities need to be covered and integrated into the various layers of 5G systems (physical, network, and application) and the various components of the E2E5G structure inside the threat administration agenda that looks at the occurrence of security threats. Scott et al. [38] highlights the effect of new 5G technologies, namely Software Network Defined Networking (SDN), Network Function Virtualization (NFV), and Cloud Computing on surviving Safety and Confidentiality Rules.

7.3 IoT Background and Requirements

To design a successful 5G mobile network in collaboration with IoT, an understanding the basic features of IoT is required so that the infrastructure is designed and works properly to satisfied user needs. Section 7.3.1 discusses the background and all the requirements of IoT and layers of IoT characteristics, main features of IoT infrastructure, and characteristics of IoT applications, are explained.

7.3.1 IoT and Its Characteristics

IoT research is still in its infancy, and a general description of IoT is not yet obtainable. IoT can be observed in three ways: Internet-focus orientation, things oriented objects (senses or intellectual objects), and semantic (informative) information. Also, IoT can be considered as consumer support or industrialized applications and can be referred to as the Human's Internet of Things (HIoT) or the Industrial Internet of Things (IIoT). The first description of IoT came from the "object-centered" concept, in which RFID tags were measured objects. The RFID community says [39] IoT can be well-defined as "a global network of connected things that can be exceptionally targeted according to standard communication protocol." The meaning of "things" in the IoT concept is very broad and contains a multiplicity of physical objects. This contains personalized items such as smart phones, smart watch, tablets, and digital cameras. It also consists of items (home, car, or work) in our premises, businesses (machinery, car, robot), and tagged items (RFID), which are associated via a gateway device (smart phone). Sensor networks (SNs) as well as WSNs and wireless sensors and actuator networks (WSANs), RFID, M2M connections, and SCADA, are key IoT components [40].

7.3.2 Characteristics of IoT Infrastructure

Some characteristics of IoT infrastructure are now explained for better arrangement of IoT devices in co-relation with the 5G network:

- Dissimilar devices: For the surrounded state of the computer for most IoT devices, low-cost computer platforms may be used. In detail, to reduce the effect of these devices on the surroundings and power ingestion, less powerful radios may be used to connect to the Internet. These low-power radios do not use Wi-Fi or fixed mobile network technology [41].

- Restricted resources: Computers and sensors require a minor device for influence, which confines its functionality, memory, and communication capability. For instance, RFID devices or tags may not have the processing power or battery power.
- Automatic communication: In IoT applications, rapid connections can occur when things are moving, and they enter into the intermediate range of other objects, prominent to the automatic creation of actions. For example, a smart phone user can interact more with a TV/refrigerator/home laundry machine and can produce actions lacking physical participation [42].
- Large-scale network and large number of actions: In the case of IoT, many thousands of strategies or objects can communicate on their own or in one place (e.g., building, store, and university), greatly higher than most common communication systems. Universally, IoT will be the largest network consisting of nodes at a billion and even trillions of nodes. It was projected that there were approximately 26 billion IoT devices in 2020 [43].
- No infrastructure and dynamic network: IoT will assimilate devices, most of which will be mobile, wireless, and have limited resources. Mobile locations within the network are moving or joining whenever they want. Also, locations may be disconnected due to wireless connectors or lack of battery. These features will mark the IoT network as more powerful. Within such ad hoc surroundings, where there is partial amount connection or sometimes there is no connection provided for fixed set-up, it will be problematic to preserve a constant network to sustain multiple IoT-dependent operating systems [44].

7.3.3 Characteristics of IoT Applications

Some main features of IoT application are now enlightened upon that are essential to understand the better employment of IoT devices:

- Various applications: IoT can provide its facilities to a great number of applications across multiple dominions and locations. These dominions and areas can be subdivided into (non-functional) dominion groups such as: Transport and operational, Healthcare, Intelligent atmosphere (home and plant), Industrialized and personal or social background [45].
- Real-time: Applications by means of IoT can be largely categorized as real-time and non-real-time. For example, IoT healthcare and transport will need to be delivered on time for their statistics or facility. Overdue transport of data can render an application or facility unusable and risky for sensitive applications [46].
- Everything-as-a-service (X-a-a-S): This application provides facility in a very efficient, awesome, and ease manner. The X-a-a-S model promoted sensitivity as a function in WSNs and this could lead of IoT in the X-a-a-S model. As an additional connection, the gathering of facilities is also growing and turns out to be available online; it will be obtainable for use again and again.
- Increased security attacks: Whereas there is great prospective for IoT in various dominions, there is also worries about the safety of applications and systems. IoT requires worldwide association and openness, which says that anybody can gain admittance at anytime,

anywhere, and anyhow. This dramatically expands the attack areas of applications and IoT networks [47].

- Privacy Leaks: Using IoT, applications can assemble data of people's regular activities. Since information about such activities (traveling, shopping, and everyday expenses) is considered by many to be confidential, the disclosure of these statistics may affect the privacy of those persons. The use of cloud computing creates the difficulties in private leakage even more adversely [48].

7.3.4 Expected Benefits of IoT Adoption for Organization

The impression of IoT acceptance on administrations is mainly correlated to the statistics produces by IoT. IoT takes three features: "Big," "Open," and "Linked" (BOLD). Initially, IoT creates large volumes of information which is often of healthier quality than information produced by traditional resources: existence of advanced granularity and frequently better correctness, presence of larger heterogeneity, upcoming from a crowd of foundations, being more appropriate than traditional information; regularly being real or near real-time; and taking significantly higher dimensions [49].

7.3.4.1 Benefits Correlated to Big Data Created by IoT

A key factor in the implementation of IoT is the combined integration of multiple technologies and communication explanations such as diagnostic and tracing technologies, cabled and wireless sensor device and actuator systems, advanced communication protocols, and the dissemination of intelligent intellectual property, radio data rate identification technology, electric merchandize coding, and ZigBee expertise [50]. The IoT diversity means that, for instance, numerous kinds of sensors from several resources can be used to enhance public security and compliance with guidelines for illustration, which may provide better control mechanisms than the traditional methods discussed by researchers. Therefore, Big Data analytics can play a significant part in empowering intelligent management, which facilitates collaboration between collaborative organizations suggesting timely data from linked materials to improve decision-making, thus allowing for improved analysis in terms of chasing or status consciousness. IoT applications permit more data collection side-by-side and through mechanization they also capture new data with advanced easiness about effective procedures and on-going jobs [51]. Wise administration of the traffic network through the facility of real-time data material to the civilian, depending on the recent state of the traffic, has a significant influence on the health of the civilian and improves the presentation of urban authorities. It also rises to the amounts of data generated by IoT, and reduces the common definition error in data analysis and can lead to superior confidence in the results delivered. suggesting that large data acceptance can have a significant impact on data quality [52].

7.3.4.2 Benefits Interrelated to the Openness of IoT

IoT applications are realized as permitting automatic scanning of data records, thus building physical manipulation of data capture in excessive amount. IoT delivers great information that can be accessed for general open use [53]. Making information accessible to the community can increase the visibility of the organization, helping to expand business processes and decrease waste. Empowering customer self-employment, even though IoT can authorize people and corporations with enhanced admittance to evidence and have faith in that business value, can be based on IoT. This can be achieved by refining real-world perceptibility, and slowing down business procedures as IoT empowers administrations to observe real-world realities, increasing tractability of service, and better performance of services, which allows for better decision-making, and repeatedly leads to fresh profits streams. Finally, the IoT's ability to enlighten and systematize can lead to a change in surviving business procedures [54].

7.3.4.3 Benefits Related to the Linked Aspect of IoT

The connected IoT feature can decrease labor costs and equip the community by empowering consumer self-employment, such as self-supply testing in superstores. Combined information can be composed leading to an understanding of product demand, whereby serving superstores expand the value of their collection and recover client consummation [55]. It is believed that being capable to associate information from diverse sources by means of IoT can allow for fraudulent detection, reduce fraud-related expenditures, and increase consumer confidence. In addition, the visions increased from connecting information from a variety of resources sanction administrations to connect more efficiently with their customers, offer innovative openings to interconnect and support service income, actual analysis, and use of big data, which are mandatory to achievement in numerous commercial and facility dominions. This includes the ability of IoT technologies to efficiently gather information about work procedures, avoiding wasting time on physical calculations so that already generated and examined data can be used to improve efficiency and consistency [56].

7.4 Non 5G Standards Supporting IoT

There are already many communication technologies for IoT integration devices, certain narratives are also planned for IoT use-cases, and various technologies are previously well-used in some cases, such as Bluetooth 5.0 Low Energy (BLE), IEEE 802.15.4, LoRa, and Sigfox. However, the technology is not old enough or does not meet all the requirements to be used on a large scale [57].

7.4.1 Bluetooth Low Energy

Bluetooth technology is often established in smart phones, laptop, air pods, cars, wireless mouses, and keyboards. This communication protocol defines a Personal Area Network (PAN), and is deliberately used to deliver over a small distance range, while communicating (nearly 10 m) between devices. Conversely, to extend its position in IoT mode, the standard is required to decrease power ingestion and so be capable to run with batteries that are similar to the size of a coin. Therefore, Bluetooth Special Interest Group (SIG) inherits low power (BLE) when releasing Bluetooth version 4.0 to identify this market segment [58]. Unlike standard Bluetooth, BLE is designed to transfer small packets of data periodically. The main difference between standard Bluetooth and its low-power configuration is shown in the Physical layer (PHY). Whereas there are typically 79 standard channels with a bandwidth of 1 MHz, BLE offers 40 standard channels with a bandwidth of 2 MHz. In both Bluetooth standard and BLE standard channels, these RF are categorized into two groups, publicity RF channels, and data RF channels. Advertising is used for 5G device acquisition, streaming, and connectivity, while data channels are used for data transference between linked Bluetooth devices [59]. Also, both activate on an unregistered Industrial, Scientific, and Medical (ISM) bandwidth (2.4 GHz). In standard Bluetooth, the switching process can vary from Gaussian Frequency Shift Keying (GFSK) to four phase shift keys 4-PSK and 8-PSK phases, while in BLE only GFSK is optimized. The GFSK produces a very low Peak-to-Average Power Ratio (PAPR), which interprets to low-power ingestion, because of high magnification power working near a pressure point of 1 dB can be monitored [60].

7.4.2 IEEE 802.15.4

IEEE 802.15.4 describes Moderate Access Control (MAC) and PHY layers. The physical layer functions in dissimilar ISM groups permitted to the region in which it operates. The 2.4 GHz band is worldwide, with other data rate bands like 868 MHz in Europe and 915 MHz in the USA [61]. IEEE 802.15.4 is intended for PANs, and is primarily integrated into embedded structures for agriculture, ecological, and engineering observing organizations. Unlike the IEEE 802.11 family, IEEE 802.15.4 does not place emphasis on higher data ranges, and unlike Bluetooth, does not concentrate attention on linking their own devices. It was suggested that low-cost and low-power wireless networks could be depleted of resources in large parts of the world. For instance, Zigbee successfully used IEEE 802.15.4. The biggest benefit, as compared to others, is that Zigbee is capable to work with multi-hop structures, permitting for network failures [40]. However, as mentioned above, the delay may not reach the required values due to the network status and configuration. IEEE 802.15.4 has defined seven dissimilar working modes. After the IoT standpoint, methods that prime to low-energy ingesting are Offset Quadrature Phase Shift Keying with Direct Sequence Spread Spectrum (O-QPSK-DSSS), QPSK variation by Chirp Spread Spectrum (DQPSK-CSS), and Gaussian Frequency Shift Keying (GFSK) with no virtual distribution process used [62].

7.4.3 LoRa

LoRa is categorized as Low Power Wide Area Network (LPWAN) machinery established by Semtech in 2015, so it has recently been made available with Zigbee and Bluetooth. Likewise, it works at a low level and suggests low energy ingestion. On the other hand, in contrast with the aforementioned technologies, LoRa (and LPWANs as a whole) is useful to cover a wide area. LPWANs have more coverage area than PANs but smaller than mobile networks [63]. The cover of LoRa is 20 miles. This functionality marks LoRa as an exciting choice for accurate applications associated to smart cities and agriculture. Applications requiring network override can be excluded, as LoRa affords, at max, is 50 kbs. In addition, the visible results have shown that a few billion devices can be accessed by a solitary primary channel. Conversely, the exposure area is greatly reduced as the number of devices rises. For physical lexicons, LoRa uses a measurement rotation process based on Chirp

Spread Spectrum (CSS) in conjunction with GFSK. CSS permits for easy synchronization of time and frequency range, so inexpensive items can be used. Additionally, the basic flexibility of CSS reveals internal strength duplicate channels. Strength of LoRa PHY is that evolutionary schemes have used an endless envelope. Therefore, more efficient amplifiers can be used, and power consumption is improved [64].

7.4.4 Sigfox

Sigfox is also categorized as LPWAN and works with small bandwidth fixes. In distinction to LoRa, Sigfox is a related system, consequently, with a sealed certification. As a result, a very tiny amount of data is accessible to the public. However, more details on its PHY are not yet known [65]. The functioning rate of recurrence band range is 868 MHz in Europe and 915 MHz in the United State. Most of the admission process is asynchronous, so no money is spent on syncing devices to RBS. To create a link, the device conveys three successive messages. Every message is conveyed at varying times and randomly selected frequencies. This is an instance of a standard shift in multi-access technology, and has prepared Sigfox with a popular solution for IoT connectivity. An additional important point is that acceptance is grounded on a collaborative process. RBS can accept messages from communicating devices [66]. This diversity of response interprets to better service quality. Sigfox has shaped a transportations protocol centered completely on IoT necessities, and it has delivered to be one step forward to further identifying the visual IoT paradigm [67].

7.4.5 Wi-Fi HaLow

Wi-Fi HaLow is a portion of the Wi-Fi family and is standardized by standard IEEE 802.11ah. Likened to the technology mentioned above, it suggests a lengthy space, low power, low energy, and low-cost solution to attach a great amount of devices [68]. Differently, IEEE 802.11ah offers a composite PHY. It maintains voice fluctuations up to 256-QAM transmissions using Orthogonal Frequency Division Multiplexing (OFDM), and Multi-Input Multi-Output (MIMO) is available in down link and up link. Hence, the full data rate is 78 Mbps but the channel provides a bandwidth of 16 MHz and the frequency indicator is fixed at 256. Due to it operating on unlicensed waves below 1 GHz, the coverage range is from 100 m to 1 km [69]. A comparison of non-5G standards of IoT that based on various attributes is shown in Table 7.2 [70].

7.5 5G Advanced Security Model

Wireless communication organizations are restricted to distinctive telephonic audial and audio-visual phone calls. Similarly, they sustain a quantity of uses including play station, transporting, market, wireless power transmission, Wi-Fi, social networking, home application Bring Your Own Device (BYOD), fog computing, and cloud machineries, which must free up widespread study experiments to inventors [71]. For providing security in the 5G model, the constraints shown in Figure 7.2 should be accomplished.

Table 7.2 Comparison of non 5G standards of IoT.

Non 5G standards of IoT	Frequency	Sensitivity	Spectrum Strategy	Range	Data rate	Modulation	1-Hop Latency	Power	Cost
Bluetooth Low Energy	2.4 GHz	−95 dBm	Wide Band	100 m	1 Mb/s	GFSK	3 ms	Low	Low
IEEE 802.15.4	2.4 GHz	−97 dBm	Wide Band	100 m	0.250 Mb/s	O-QPSK CCK/DSSS	1.5/10/20 ms	Low	Low
LoRa	<1 GHz	−149 dBm	Wide Band	2–5 km	18 b/s −37.5 kb/s	LoRa	500 ms	Low	Medium
Sigfox [70]	<1 GHz	−126 dBm	Ultra Narrow Band	Several km	100 b/s	DBPSK	2 s	Low	Medium
Wi-Fi HaLow	<1, 2.4, 5 GHz	−95 dBm	Wide Band	Several km	1–54 Mb/s	256-QAM	N/A	Medium	Low

Source: Based on Morin, E. et al. [70].

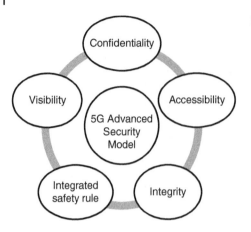

Figure 7.2 5G security model.

7.5.1 Confidentiality

In 5G security prototype, data confidentiality can defend data transference from exposure to illegalities such as Submissive attacks (i.e., Snooping). Seeing the 4G-LTE and 5G structural designs, any consumer information needs to be trusted and secure from unlawful consumers. Ordinary information encryption procedures have been extensively accepted to understand the data confidentiality in 5G networks [72].

7.5.2 Integrity

Integrity is used to retain moderation and to protect data from damage for the period of transmission from start point to end point. Integrity of 5G standard New Radio (NR) stream of circulation flow is sheltered as with 4G standards at the Packet Data Convergence Protocol (PDCP) layer. In 4G LTE, integrity security is delivered individually for Non-Access Stratum (NAS) and Access Stratum (AS). On the other hand, leading of key improvement in 5G integrity security involves that 5G NR deals with the integrity guard of the user as well [73].

7.5.3 Accessibility

In 5G dominions, systems accessibility is to guarantee that the system properties can be available each and every time they are required by authentic users, meaning that accessibility belongs to the status of the facility supplier. In alternative arrangements, the accessibility guarantees the great likelihood efficiency of the web substructure. It also processes the sustainability of a network in contradiction of vigorous attacks, e.g., DDoS attack, flooding attacks, etc. [74].

7.5.4 Integrated Safety Rule

In the 5G web, the existing 3GPP 4G safety designs cannot remain practical to the innovative 5G standard used scenario by way of their commitment to their outdated dealings with the supported dependence model. Hence, to maintain innovative ideas (NFV and SDN), it is required for central safety rules controlling organization that offers suitability for manipulators admittance of the uses and services [75].

7.5.5 Visibility

Visibility allows End-to-End consciousness of mobile systems to the controller's plane. This can professionally challenge the simple system topics to confirm safe surroundings. The 5G systems engaged wide-ranging end-to-end safety policies, and covered all network layers including physical, application, transport, signaling, and data planes. To execute such inclusive protection methods, 5G operatives must have a comprehensive visibility, review, and command entire layers in the 5G system. Now, the 5G machineries would be combined with exposed Application Programming Interface (APIs) to succeed in the safety strategies. In this a manner, the 5G web can have dependable safety strategies in software and hardware of the set-up. The heightened visibility through the system and safety strategies will support the appliance of appropriate safety machineries which is appropriate for innovative 5G facilities [76].

7.6 Safety Challenges and Resolution of Three-Tiers Structure of 5G Networks

To appropriately examine the safety viewpoints of the whole system in a regular mode, safety in the web structural design is labeled in three tiers. The first one is the Access network, the next is the Backhaul network, and last is the Core network. For clarity, we need to emphasize the safety experiments and then offer the possible clarifications.

7.6.1 Heterogeneous Access Networks

Heterogeneous Access Networks are explained in two main parts, The first is security challenges that come under a heterogeneous network and the second one is the required solutions to achieve security in this network.

7.6.1.1 Safety Challenges

Due to high demands of 5G services, it is required to have very great data rates with omnipresent accessibility and very low potential. The fresh operating circumstances of MTC, blockchain, IoT, and Vehicle-to-everything (V2X), etc. will place different necessities on the Web. For example, V2X and MTC life-threatening applications will require potentials in the direction of 1 ms or fewer. In addition to such necessities, the required consistency and accessibility of facilities will be of a much larger order than existing systems. Existing networks, on the other hand, were previously inclined to various Net-based intimidations that could aim admittance opinions such. as eNBs on LTE and low-power access points, as identified by investigators. With the integration of various IP devices into 5G, safety threats will also growth. With the speedy development of a large number of new devices and facilities, the claim for network power is increasing faster than ever. In addition to improving link and distribution budget, Het-Nets will hold system nodes with various features such as broadcast energy, transistor rate of recurrence, small electrical node and macro node control, all under the similar operative control [77].

7.6.1.2 Safety Resolutions

In an idyllic case, the security of the access system must defend the user, the system substructure, and amenities from all potential threats to the radio network components. Various technologies are used by 5G network standard for access to lengthy exposure, high penetration, and lesser distances, therefore 5G network will use virtualization technology, SDN, NFC, and cloud technology to synchronize performance and precise facilities with the construction and robustness of admittance to various network places. By way of such abilities, 5G network will recover organization toughness in contradiction to a variety of security challenges ranging from various access technologies. Communication safety can be delivered in a variety of ways, regularly in the superior layers of the network. Physical layer safety, conversely, tends to reduce the difficulty of cohesion without guaranteeing data security. The security of the physical layers of wireless communication does not place trust on high-level security systems or encryption. The basic goal of physical layer protection is to use random sound and communication channels to limit the amount of data that can be exposed by illegal recipients. Therefore, the use of physical layer protection outlines makes it very challenging for hackers to determine or operate the information conveyed [78].

7.6.2 Backhaul Networks

Backhaul network is explained in two main parts. The first one is the security challenges that generally are uncovered in backhaul network and the second one is the required solutions to secure this network.

7.6.2.1 Safety Challenges

The backhaul system contains network and communication resources between the sub-channel and the main Web. For that reason, the retrieval (backhaul) system is exposed to the safety risk access of network. Transmission of data in the network can be wireless, microwave frequency linked, and irregular connection links of satellite, wire cable using committed copper, or fiber optics, or using both in integration. Variations in RAN technology have also emerged in backhaul networks, most notably LTE. Backhaul security differs in the logic that it includes the radio and the main parts of the system. In some cases, security issues are common in parts of an admittance network, like the eNB and FAU in the main core system. With the Internet or Internet connection, eNB provides the traffic flow to the main server gateway and hub via the GPRS Tunnel Protocol (GTP). By using this gateway, data traffic is guided straight to the Public Data Network (PDN) gateway that connects to external systems or to the main net-web. GTP is also used above the S5/S8 virtual line among the operating gateway and the required PDN gateway. The LTE backhaul network improved safety by announcing Internet Protocol Security (IPSec) constructed on the GTP routes of the X2 cross-section among eNBs, and the S1 link in the middle of eNBs and FAUs. Another humble test for such channels is that the channel should be designed once UE goes live, or smooth start-up of the gathering by means of a new structure. In the event of a wireless backhaul standard, for increased traffic flow request, large MIMO is used for its safety tests [79].

7.6.2.2 Safety Resolutions

In a backhaul network, limited safety experiments are present as it works as a medium between two secure nodes, because of the close nature of nodes. GTP has numerous loopholes with several benefits regarding safety. Meanwhile, GTP activates on an upper layer of User Datagram Protocol (UDP), which runs fine with firewalls. Even if GTP is used in an internal worker system, firewall feeds may be required to prevent malware attacks at GTP end points. In addition, the traffic we experienced due to a tiny traffic break as of active state modification of devices, or growth of IoT devices, can be decreased by combining the identical UE channel with various UE using the similar service necessities. In addition, the backhaul network is converted in computer generated data logs with SDN, and maximum control data logs roles will be transferred to the main core network. The safety fears on the backhaul network will be greatly reduced by it working as simple dispatching device [80].

7.6.3 Core Network

To understand the security of the core network properly, two parameters are covered. The first one is the security challenges that arrived in the core network and the other is the security solutions to provide safety.

7.6.3.1 Safety Challenges

The main LTE or 4G standard networks, named EPC, consisted of various organizations such as FAU, router, gateway, firewall, PDN gateway, hub, and Hollow Structural Sections (HSS). In 5G network, the main system essentials are characterized by system roles. Comprehensive configurations of the main 5G standard network and an explanation of the set-up jobs are accessible in the newest statement of 3GPP 15. The core system is totally constructed on IP and guarantees endwise, safety, and Quality of Service (QoS) facility distribution and preserves subscriber statistics. The main 5G system is extra commanding, parallel to earlier peers using NFV, SDN, and cloud technology as defined previously. However, it has the most beleaguered safety fears and is also susceptible to safety threats. The enormous influx of IP protocols to manage and operator flights of dissimilar system tasks makes the main 5G standard core network more susceptible. The system requisite needs to be extremely patient and confirm obtainability with growing display capacity. Growing types of communication facilities and devices can prime to a large number of vehicles for display purposes. Signing/Export signature processes, carrier installation, location renewal, and authentication that occur in the Non-Access Stratum (NAS), 3GPP network proprieties can create NAS tempests. This is going to be very difficult [81].

7.6.3.2 Safety Resolutions

The 5G standard core network presents fresh structures and adopts innovative technical ideas paralleled to the EPC. Allowing to the release 15 of 3GPP, the best and clear variations are user control partitioning, network cutting, service building, and elastic non-3GPP admitted bury set-up. Complete variations must remain prepared and imaginable by technical notions like as SDN, NFV, and virtualization, as well as cloud computing or movable edge computing and their robustness to wireless system surroundings to achieve cost-efficacy and ease of set-up administration. This section defines the safety

Table 7.3 Summary of safety challenges and resolutions of three-tier structure of 5G networks.

3-tiers of 5G networks	Safety challenges	Safety resolutions
Heterogeneous Access Networks [82]	3GPP endorses using a variety of safety facilities like nodule certification, payment-encoding, and IP attack protection to offer more powerful heterogeneous access network.	Initiate reliable configurations performed on a specific domain, offer the appropriate level of security for a heterogeneous access network.
Backhaul Networks	Safety measures, Deep Packet Inspection. (DPI), IPsec transport, and firewall reconstruction strategies should be used on backhaul networks to protect against numerous harmful outbreaks.	Backhaul structures designed for IPsec VPN must be united with new safety systems to provide a fully secure backhaul network.
Core Network [82]	This set-up is visible to millions of campaigns on the net and additional unreliable IP networks standards. Therefore, it can easily be affected by net-centered fears containing malicious, DDoS attacks, botnets, etc.	Protect traffic on the S1 interface. For safety, gateway must be used at the end of the core network by looking at exterior IP networks.

Source: Based on Liyanage, M. et al. [82].

explanations that view a core network as a complete IP based on a logical core network. For a well-designed central network, the largest task can be overcrowding due to the large number of different devices. Conversely, another major change to 5G is that the main features of the network like FAU is characterized by system jobs, likewise the Admittance and Flexibility Administration Function (AFA). Therefore, to successfully manage the challenge of over-signing, the two methods are deliberated in 5GPPP. Firstly, it uses simple certification and key protocols for large IoT communications. Secondly, use of protocols allow devices to be integrated together with different types of AKA-based assembly protocols [81]. A summary of the safety challenges discussed above, and required possible resolution, is presented in Table 7.3 [82].

7.7 Conclusion and Future Research Directions

Wireless communication networks have grown from connecting basic 1G phone to connecting every phase of life to 5G. Through this time of progression, the security environment has evolved equally from the light touch of the phone to various security attacks on movable equipment, system tools, and facilities. By incorporating different features (IoT) and resources hooked on to the system, 5G systems will use innovative machineries like progressive cloud computing ideas such as SDN and NFV, etc. All these novel machineries have their individual safety tasks that can also compromise system safety. Thus, this chapter discusses the standard mobile networks with its evaluation from 1G to 5G and

Non-IP networks. Additionally, brief details of 5G networks along with its importance, functionality, applications, and why we found the need to provide security in it are explored. It is stated how 5G networks is co-relating with IoT by uncovering its background and requirements with its basic characteristics of IoT infrastructure (Heterogeneous device, Restricted resources, Automatic communication, large-scale network with large number of events, and no infrastructure and dynamic network) and characteristics of IoT applications (various application, real-time application, X-a-a-S, increase security attack surface, and privacy leaks). Furthermore, some additional benefits adopted by organizations, such as related to big data, openness of IoT, and link aspect of IoT, are covered. Some non-5G technologies are also elaborated upon, that are addressing IoT environments such as BLE IEEE 802.15.4, LoRa, Sigfox, and Wi-Fi haLow. Additionally, it is explained how we can secure this mobile network by describing 5G advanced security model. Finally, the chapter explains security of three tiers (Heterogeneous access networks, Backhaul network, and Core network) in 5G networks with its safety challenges and possible solutions. This very informative study leads to providing security in 5G network by co-relating it to IoT.

Extra research is vital into the impending and long-lasting significances of IoT implementation with 5G networks. Experts requisite to create the needed administrative fluctuations to achieve demonstratively earnings from IoT as well as the 5G network. In this regard, 5G safety can be the main evolutionary stream of research relating 5G with required IoT connecting areas. In addition, 5G network safety issues can be solved by handling the safe uniqueness, common certification of the devices, suitable documentation and requirements, and powerful effective processes and guidelines. In accumulation, specific key administrations can be executed at the edge line of the 5G grid system by adopting MEC aptitudes. It can decrease the safety allied communication in the cloud over 5G backhaul grid. This methodology will cement the approach for the interruption in IoT and 5G requests as well. As a future research direction, we need to study more in this domain and will provide enhanced versions of security with proper collaboration and adoption of new technologies along with ease.

References

1 Zhang, P., Yang, X., Chen, J. et al. (2019). A survey of testing for 5G: solutions, opportunities, and challenges, *China Communications* 16 (1): 69–85. doi: 10.12676/j.cc.2019.01.007

2 Liyanage, M., Ahmad, I., Abro, A.B. et al. (2018). *A Comprehensive Guide to 5G Security*. Wiley Publishing. doi:10.1002/9781119293071.

3 Maier, G. and Reisslein, M. (2019). Transport SDN at the dawn of the 5G ERA. *Optical Switching and Networking* 33: 34–40. doi:10.1016/j.osn.2019.02.001.

4 Bonfim, M.S., Dias, K L., and Fernandes, S.F. (2019). Integrated NFV/SDN architectures. *ACM Computing Surveys* 51 (6): 1–39. doi:10.1145/3172866

5 Kitanov, S., Popovski, B., and Janevski, T. (2021). Quality evaluation of cloud and fog computing services in 5G networks. *Research Anthology on Developing and Optimizing 5G Networks and the Impact on Society* 2: 240–275. doi:10.4018/978-1-7998-7708-0.ch012

6 Han, B., Wong, S., Mannweiler, C. et al. (2019). Context-awareness enhances 5G multi-access edge computing reliability. *IEEE Access* 7: 21290–21299. doi:10.1109/access.2019.2898316

7 Zhang, S. (2019). An overview of network slicing for 5G. *IEEE Wireless Communications* 26 (3): 111–117. doi:10.1109/mwc.2019.1800234

8 Sarraf, S. (2019). 5G emerging technology and affected industries: quick survey, *American Scientific Research Journal for Engineering, Technology, and Sciences (ASRJETS)* 55 (1): 75–82. https://www.researchgate.net/publication/334282546_5G_Emerging_Technology_and_Affected_Industries_Quick_Survey.

9 Han, B., Gopalakrishnan, V., Ji, L. et al. (2015). Network function virtualization: challenges and opportunities for innovations. *IEEE Communications Magazine* 53 (2): 90–97. doi:10.1109/mcom.2015.7045396

10 Pranto, T.H., Noman, A.A., Mahmud, A. et al. (2021). Blockchain and smart contract for IoT enabled smart agriculture. *Peer Journal of Computer Science* 7: doi:10.7717/peerj-cs.407

11 Malik, A., Gautam, S., Abidin, S. et al. (2019). Blockchain technology – Future of IoT: Including structure, limitations and various possible attacks, Proceedings of the 2nd International Conference on Intelligent Computing, Instrumentation and Control Technologies (ICICICT), Kannur, India 1100–1104. doi: 10.1109/ICICICT46008.2019.8993144.

12 Gautam, S., Malik, A., Singh, N. et al. (2019). Recent advances and countermeasures against various attacks in IoT environment. Proceedings of the 2nd International Conference on Signal Processing and Communication (ICSPC). https://doi.org/10.1109/icspc46172.2019.8976527

13 Netra, K., Manjunath, K.G., and Shankar, A. (2019). An effective vehicular Ad hoc network using cloud computing: a review. Proceedings of the 9th International Conference on Cloud Computing, Data Science & Engineering (Confluence), Noida, India. 69–74. doi: 10.1109/CONFLUENCE.2019.8776992.

14 Saxena, S., Bhushan, B., and Ahad, M. A. (2021). Blockchain based solutions to secure Iot: background, integration trends and a way forward. *Journal of Network and Computer Applications* 181: 103050. doi:10.1016/j.jnca.2021.103050

15 Bhushan, B., Sahoo, C., Sinha, P. et al. (2020). Unification of Blockchain and Internet of Things (BIoT): requirements, working model, challenges and future directions. *Wireless Networks* 27: 55–90. doi:10.1007/s11276-020-02445-6

16 Shahabuddin, S., Rahaman, S., Rehman, F. et al. (2018). Evolution of cellular systems. *A Comprehensive Guide to 5G Security* 32: 1–29. doi:10.1002/9781119293071.ch1

17 Conti, M., Dragoni, N., and Lesyk, V. (2016). A survey of man in the middle attacks. *IEEE Communication Surveys and Tutorials* 18: 2027–2051. https://doi.org/10.1109/COMST.2016.2548426

18 3GPP (2001–2003) Security Objectives and Principles, 3rd Generation Partnership Project (3GPP), TS 33.120. http://www.3gpp.org/ftp/Specs/html-info/33120.html.

19 Garg, V.K. and Rappaport, T.S. (2001). Wireless Network Evolution: 2G to 3G. Prentice Hall PTR. https://dl.acm.org/doi/book/10.5555/559259.

20 Yu, W. (2010). The network security issue of 3G mobile communication system research. International Conference on Machine Vision and Human-machine Interface, Kaifeng, China. 373–376. doi: 10.1109/MVHI.2010.47.

21 Pattekar, A., Dhanvijay, M., Jadhav, V. et al. (2017). Dual-band compact MIMO antenna for 3G, Wi-MAX and WLAN applications. Third International Conference on Sensing, Signal Processing and Security (ICSSS), Chennai, India. 357–360. doi: 10.1109/SSPS.2017.8071620.

22 Li, H., Guo, S., Zheng, K. et al. et al. (2009). Security analysis and defence strategy on access domain in 3G. In Proceedings of the 2009 First IEEE International Conference on Information Science and Engineering ICISE '09). IEEE Computer Society, USA, 1851–1854. doi: https://doi.org/10.1109/ICISE.2009.1050

23 3GPP (2014). Technical Specification Group Services and System Aspects; Study on security aspects of Machine-Type Communications (MTC) and other mobile data applications communications enhancements (Release 12), 3rd Generation Partnership Project (3GPP), Tech. Rep. 33.868, https://patents.google.com/patent/WO2016004102A3/un.

24 3GPP (2009). Technical Specification Group Services and System Aspects; Security of Home Node B (HNB) / Home evolved Node B (HeNB) (Release 11), 3rd Generation Partnership Project (3GPP), Tech. Rep.33.320https://http://portal.3gpp.org/desktopmodules/Specifications/SpecificationDetails.aspx?specificationId=910.

25 Yarali, A. (2020). *Issues and Challenges of 4G and 5G for PS, Public Safety Networks from LTE to 5G*. Wiley. 189–194. doi: 10.1002/9781119580157.ch11.

26 Bikos, A.N and Sklavos, N. (2018). Architecture Design of an Area Efficient High Speed Crypto Processor for 4G LTE. *IEEE Transactions on Dependable and Secure Computing* 15 (5): 729–741. doi: 10.1109/TDSC.2016.2620437.

27 Zeqiri, R., Idrizi, F. and Halimi, H. (2019). Comparison of Algorithms and Technologies 2G, 3G, 4G and 5G. Proceeings of the 3rd International Symposium on Multidisciplinary Studies and Innovative Technologies (ISMSIT), Ankara, Turkey 1– 4, doi: 10.1109/ISMSIT.2019.8932896.

28 Telecommunication Standardization Sector of ITU, Security architecture for systems providing end-to-end communications, International Telecommunication Union, Tech. Rep. X.805, 2003.

29 3rd Generation Partnership Project (3GPP), Technical Specification Group Services and System Aspects (SA3);TS 33.501: Security Archi-tecture and Procedures for 5G System, Release 15, 3rd Generation Partnership Project (3GPP), Tech. Rep. 33.501, 2018.

30 Hashim, F., Munasinghe, K.S., and Jamalipour, A. (2010). A biologically inspired framework for mitigating epidemic and pandemic attacks in the NGMN. Proceedings of the IEEE Wireless Communication and Networking Conference, Sydney, NSW, Australia 1–6. doi: 10.1109/WCNC.2010.5506116.

31 Le, T.K., Salim, U., and Kaltenberger, F. (2021). An overview of physical layer design for ultra-reliable low-latency communications in 3GPP releases 15, 16, and 17. *IEEE Access* 9: 433–444. doi: 10.1109/ACCESS.2020.3046773.

32 Alliance (2015). NGMN, NGMN 5G white paper, Next Generation Mobile Networks. White paper, https://www.ngmn.org/wp-content/uploads/NGMN_5G_White_Paper_V1_0.pdf.

33 Khan. R., Kumar. P., Nalin. D. et al. (2019). A survey on security and privacy of 5G technologies: potential solutions. *Recent Advancements and Future Directions IEEE Communications Surveys & Tutorials.* doi: 10.1109/COMST.2019.2933899

34 https://www.5gradar.com/features/5g-security-5g-networks-contain-security-flaws-from-day-one

35 Anamalamudi, S., Sangi, A.R., Alkatheiri, M. et al. (2018). 5G-Wlan security. *A Comprehensive Guide to 5G Security* 143–163. doi:10.1002/9781119293071.ch7

36 Atzori. L., Iera, A., and Morabito. G. (2010). The Internet of Things: a survey. *Computer Networks* 54 (15): 2787–2805, https://www.cs.mun.ca/courses/cs6910/IoT-Survey-Atzori-2010.pdf.

37 Dutta, A. and Hammad, E. (2020). 5G Security Challenges and Opportunities: A System Approach. Proceedings of the 2020 IEEE 3rd 5G World Forum (5GWF), Bangalore, India. 109–114. doi: 10.1109/5GWF49715.2020.9221122.

38 Scott, M. and Whitney, R. (2014). The Industrial Internet of Things. http://www.mcrockcapital.com/uploads/1/0/9/6/10961847/mcrock_industrial_internet_of_things_report_214.pdf.

39 Ahmad, I., Kumar, T., Liyanage, M. et al. (2017). 5G security: analysis of threats and solutions. Proceedings of the 2017 IEEE Conference on Standards for Communications and Networking (CSCN). doi:10.1109/cscn.2017.8088621

40 TE Commission (2008). Internet of Things in 2020. http://www.caba.org/resources/Documents/IS-2008-93.pdf

41 Bellavista, P., Cardone, G., Corradi, A. et al. (2013). Convergence of MANET and WSN in IoT urban scenarios. *IEEE Sensors Journal* 13 (10): 3558–3567. doi:10.1109/jsen.2013.2272099

42 IT, ITU Internet report Union (2005). The Internet of Things. https://dl.acm.org/doi/10.1155/2017/9324035.

43 Greg, C.P. and Chatha, A. (2014). Planning for the Industrial Internet of Things. http://www.arcweb.com brochures/planning-for-the-industrial-internet-of-things.pdf

44 IIS Group (2014). The Internet of Things starts with intelligence inside. http://www.intel.com/newsroom/kits internetofthings/pdfs/IoT_Day_presentation.pdf

45 Gartner (2013) Gartner says the Internet of Things installed base will grow to 26 billion units by 2020. https://www.ageinplacetech.com/pressrelease/gartner-says-internet-things-installed-base-will-grow-26-billion-units-2020.

46 Bay. O. (2013). More than 30 billion devices will wirelessly connect to the Internet of Everything in 2020. https://www.abiresearch.com press/more-than-30-billion-devices-will-wirelessly-conne.

47 Banerjee, P., Friedrich, R., Bash, C. et al. (2011). Everything as a service: powering the new information economy. *Computer* 44 (3): 36–43. doi:10.1109/mc.2011.67

48 Zaslavsky, A., Perera, C., and Georgakopoulos, D. (2013). Sensing as a service and big data. doi.arXiv:1301.0159

49 Tracey, D. and Sreenan, C. (2013). A holistic architecture for the Internet of things: Sensing services and big data. Proceedings of the 2013 13th IEEE/ACM International Symposium on Cluster, Cloud, and Grid Computing. doi: 10.1109/ccgrid.2013.100

50 TE Commission (2012). Protection of personal data. http://ec.europa.eu/justice/data-protection

51 Palattella, M.R., Dohler, M., Grieco, A. et al. (2016). Internet of things in the 5G ERA: enablers, architecture, and business models. *IEEE Journal on Selected Areas in Communications* 34 (3): 510–527. doi:10.1109/jsac.2016.2525418

52 Bluetooth SIG (2016). Bluetooth Core Specification Version 5.0. https://www.mouser.it/pdfdocs/bluetooth-Core-v50.pdf.

53 Chang, KH. (2014). Bluetooth: a viable solution for Iot? [industry perspectives]. *IEEE Wireless Communications* 21 (6): 6–7. doi:10.1109/mwc.2014.7000963

54 Castro, P., Afonso, J.L., and Afonso, J.A. (2016). A low-cost zigbee-based wireless industrial automation system. *Lecture Notes in Electrical Engineering* 402: 739–749. doi:10.1007/978-3-319-43671-5_62

55 IEEE Computer Society LAN/MAN Standards Committee (2011). IEEE Stan-dard for Local and Metropolitan Area Networks – Part 15.4: Low-Rate Wireless Personal Area Networks (LR-WPANs). http://ecee.colorado.edu/~liue/teaching/comm_standards/2015S_zigbee/802.15.4-2011.pdf.

56 LoRa Alliance Technical Committee (2017). LoRaWAN 1.1 Specification. https://lora-alliance.org/resource_hub/lorawan-specification-v1-1.

57 Petäjäjärvi, J., Mikhaylov, K., Pettissalo, M. et al. (2017). Performance of a low-power wide-area network based on lora technology: Doppler robustness, scalability, and coverage. *International Journal of Distributed Sensor Networks* 13 (3): 155014771769941. doi:10.1177/1550147717699412

58 Centenaro, M., Vangelista, L., Zanella, A. et al. (2016). Long-range communications in UNLICENSED bands: the rising stars in the IoT and smart city scenarios. *IEEE Wireless Communications* 23 (5): 60–67. doi:10.1109/mwc.2016.7721743

59 IEEE Computer Society LAN/MAN Standards Committee (2016). Part 11: Wireless LAN Medium Access Control (MAC) and Physical Layer (PHY) Specifications Amendment 2: Sub 1 GHz License Exempt Operation. https://ieeexplore.ieee.org/document/8624660.

60 Liyanage, M., Braeken, A., Jurcut, A.D. et al. (2017). Secure communication channel architecture for software defined mobile networks. *Computer Networks* 114: 32–50. doi:10.1016/j.comnet.2017.01.007

61 Franco de Almeida, I.B. and Leonel Mendes, L. (2018). Linear GFDM: a low out-of-band emission configuration for 5G air interface. Proceedings of the 2018 IEEE 5G World Forum (5GWF). doi:10.1109/5gwf.2018.8516993

62 Gerzaguet, R., Bartzoudis, N., Baltar, L.G. et al. (2017). The 5G candidate waveform race: a comparison of complexity and performance. *EURASIP Journal on Wireless Communications and Networking* 2017 (1): 53–78. https://doi.org/10.1186/s13638-016-0792-0

63 Granjal, J., Monteiro, E., and Sa Silva, J. (2015). Security for the internet of things: a survey of existing protocols and open research issues. *IEEE Communication Surveys and Tutorials* 17 (3): 1294–1312. https://doi.org/10.1109/comst.2015.2388550

64 Thanh, T Q., Covaci, S., and Magedanz, T. (2019). VISECO: An Annotated Security Management Framework for 5G. Mobile, Secure, and Programmable Networking 6: 251–269. https://doi.org/10.1007/978-3-030-03101-5_21.

65 Zhang, X., Kunz, A., and Schroder, S. (2017). Overview of 5G security in 3GPP. Proceedings of the 2017 IEEE Conference on Standards for Communications and Networking (CSCN). https://doi.org/10.1109/cscn.2017.8088619

66 Goyal, S., Sharma, N., Kaushik, I. et al. (2021). A green 6g network era: architecture and propitious technologies. *Data Analytics and Management* 54: 59–75. doi:10.1007/978-981-15-8335-3_7

67 Bhushan, B. and Sahoo, G. (2017). Recent advances in attacks, technical challenges, vulnerabilities and their countermeasures in wireless sensor networks. *Wireless Personal Communications* 98 (2): 2037–2077. doi:10.1007/s11277-017-4962-0

68 Bhushan, B. and Sahoo, G. (2020). Requirements, protocols, and security challenges in wireless sensor networks: an industrial perspective. *Handbook of Computer Networks and Cyber Security*. 3: 683–713. doi:10.1007/978-3-030-22277-2_27

69 Bhushan, B. and Sahoo, G. (2018). Routing protocols in wireless sensor networks. Computational intelligence in sensor networks studies in *Computational Intelligence* 6: 215–248. doi:10.1007/978-3-662-57277-1_10

70 Morin, E., Maman, M., Guizzetti, R. et al. (2017). Comparison of the device lifetime in wireless networks for the internet of things. *IEEE Access* 5: 7097–7114. https://doi.org/10.1109/access.2017.2688279

71 Ahmad, A., Bhushan, B., Sharma, N. et al. (2020). Importunity & Evolution of IoT for 5G. Proceedings if the 2020 IEEE 9th International Conference on Communication Systems and Network Technologies (CSNT). doi: 10.1109/csnt48778.2020.9115768

72 Arora, S., Sharma, N., Bhushan, B. et al. (2020). Evolution of 5G Wireless Network in IoT. Proceedings of the 2020 IEEE 9th International Conference on Communication Systems and Network Technologies (CSNT). doi: 10.1109/csnt48778.2020.9115773

73 Malik, H., Pervaiz, H., Mahtab Alam, M. et al. (2018). Radio resource management scheme in NB-IoT systems. *IEEE Access* 6: 15051–15064. https://doi.org/10.1109/access.2018.2812299

74 Atakora, M. and Chenji, H. (2018). A multicast technique for fixed and mobile optical wireless backhaul in 5G networks. *IEEE Access* 6: 27491–27506. https://doi.org/10.1109/access.2018.2832980

75 3rd Generation Partnership Project (3GPP) (2018). Technical Specification Group Services and System Aspects; Network architecture, Release 8, 3rd Generation Partnership Project (3GPP), Tech. Rep. 3GPP TS 23.002. https://dl.acm.org/doi/abs/10.1109/COMST.2019.2916180.

76 Ijaz, M., Alfian, G., Syafrudin, M. et al. (2018). Hybrid prediction model for type 2 diabetes and hypertension USING DBSCAN-BASED outlier DETECTION, SYNTHETIC minority over sampling technique (SMOTE), and random forest. *Applied Sciences* 8 (8): 1325. doi:10.3390/app8081325

77 Sharma, P., Shankar, A., and Cheng, X. (2021). Reduced paper model predictive control based Fbmc/oqam signal for NB-IoT paradigm. *International Journal of Machine Learning and Cybernetics* 5: doi:10.1007/s13042-020-01263-8

78 Khapre, S.P., Chopra, S., Khan, S. et al. (2020). Optimized routing method for wireless sensor networks based on improved ant colony algorithm. Proceedings of the 2020 10th International Conference on Cloud Computing, Data Science & Engineering (Confluence), Noida, India. 455–458. doi: 10.1109/Confluence47617.2020.9058312.

79 Alfian, G., Syafrudin, M., Ijaz, M. et al. (2018). A personalized HEALTHCARE monitoring system for diabetic patients by UTILIZING Ble-based sensors and real-time data processing. *Sensors* 18 (7): 2183. doi:10.3390/s18072183

80 Nie, X., Fan, T., Wang, B. et al. (2020). Big data analytics and IoT in OPERATION safety management in under water management. *Computer Communications* 154: 188–196. doi:10.1016/j.comcom.2020.02.052

81 Latif, G., Shankar, A., Alghazo, J.M. et al. (2019). I-cares: advancing health diagnosis and medication through IoT. *Wireless Networks* 26 (4): 2375–2389. doi:10.1007/s11276-019-02165-6

82 Liyanage, M., Ylianttila, M., and Gurtov, A. (2014). A case study on security issues in LTE backhaul and core networks. *Case Studies in Secure Computing* 3: 190–211. https://doi.org/10.1201/b17352-3

8

Blockchain Assisted Secure Data Sharing in Intelligent Transportation Systems

Gunjan Madaan[1], Avinash Kumar[2], and Bharat Bhushan[2]

[1]*Deloitte Consulting India Private Limited, Bengaluru, India*
[2]*School of Engineering and Technology (SET), Sharda University, Greater Noida, Uttar Pradesh, India*

8.1 Introduction

Blockchain initially came into view with the concept of bitcoin (also known as Cryptocurrency) [1–3]. People nowadays consider blockchain as the "Topic of Interest," as it is immutable and transparent between stakeholders and trusted parties. The introduction of blockchain has inspired a transfer from centralized to decentralized dynamic architecture. Blockchain's decentralized architecture is used by several distributed nodes and each of these nodes contain some cryptographically-chained transaction records. These transactions are set up in blocks and agree upon some of the common consensus protocols. Because of the chaining of blocks with the distributed protocol, the immutability of the blockchain is ensured. The transparency and immutability make blockchain desirable. Any user can verify the history of the bitcoin transaction and any tampering with it. The work finally states that once the data is added to the chain it cannot be modified. The three main characteristics (transparency, decentralization, and genuine ability) make it a publicly verifiable architecture. Many applications have been planned for blockchain, and many financial experts and scientists are expecting great innovations to center on blockchain. "Blocks" in blockchain is digital information which has three parts. Firstly, blocks store information regarding transactions like the recent purchase information, the date, and time of the purchase. Secondly, blocks store information related to all the participants of transactions as it stores the user's details as well. For example, if a person shops from Amazon, it stores the name and other details of the person and links it with Amazon.inc but instead of storing the name, user's details regarding the purchase are recorded in the form of digital signatures. Thirdly, blocks also store information that distinguishes it from other blocks. Humans have many features that distinguish them from each other; a similar concept is used within the blocks. Each block contains a unique code called "Hash," which helps to distinguish one block from another. Hash is derived from a cryptographic code that is derived from a special algorithm which contains the values that will store all the changes, which helps us to depict whether the data is tampered with or not.

Smart and Sustainable Approaches for Optimizing Performance of Wireless Networks: Real-time Applications, First Edition.
Edited by Sherin Zafar, Mohd Abdul Ahad, Syed Imran Ali, Deepa Mehta, and M. Afshar Alam.
© 2022 John Wiley & Sons Ltd. Published 2022 by John Wiley & Sons Ltd.

Blockchain has grown itself into a complex system. Blockchain is used in many applications such as vehicular technology, satellite application, and cyber-physical systems [4]. It is also used in smart solar power plants [5]. It has given rise to many concepts such as the Internet of Vehicle which has further supported the Intelligent Transportation System (ITS). ITS targets for providing urban mobility, which is ecological as well as economical and is done by connecting the vehicle with the network interface. There are five kinds of services which are delivered by an ITS. The first one is the intelligent routing, which provides planning and navigation of an optimal path taking into consideration time and cost. Moreover, the route should also encounter minimum traffic congestion. Second is the safe driving that should be have multiple alarms relating to obstacles. It coordinates with these sensors and devices can make the system automated. The third is that vehicle or driver related services may include fines or automatic maintenance of the vehicle. Fourth, it may also include information on the people on board the vehicle. Finally, the fifth is the ITS that also provides environmental data, including traffic data, etc.

From the above discussion, it is clear that ITS provides us with all the above five mentioned benefits that is needed data from the vehicle, but a problem that arises is the privacy of the sensor on the vehicle that gives us information such as location activity, model of the car, and any traffic rule violation, which may give rise to security risks too. There are so many services provided by the ITS that may be providing convenience to the user, which in turn may come at risk of privacy exposure. Moreover, the automated vehicle's decisions depend on data that is obtained by the network in which security problems are present. This happens because the ITS is based on existing infrastructure and does not consider privacy and security at the design stage. There are various researches and studies done on ITS, but this work is distinct as it tries to insight all of the prominent features of blockchain that could make the ITS a more efficient system to be executed in a real life scenario. To summarize briefly, the major contributions of this work is enumerated as follows:

- This work covers the drawback of the traditional transport system by considering human loss in the form of accidents due to communication mismatch or hackers' involvement.
- This work emphasizes the two vital entities, that is, privacy and security of the driver for driving the vehicles in ITS.
- The work highlights the vitality of blockchain for securing the identify theft of drivers and vehicle done by hacker.
- The work compares the various constraints that make the existing system undesirable for the smart cities.
- The work explains the criticality of using blockchain for Vehicle of Things (VoT) to avoid compromise of the vehicle by the adversary.

The remainder of this chapter is organized as follows: Section 8.2 presents the background of ITS. Section 8.3 explores the basics of blockchain technology and discusses the features and major role of blockchain in securing data and information. Section 8.4 presents blockchain-assisted ITS for enhanced security and privacy. Section 8.5 presents future research prospective for the use of blockchain technology, followed by the conclusion in Section 8.6.

8.2   Intelligent Transport System

ITS plays a vital role in the management of traffic as well vehicles in a transportation system. The most important features of the ITS is to exchange the information between the vehicles gathering information regarding road congestion and accidents on another roadway, entities that could help to generate more traffic awareness. The ITS communication could range from smaller distances of a few meters to longer distances of a few kilometers. This shorter distanced communication would help in automation decision-making by VoT, while the longer communication would help regarding knowledge of traffic congestion [6]. The previous one will help in vehicles navigation. while the latter will help in making decisions regarding choosing the most suitable route to reach the desired destination. ITS is also very useful in the smart city where smart healthcare could be one of the components and also where public data is used, such as where Europe GDPR (General Data Protection Regulation) is considered [7–11]. ITS will drive human life for mobility in a more convenient and safer manner. The further working and features inspection of ITS are traversed below.

8.2.1   ITS Overview

The ITS has evolved from an archaic system to an intelligent one by overcoming various real-life problems that occur while driving vehicles. ITS is based on the communication between vehicles, traffic signals, Closed-Circuit Television (CCTV) cameras, sensors, and other traffic-related devices working through interconnected communication, either through wired or wireless media. The traditional system does not have an efficient mechanism to manage the huge amounts of traffic during peak hours. Office hours are usually flooded with a huge number of vehicles, with everyone wanting to reach their destinations in the most optimal time. Thus, this situation results in more accidents and traffic jams. ITS could tackle this situation in a more efficient manner by suggesting the vehicles, the most relevant route, and the duration which the driver could incur by traveling on the suggested route. Also, if there is any accident on a certain route, the preceding vehicle on the same route cannot be informed in advance. So, the driver has to manually decide on a new route. But in the case of ITS, this information is automatically directed to the driver on the vehicle's screen. Apart from these concerns, the most vital issue is keeping a route open for the emergency vehicles. Emergency vehicles such as ambulances, fire fighting vehicles, and police cars need special access to deliver emergency services. The modern traffic system or ITS considers these situations while tackling the effect that moving vehicles have on the traffic.

Another important role that ITS could perform is to save human lives by providing navigational help during foggy winter. Lack of visibility on the road occurs bad weather and other vehicles become almost invisible to other drivers. Here the ITS, with the help of sensors on the road, as well as sensors on VoT, navigates with each other in real time to guide the drivers, thus helping to avoid any collision during times of poor visibility on the road. This fog management of ITS could also be implemented for managing other public transportation systems such as trains, buses, metros, cargos, etc. City buses are one of the most used public transportation systems. The city bus transportation system consists of regular routine journeys from source to destination. ITS could easily map the route and suggest the average speed that the driver needs to keep to in order to adapt to the most optimal working that includes bus stops, passenger boarding and deboarding, ticket collection, and other related activities. Similarly, ITS could help train systems, including metro systems, in track navigation and also alert them to any situations ahead on the track that could lead to disaster. The real-time monitoring feature of ITS is also very useful for avoiding accidents that arise due to human error.

Safety is the highest priority for humans while on the road. Therefore, the ITS also uses the road symbols for alerting moving vehicles. Roads can be damaged at any time because of weather or other natural disasters. The ITS could alert the vehicles or drivers about these incidences by sending them information before they reach the problem area. Often, road workers have to work during the daytime when traffic is still on the move, so the ITS could safeguard these workers as well. ITS could suggest that drivers slow their speed to avoid any accidents due to roadworks. Moreover, the ITS could also help in cargo navigation while entering and leaving a sea port. Sea ports are not very easy to manage because here the medium is water rather than road. Therefore, the use of ITS could give an edge to managing huge cargos and ships. Finally, the ITS could also guide efficient take-off and landing of airplanes. Thus, ITS is very useful for every type of transportation system, whether it is private, public, small, or larger. Though ITS seems useful, it has some issues that are discussed in next section. Figure 8.1 represents various ways by which decentralized blockchain-based technology can be used for ITS.

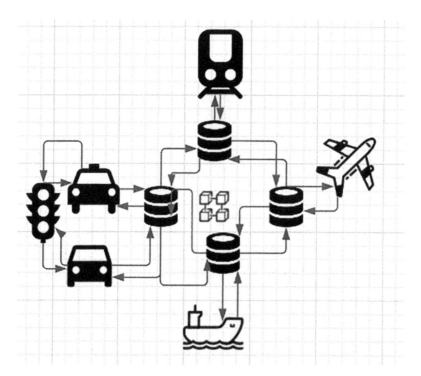

Figure 8.1 Area of implementation of blockchain based ITS.

8.2.2   Issues in ITS

Many problems are arising due to the ITS architecture related to privacy and security. Firstly, the server is centralized. ITS services are developed on centralized clouds, and due to this centralization the user's identification, authorization, and authentication services are provided centrally, or we can say from the same slot or at the same point, which means that ITS become vulnerable to security attacks and privacy breaches. Secondly, as we have seen, the current ITS faces many privacy and security concerns, as the data being transferred to the ITS for its working contains private information. The environmental monitoring of data as private data is very difficult to discriminate which may lead to privacy issue. Thirdly, the ITS incurs higher extra costs for provisioning security and privacy to protect services. Some services, such as network infrastructure, complex management, authorization, authentication, and the high-cost, restrain people from availing these services. Finally, is the difficulty of tracing malicious devices. Nowadays, vehicles rely more on software and control functions. Any malicious attack on the software may result in a threat to the safety of the driver and passenger. However, the problem becomes worst when this malicious behavior cannot be traced in time.

8.2.3   ITS Role in IoT

With the advancement in technology, the ITS has become one of the indispensable components in transforming smart cities into digital societies, to make mobility more ecological and economical. The key concern of any city is our mobility as we are travel to schools, colleges, offices, etc. The ITS is achieving this by minimizing traffic problems and gaining traffic efficiency. It provides information to the user regarding the running time with the real conveyance, traffic, allocation of seats, etc., which ensures the privacy and safety of commuters and also makes traveling hassle-free.

The detecting and controlling of the system are done in parallel with a framework termed as the Cyber Physical System (CPS). The CPS is typically associated with the Internet and is advanced to make an extension between the system and physical world. Intelligent vehicle systems are based on blockchain technology. The services could be tested for vulnerability using penetration testing to make the system more resilient [12].

ITS provides five services: Intelligent Transportation (IT), Vehicle-related or Driver services, Safe driving, Board vehicles infotainment for people, and environmental data provider. The ITS provides the different innovative services such as management of traffic, planning, and navigation through intelligent routing and also optimizes the route in terms of time and economy. The vehicle-related and driver-related services give an edge to users or passengers while booking and riding on a commercial vehicle, as it provides details of driver and vehicle before they enter the vehicle and also provides live route, distance, or times of vehicle from pick-up point to destination. Safe driving is the most important factor for successful driving, regarding the safety of the driver and the passengers. So, ITS vehicles have inbuilt sensors that can detect surrounding vehicles and the distances between them. This information helps the system to avoid collisions by slowing down vehicles when necessary and by activating an alarm which will start when a vehicles come too close. Therefore, ITS offers a safe driving mechanism. Onboard vehicle infotainment for people

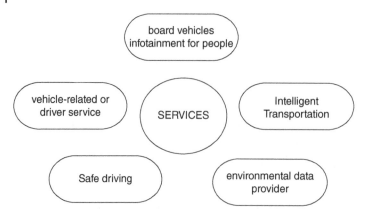

Figure 8.2 Services provided by ITS.

is used to provide audio and video entertainments in the vehicles. The Environmental Data Provider (EDP) is one of the important aspects of ITS, as it helps the ITS structure to gather information from its surroundings such as location, number of vehicles on the road, etc., which help the ITS system to provide accurate details about traffic and the route which also helps the drivers by suggesting alternative routes where traffic is lower. This all helps drivers to reach their destination in less time.

The development of technology in computation, communication, sensing, and analysis has increased ITS growth, which enables the facilities of transportation system to become more convenient. ITS tendency is a critical security risk toward centralization. The ITS came with the idea of blockchain-based ITS, also termed B2-ITS, to tackle the above-discussed challenges [4]. It is known as a trusted, secured, and decentralized system based on blockchain for an ITS autonomous ecosystem. Figure 8.2 represents the services provided by ITS.

8.3 Blockchain Technology

Blockchain is very much needed in the modern era of technology. In blockchain, the database is immutable and is distributed peer-to-peer. The data in blockchain is secured cryptographically [13]. Blockchain technology is made up of blocks which consist of different data stored within it and these blocks are linked with each other using certain cryptographic techniques to build a chain. This chain is added with the new blocks in a digital ledger, hence the blockchain is also known as Distributed Ledger Technology (DLT). The chain formed is immutable in nature and known as an immutable ledger. It permits executing and gathering information between the partner to set up trust among untrusted circumstances in a decentralized manner. Because of the enormous achievement of Bitcoin, blockchain has taken in account extraordinary consideration of the research network as a developing innovation. The four techniques on which blockchain rely are Smart contracts, Cryptography, Structure, and Consensus.

8.3.1 Overview

Blockchain is the system that does not permit the transaction to be checked by an unreliable actor. It gives a permanent, dispersed, auditable, secure, and transparent ledger. It is a conventional structure of data in blocks which are connected in the form of a chain, where the connection of blocks with each other is made with reference to the previous block. Network peers provide the following functionality to operate and support the blockchain. Firstly, routing is by the propagation of blocks, and transactions are necessary for P2P networks to participate. Secondly, in storage, the node keeps a copy of the chain. Thirdly, in the "Wallet," the security key is provided to the user which allows the order to complete the transaction. Fourthly, by solving a proof of work, new blocks are generated, which is known as Mining.

The newly-generated bitcoins are received by miners as a reward for mining. The new blocks are published by the miner once the Proof of Work (PoW) is completed and the validity is verified by the network before the addition of the block to the chain. In the network, the blocks are generated concurrently and may fork the blocks of the blockchain into many different branches. This issue is solved by considering the longest chain of blocks as the valid one. The process is too expensive for the attacker to corrupt and modify the blocks inside the blockchain. Altogether, for a controlled square to be effectively added to the chain, it is important to unravel the PoW more quickly than the remainder of the network, which has excessive computational loss. Because of the enormous computational limit needed to alter the blockchain, the debasement of its blocks is outlandish. This implies, regardless of whether the members are genuine or not about the utilization of Bitcoin, an accord is constantly reached in the arrangement as long as a large portion of the system is framed by legitimate members. Insights concerning the blockchain design can be found in [14, 15].

8.3.2 Types of Blockchain

Blockchains are of three types based on the mechanism by which they control the node and its validation. These are called public, private, and federated. The blockchain here differs in the sense of ownership for node acquisition.

8.3.2.1 Public Blockchain

A public blockchain is available for the entire user, as drivers on the concept of open source. The participating node can perform desirable operations that could be creating of node, validating of nodes, and other blockchain-related activities. The time required for the PoW in the case of public blockchain is very high because there is no dedicated node that is responsible for validating the node. Public blockchain could only be overtaken by a hacker when it has 51% of total available mining power [16]. Moreover, public blockchain consists of unknown nodes before the start of mining and this makes the public blockchain more prone to Sybil attack.

8.3.2.2 Private Blockchain

Private blockchain is different from the public blockchain in the sense of ownership. Private blockchain does all the inspection in blockchain with their own nodes and does not allow an unknown node to validate and share of data. The organization having private blockchain shares data and information among themselves and it not open to the public. The new user could only become part of the private blockchain if an authorized node sent a request to a new user to participate in it. Moreover, here only the controlling node could perform the transaction within the group. The private node is more efficient because here the PoW is carried out in less time in comparison to public blockchain.

8.3.2.3 Federated Blockchain

Federated blockchain is somewhat similar to private blockchain, but the only difference is that the federated blockchain has no complete influence over the organization. Federated blockchain is also known as consortium blockchain. This blockchain also works on the ledger system, where permission must be gained before the actions are being taken. Here, multiple nodes make use of the network to form a decentralized system. The federated blockchain is maintained by the community and hence multiple organizations are involved. A practical implementation of federated blockchain is Enterprise Ethereum Alliance (EEA), which includes Microsoft, J.P. Morgan Chase Bank, Accenture, etc.

8.3.3 Consensus Mechanism

These systems aim to empower substances to concur over one single variant of a substantial block to guarantee a transparent and reliable viewpoint, which finally tackles forks and clashes inside the network. Every component has one-of-a-kind standards and calculations which structure the necessity to be trailed by the nodes in order to incorporate new squares to the chain. A portion of the remarkable agreement components is Proof of Stake (PoS) [17], PoW, [18] and Practical Byzantine Fault Tolerance (PBFT) [19]. The different consensus protocols are discussed below.

8.3.3.1 Proof of Work

In the blockchain system, PoW is the most the widely-adopted and the first consensus protocol. The work done by you is confirmed by the PoW. In this concept, PoW is sent prior to the main mail message. The PoW aims to solve problem mathematically by meeting the following conditions:

- It should be correlated to the message in order to prevent replay attack.
- The message should be difficult to be revealed by the adversary.
- The verification of the recipient should be implemented with easier computation.

The biggest advantage of PoW is it has the same possibility of corresponding blocks to be generated and gets paid as if the p% is the computing power of the miner to that of the total computing power of the network. Hence the protocol fairness is reflected by the PoW [20]. So, it creates the difficulty of an attack as the computing power of the attacker must be equal to the power of other nodes so that the blocks are generated to benefit them. So, the algorithm of PoW in the bitcoin network guarantees safety.

8.3.3.2 Proof of Stake

PoS reduces the problem of waste of computer power in mining PoW. "Coin Age" is the name of a PoS early version, i.e., period from which an amount of currency has been held. In contrast to this, where in PoW the data is recorded as nodes ability, in PoS the data recording ability is more because nodes have more stakes. With this ability, there is a relatively low difficulty for longer coin age nodes. PoS is not perfect due to the following problems:

- Initial currency distribution
- Encourages boarding behavior
- Accumulate the age of the coin when the node is offline
- Costless simulation

8.3.3.3 Delegated Proof of Stake

For solving the problem of security and speeding up the transaction process that is created offline of nodes in PoS, which accumulate the age of the coin, a new consensus was introduced. In April 2014, [19] Delegated Proof of Stake (DPoS) was proposed by Danny Larimer, which is used in Bitcoin as a consensus mechanism [21], and the Decentralized Platform (DP) [22]. Two roles are introduced by DPoS in system delegates and witnesses. In DPoS, to become a delegate, the node has to pay a deposit in the form of security money.

8.3.3.4 Practical Byzantine Fault Tolerance

Practical Byzantine Fault Tolerance (PBFT) is a consensus protocol that can determine the number of participants. It is permission less protocol. PBFT nodes are divided into two types. The first is that the primary node are those nodes where the client's requests are sorted. The second is known as the secondary node, which executes the request as the primary node provides. There are three basic protocols provided by this algorithm. First is the agreement in which the client requests are executed in each server in a fixed order. This order consists of five stages which are Request, Pre-Prepare, Prepare, Commit, and Reply. Second is the checkpoint which is used to handle the status of the correct node where log of this protocol has been executed. Finally, the change of viewpoint is used when failure of the primary node occurs, where it replaces the node with the node having backup and also ensures not to tamper with the node which the normal node executes. For use in consortium and private blockchain, PBFT is suitable where the source of the node is relatively reliable.

8.3.3.5 Casper

PoS protocol which is used for an operating system and distributed computing platform in Blockchain and Ethereum V2.0, [23] which is a security deposit protocol, is known as Casper. The process of Casper is implemented in such a way that all malicious elements can be removed. Casper works under PoS as a stake of Ether, where some parts are taken by validators, and blocks are validated. Validators get the reward if the block is appended, which is proportional to a stake. If nothing is performing at stake, then validator performance is malicious. Casper is more Byzantine Fault Tolerance (BFT), as it works with a Trust-less system. Network security is more critical in Casper.

8.3.3.6 Ripple

Ripple is an open-source Internet-based payment protocol which allows decentralization for the exchange of currency, clearing functions, and payments. In a network created by Ripple, the client initiates the transactions and broadcasts them to the entire network through validating nodes or tracking nodes. The tracking node's primary duty is to disseminate transaction data and respond to ledger requests from clients. The consensus protocol allows the validating node to add new data to the ledger in between the validating nodes and finally Ripple's consensus is achieved. Each validating node is pre-configured with a list of trusted nodes called the Unique Node List (UNL).

8.3.3.7 Proof of Activity

The Proof of Activity (PoA) incorporates the features of PoW and PoS The power of computing is centralized which contributes to PoW. Due to the effect of stakes, an oligarchy of stakes is formed by PoS/DPoS. Since the PoA consensus algorithm is based on the importance of identities, block validators stake their own credibility rather than coins. As a result, validating nodes, which are chosen at random as trustworthy individuals, make PoA blockchain a stable entity. Table 8.1 summarize the features of various consensus protocols.

8.3.4 Cryptography

Communication in plain form could lead to loss of data and breach of privacy. The hackers could easily manipulate and use the plain form of data for illegal purpose. Cryptography helps in protecting private and sensitive information from being captured by adversary or hackers. This is a process whereby the ordinary plain text gets converted into an unintelligible readable format, and vice-versa. Cryptographic methods are utilized to guarantee the integrity and security of information within the advisory. The cryptography does not only guarantee the protection of data but also helps authenticate the legitimate user who is intended to either send or receive the data. Blockchain sustains the privacy and security of data, either by using asymmetric key cryptography or via hash function.

A blockchain asymmetric key is used for securing exchange of data and information between the parties involved in the chain. The authentication is performed by digital signature. The initiating user signs using the private key and the verification of the signature is done using the public key, which is available publicly. Hence, the algorithm used to perform the digital signature process gets authenticated because only the authorized user will have generated or created the digital signature. Moreover, the public key cryptography is vital for

Table 8.1 Comparison of consensus protocols.

Consensus Nodes		Permission	Scalability	Consumption of energy	Throughput
PoW	Highly expandable	Permissionless	Highly scalable	High usage of energy	Low
PoS	Highly expandable	Permissionless	Highly scalable	Low usage of energy	Low
DPoS	Highly expandable	Permissionless	Highly scalable	Low usage of energy	High
PBFT	Restricted expansion	Permissioned	Low scalability	Low usage of energy	High
Casper	Highly Expandable	Permissionless	Highly scalable	Low usage of energy	High
Ripple	Restricted expansion	Permissioned	Highly scalable	Low usage of energy	High
PoA	Highly expandable	Permission less	Highly scalable	Low usage of energy	Low

the wallet that contains various stored files as well as normal data. Therefore, each user in the blockchain consists of a wallet which is related to a minimum of one public address and one private key that is needed for signing the transactions. The use of public key cryptography does require a bit of more computation but it is worth it as it give more surety about data integrity and authenticity.

Another concept used for sustaining privacy and security in blockchain is done by using hash function. The hashes, such as SHA-0, SHA-224, SHA-256, and SHA-384, are examples of hash function. These are used because of their less complexity in implementation and processing. The hash function is very useful in blockchain for making links between the various block that are designated for transactions. The blocks become linked in a sequential manner, where the successive block contains the hash of the preceding block. Moreover, the hash reduces the size for using a public address, which is used in blockchain for authentication of both user and data. The hash value makes the blockchain tamper-proof. In order to improve the privacy of the information in the block, zero knowledge of proof is utilized in the blockchain.

8.3.5 Data Management and Its Structure

Blockchain became popular in cryptocurrency and is well-known among different applications due to its one-of-a-kind data structure. The linear structure is very normal also, and another sort of structure is accessible in a likewise manner, which is known as Directed Acyclic Graph (DAG). In contrast to the linear structure, blocks are associated in a DAG with their past various blocks. Aside from data structure, there are three sorts of information for the board methods in blockchain, which are an on-chain method, off-chain method, and side-chain method. In on-chain, the content inside the blocks is recorded in the blockchain, which is obvious to all elements of the organization, though, in off-chain, a few instances are recorded also, handled outside of the system for block execution typically confided by the element. Then again, side-chain [24] is an autonomous blockchain that runs concurrently to the primary chain and keeps up a connection with the principle chain. It permits clients to move and utilize their digital currencies or resources to the supplementary chain and finally return to the primary chain.

The database management has been recently adopted by many technologies. The blockchain could be easily adopted for distributed management [25]. Febriyanto et al. [26] implemented blockchain to maintain the database in an e-voting system. Blockchain-based data management could also be used for managing databases for power systems [27]. Kuperberg et al. [28] used the blockchain-based data management concept for growth management in Europe. Paik et al. [29] utilized blockchain-based data management for governance purposes. This mechanism database management using blockchain could also sustain framework management [30]. The proxy management to guard the extra layer of security is also managed by blockchain data management using encryption mechanism [31]. The management of data is crucial in agriculture and this could be achieved using blockchain [32, 33]. As the data are increasing rapidly, big data has become more prevalent in any technology. The data management in IoT, where big data is involved, could be managed using blockchain technology [34].

8.4 Blockchain Assisted Intelligent Transportation System

Blockchain has many useful features for security, privacy, and authentication of data. The features and mechanism discussed above confirm that blockchain is very efficient for ITS.

8.4.1 Security and Privacy

Security as well as protection have been recognized as one of the most relevant necessities in vehicular systems and it is considered as the essential foundation of ITS. Different conventions had been proposed to keep vehicles and clients from being perceived from the traded message, even though an unknown validation is acknowledged. It depends on the mysterious authentication over the top computational expense that can be decreased inside the certificate revocation Checking Process List (CRL), also inside the testaments, and mark confirmation method could be used. For all these basic operations, the considered authority is required and is responsible for keeping up the whole vehicular community. A set mark is produced for demonstrating confirmation to a lot of cars, yet when this technique is used, each vehicle wants to store the announcement intending to maintain a strategic distance from verbal trade with denied autos. Therefore, the check strategy develops directly with the development of the number of repudiated vehicles inside the posting [35]. Across the board, the leading edge techniques utilize a transfer on the validation system to spare us, clients, from being effortlessly perceived from the traded messages and attempts to blast the effectiveness in the registered forms, since the basic framework does not offer any instruments for security and protection well-being. We endeavor to determine the security and privations issues characteristically from the viewpoint of ITS structure, utilizing the use of the blockchain period. We present a blockchain overlay on the zenith of the leading-edge ITS structure to adapt to the well-being and security assurance [36]. Along these lines, the data dispatched with the guide of clients and autos is mysterious, and the respectability of the realities might be ensured all through transmission in systems.

8.4.2 Blockchain and Its Applications for Improving Security and Privacy

In recent years, blockchain has become a hot topic for research. Typically, the blockchain has the inherent various features [37]. A blockchain uses decentralizing of the control and storage of data. Throughout the network of the blockchain, the blocks are stored in unique nodes that are adjoined effectively. No polices are dictated by central authorities. Also, the auditability and transparency of data is maintained by blockchain. A complete replica of each transaction achieved within the blockchain is recorded in the blocks and is public to all of the peers in the equal chain. Inside the network, all the transactions can be traced. A consensus decentralization feature of blockchain is very essential, because a new block appends the chain by the members whenever a new transaction consensus is completed within the blockchain, which was not the case with a traditional database system.

The anonymity and encryption in blockchain is handled using public and private key cryptography. The supply and destination of a transaction are identified by way of users' public keys. In this scenario, customers' anonymity may be assured. Besides, the transactions are signed with the aid of the private keys of users, and the content is hashed, which additionally assures a sure level of safety. Finally, the independent program execution is one of the features of blockchain that enables with or without the usage of any saved information and the program execution is done automatically by fulfilling the predefined conditions.

8.4.3 ITS Based on Blockchain

Some work related to the vehicular network and ITS network is introduced in blockchain due to its unique features. For establishing the Vehicular Ad hoc Network (VANET) trusted model, the author has proposed the Blockchain-based Anonymous Reputation System (BARS) for protecting vehicle identity so that forged message distribution is prevented [38]. The certificate authority operates to protect the vehicle identity to dispose of the connection between the vehicle identity and public key. Singh et al. [39] proposed a concept for the communication network where trust is created by an authorized dealer or the vehicle seller issues an "Intelligent Vehicular Trust Node." An approved foundation creates a remarkable ID and circulates it to a vehicle. By utilizing blockchain, the total history of the node can be recorded, including the uprightness worth and criminal history. The utilization of charging through smart contracts via autonomous vehicles is proposed, which is also very efficient [40]. Table 8.2 compares and contrasts the feature of traditional ITS with the ITS based on the blockchain.

In contrast to the above works, the work purpose blockchain helps with an ITSs design, giving inborn security and privacy insurance for transmissions of information and clients. Simultaneously, new administrations can be presented deftly and quickly. Moreover, we

Table 8.2 Traditional traffic system vs. ITS.

Categories	Traditional ITS	ITS based on blockchain
Privacy	The information of the user can be accessed by other authorities.	The user information is secured and no middle party is involved in transactions or information sharing.
Attack Vulnerability	The traditional ITS is more susceptible to outside interventions.	User information is safe and cannot be intervened by any attacker
Access Control	It has a centralized controlled network, which means a single entity controls the entire network.	The network is decentralized, which means no particular node has control over the entire network
Trust	Users have to trust a particular organization for its privacy and security.	Users trust the network as it is not controlled by any single organization or entity, as it has a decentralized network.
Authentication	Every user has its authentication information and is unique, so users can only access their information.	Every user has its authentication information and is unique, so users can only access their information.
Implementation Limitations	It requires real-time sharing of information about vehicles to provide real-time solutions.	It requires a large amount of memory space to store real-time data in different nodes.

Source: Based on Jeong, S. et al. [40].

utilize various leveled chains in the design, which makes the engineering versatile and reasonable to vehicles with various abilities. The interoperability between the neighbor layers chains is also investigated.

8.4.4 Recent Advancement

ITS has evolved in various forms and its implementations are increasing drastically using blockchain as one of the main components. Baldini et al. [41] has used the blockchain for managing vehicular networks. Also, Baldini et al. [42] has used blockchain for road transport where significance of ledger is discussed. Bhalerao et al. [43] explains the significance of blockchain in supply chain management. The Internet of Transport (IoT) using blockchain is successfully explained in [44]. Das et al. [45] infers the importance of using blockchain for VoT. The dangerous accidents due to fog could be tackled using ITS using blockchain [46]. Hatim et al. [47] explains the significance of blockchain in smart cities. Hu et al. [48] describes the use of consensus for maintaining that the vehicle has been covered as future implementation. Security being considered as highest priority is achieved using blockchain for vehicles [49]. Jabbar et al. [50] explains the significance of blockchain technology for parking of vehicles. Management of fund transfer and economy is one of the important aspects of the industrial system and has been suitably depicted [51]. The cloud-based traffic management needs extra security, and blockchain is one of the best solutions [52, 53]. The security of ITS needs extra attention and hence it has been traversed significantly [54]. Dynamically, it is one of the important features of any system and hence maintaining dynamic keys is a crucial step in heterogeneous ITS and can be handled using blockchain [55]. The preservation of any system requires state-of-art system and blockchain follows the same principle, therefore it is best suited for preserving communication for a vehicular system [56]. The shipping ports are very difficult to manage and hence ITS along with blockchain makes ship port management easier [57]. Shwetha et al. [58] traverse the need of accountability in distributed system, which is one of the features of blockchain. ITS is one crucial part of identity management and the blockchain plays a very important role [59].

8.5 Future Research Perspectives

The use of blockchain is very prominent and will continue to flourish in the future. This technology is vital for security, privacy, and integrity of data. Some of its vital uses are discussed below.

8.5.1 Electric Vehicle Recharging

Electric vehicles have improved and efficient hardware research requires recharging of the battery after a certain running period. It also leads to lack of software efficiency and reliability. As there is an increase in electric vehicles and their charging stations, a group of motorists have built such stations for charging batteries to fulfill the requirements of their energy supply, but due to lack of software research in this area, it can cause leakage of privacy and unorganized varying prices at different stations [60]. Therefore, blockchain in this area can prove to be a reliable and secure network for both parties. Motorists can match the right demand and supply of energy by building stations at varying distances, depending upon the location and its demand. They can also vary prices to maximize their profit based on demand and location, whereas the drivers can choose the lowest price and nearest stations while traveling and can pay the bill with a secure blockchain network. The increase in research of blockchain technology in electric vehicles will give a boost to the upcoming demand for more efficient electric vehicles.

8.5.2   Smart City Enabling and Smart Vehicle Security

Technologies such as Internet connectivity, knowledgeable workers, and digital establishments make a city into a smart city. The main agenda of creating a smart city is to reduce administrative costs and to deliver reliable and improved services to citizens via technology. Many technologies, such as IoT blockchain [61–64], etc., are majorly useful in establishing a smart city and it needs secure networks for data handling. This issue can be resolved using blockchain technology making a system more reliable, secure, and effective while sharing and managing information via making it a decentralized network that does not involve government policies and other outside involvement. Smart cities in the near future will produce and run vehicles or smart vehicles, which will be connected to the Internet and can share real-time information such as position, traffic, or any unusual events that may occur such as road accidents [65]. But the main concern in vehicles will be the same as for cities, like establishing a secure network for sharing real-time information of the vehicle and its environment and management of data by a single entity or organization, making it more vulnerable to attack by an unauthorized user. Therefore, to make it reliable and secure, blockchain should be introduced in a smart vehicle system to create a secured network for real-time information sharing by drivers.

8.5.3   Deferentially-Privacy Preserving Solutions

Internet of Vehicle (IoV) technology has contributed a great deal to the development of vehicle management and smart city development. It offers live location access to vehicles, traffic, etc., which helps a driver to choose a less busy traffic way to reach their destination as soon as possible [66]. For IoV technology to work properly, it requires real-time details of the driver, its location, and behavior of its environment, which can be computed to decide and help drivers to reach their destination in less time, by using Live Maps. The police can use this technology to chase stolen vehicles or catch a criminal escaping in a vehicle, etc., but if the real-time information of vehicles is managed by a third party or if blockchain transactions are stored on a third-party database or device, this can make this technology more vulnerable to malware. The hackers or adversary can use the information to gain a real-time location of any vehicle, the identity of the driver, and also details of the vehicle. This leads to leakage of the privacy for drivers. This privacy can be preserved by using Classic Blockchain Technology, which provides cryptographic techniques that ensures the transaction is secured by cryptographic key and every transaction is assigned its cryptographic key, making it more secure and can be used for privacy.

8.5.4 Distribution of Economic Profits and Incentives

The integration of Blockchain Technology and IoV has given a boost to ITS, providing it with a secure and reliable network to share real-time information with the vehicle, driver, and the surroundings. But this requires a stable connection and large computing of resources, therefore requires a large number of contributors to make it a successful ITS. These contributors help the system to be a success in return for profit and incentives, but the challenge or conflict arises in the distribution of these profits among the contributors. As there are no certain parameters set to distribute these profits, it can cause conflict among contributors and in the system. Due to the unavailability of such parameters, it can lead to several issues. Firstly, there is no surety of trust among contributors while sharing resources, which can make it hard to identify the corrupted entity, which can deny access to certain information to other entities or may deny giving incentives while trading resources. Secondly, conflict can arise about the proper distribution of profits, depending on their contribution during a stable and unstable connection. Lastly, there is a need for proper development of connections among other entities for the stable sharing of resources.

8.6 Conclusion

In recent years, blockchain technology has been growing day-by-day and is extensively used as a key feature for providing security solutions in multiple platforms. Blockchain technology has been introduced into the ITS network to overcome vulnerabilities such as third-party interference in multiple nodes causing leakage of the driver's privacy, memory safety violation, input validation error, etc. In the ITS system, the use of blockchain is to provide data privacy and to make services trustworthy. This chapter discusses and studies the combination as well as the use of blockchain and the IoV together to form the ITS, to create a better-integrated system which offers decentralization and better security measures. This chapter also discusses the structure layout of blockchain in ITS to prevent access to users' privacy by an unauthorized system or network. It also provides users with safety, trust, and secure services, including tracing of malware or corrupted data sharing over the network between the users. This structure is based on Blockchain Assisted ITS (Ba-ITS), which is a more flexible, reliable, trustworthy, and fast network design. This study of the integration of blockchain with IoV is of great interest in the scientific world and will keep going until a better and proper framework with reliable functionalities is introduced in ITS. The privacy and weak security are the most vital elements that an adversary could exploit to enter into the ITS and could lead to huge damage. This chapter also tries to show the vitality of blockchain to eradicate such attempts by the adversary.

References

1 Wright, C.S. (2008). Bitcoin: a peer-to-peer electronic cash system. *SSRN Electronic Journal* Available at: https://ssrn.com/abstract=3440802 or http://dx.doi.org/10.2139/ssrn .3440802

2 Tschorsch, F. and Scheuermann, B. (2016). Bitcoin and beyond: a technical survey on decentralized digital currencies. *IEEE Communications Surveys & Tutorials* 18 (3): 2084–2123. doi:10.1109/comst.2016.2535718

3 Belotti, M., Bozic, N., Pujolle, et al. (2019). A vademecum on blockchain technologies: when, which, and how. *IEEE Communication Surveys and Tutorials* 21 (4): 3796–3838. doi:10.1109/comst.2019.2928178

4 Gu, A., Yin, Z., Fan, C. et al. (2019). Safety framework based on blockchain for intelligent manufacturing cyber physical system. 2019 1st International Conference on Industrial Artificial Intelligence (IAI). doi:10.1109/iai47267.2019.9085328

5 Haque, A.B., Shurid, S., Juha, A.T. et al. (2020). A novel design of gesture and voice controlled solar-powered smart wheel chair with obstacle detection. 2020 IEEE International Conference on Informatics, IoT, and Enabling Technologies (ICIoT). doi:10.1109/iciot48696.2020.9089652

6 Madaan, G., Bhushan, B., and Kumar, R. (2021). Blockchain-based Cyberthreat mitigation systems for smart vehicles and industrial automation. In: Kumar, R., Sharma, R., and Pattnaik, P.K., eds.). Studies in Big Data *Multimedia Technologies in the Internet of Things Environment*.Vol. 79, Singapore, Springer. 13–32. Singapore, Springer. doi: 10.1109/ACCESS.2021.3069877

7 Haque, A.B., Islam, A.K.M.N., Hyrynsalmi, S. et al. (2021). GDPR compliant Blockchains – a systematic literature review. *IEEE Access* 9: 50593–50606. doi:10.1109/access.2021.3069877

8 Indumathi, J., Shankar, A., Ghalib, M.R. et al. (2020). Block chain based internet of medical things for Uniterrupted, ubiquitous, user-friendly, unflappable, unblemished, unlimited health care services (BC Iomt U6 HCS). *IEEE Access* 8: 216856–216872. doi:10.1109/access.2020.3040240

9 Goyal, S., Sharma, N., Bhushan, B. et al. (2021). IoT enabled technology in secured healthcare: applications, challenges and future directions. In: Hassanien, A.E., Khamparia, A., Gupta, D. et al. eds). *Cognitive Internet of Medical Things for Smart Healthcare*, Vol. 311, 25–48. Springer, Cham. doi:10.1007/978-3-030-55833-8_2

10 Sethi, R., Bhushan, B., Sharma, N. et al. (2020). Applicability of industrial IoT in diversified sectors: evolution, applications and challenges. Studies in Big Data *Multimedia Technologies in the Internet of Things Environment*, 45–67. doi:10.1007/978-981-15-7965-3_4

11 Goyal, S., Sharma, N., Kaushik, I. et al. (2020). Blockchain as a lifesaver of IoT. *Security and Trust Issues in Internet of Things* 209–237. doi:10.1201/9781003121664-10

12 Bhardwaj, A., Shah, S.B.H., Shankar, A. et al. (2020). Penetration testing framework for smart contract Blockchain. *Peer-to-Peer Networking and Applications*. doi:10.1007/s12083-020-00991-6

13 Liotine, M. and Ginocchio, D. (2020). The supply Blockchain: integrating blockchain technology within supply chain operations. *Technology in Supply Chain Management and Logistics*, 57–89. doi:10.1016/b978-0-12-815956-9.00004-1

14 Nakamoto, N. (2017). Centralised bitcoin: a secure and high performance electronic cash system. *SSRN Electronic Journal*. Available at: https://ssrn.com/abstract=3065723 or http://dx.doi.org/10.2139/ssrn.3065723

15 Martino, P. (2019). Blockchain technology: challenges and opportunities for banks. *International Journal of Financial Innovation in Banking* 1 (1): 1. doi:10.1504/ijfib.2019.10021814

16 Bhushan, B., Khamparia, A., Sagayam, K.M. et al. (2020). Blockchain for smart cities: a review of architectures, integration trends and future research directions. *Sustainable Cities and Society* 61 : 102360. doi:10.1016/j.scs.2020.102360

17 Özsu, M.T. and Valduriez, P. (2020). Distributed query processing. In: *Principles of distributed database systems*. Springer, Cham. https://doi.org/10.1007/978-3-030-26253-2_4

18 Zhang, R., Xie, P., Wang, C. et al. (2019). Classifying transportation mode and speed from trajectory data via deep multi-scale learning. *Computer Networks* 162: 106861. doi:10.1016/j.comnet.2019.106861

19 Kashyap, R., Arora, K., Sharma, M. et al. (2019). Security-aware ga based practical byzantine fault tolerance for permissioned blockchain. 2019 4th International Conference on Control, Robotics and Cybernetics (CRC). doi:10.1109/crc.2019.00041

20 Mittal, A. and Aggarwal, S. (2020). Hyperparameter optimization using sustainable proof of work in Blockchain. *Frontiers in Blockchain* 3: 1–23. doi:10.3389/fbloc.2020.00023

21 Casino, F., Dasaklis, T.K., and Patsakis, C. (2019). A systematic literature review of blockchain-based applications: current status, classification and open issues. *Telematics and Informatics* 36: 55–81. doi:10.1016/j.tele.2018.11.006

22 Wang, Z. (2020). A decentralized prediction market platform based on blockchain and master node technologies. *China Communications* 17 (9): 25–33. doi:10.23919/jcc.2020.09.003

23 Ozturan, C. (2020). Barter machine: an autonomous, distributed barter exchange on the Ethereum blockchain. *Ledger*, 5. doi:10.5195/ledger.2020.148

24 Islam, M.N. and Kundu, S. (2019). Enabling IC traceability via Blockchain pegged to embedded puf. *ACM Transactions on Design Automation of Electronic Systems* 24 (3): 1–23. doi:10.1145/3315669

25 Baig, F. and Wang, F. (2019). Blockchain enabled distributed data management - a vision. 2019 IEEE 35th International Conference on Data Engineering Workshops (ICDEW). doi:10.1109/icdew.2019.00-39

26 Febriyanto, E., Triyono, Rahayu, N. et al. (2020). Using blockchain data security management for e-voting systems. 2020 8th International Conference on Cyber and IT Service Management (CITSM). doi:10.1109/citsm50537.2020.9268847

27 Huang, Z. and Zhu, H. (2020). Blockchain-based data security management mechanism for power terminals. 2020 International Wireless Communications and Mobile Computing (IWCMC). doi:10.1109/iwcmc48107.2020.9148050

28 Kuperberg, M. (2020). Towards enabling deletion in append-only blockchains to support data growth management and GDPR compliance. 2020 IEEE International Conference on Blockchain (Blockchain). doi:10.1109/blockchain50366.2020.00057

29 Paik, H., Xu, X., Bandara, H.M. et al. (2019). Analysis of data management in blockchain-based systems: from architecture to governance. *IEEE Access* 7: 186091–186107. doi:10.1109/access.2019.2961404

30 Salman, T., Jain, R., and Gupta, L. (2019). A reputation management framework for knowledge-based and probabilistic Blockchains. 2019 IEEE International Conference on Blockchain (Blockchain). doi:10.1109/blockchain.2019.00078

31 Su, Z., Wang, H., Wang, H. et al. (2020). A financial data security sharing solution based on blockchain technology and proxy re-encryption technology. 2020 IEEE 3rd International Conference of Safe Production and Informatization (IICSPI). doi:10.1109/iicspi51290.2020.9332363

32 Yang, C. and Sun, Z. (2020). Data management system based on blockchain technology for agricultural supply chain. 2020 International Conference on Data Mining Workshops (ICDMW). doi:10.1109/icdmw51313.2020.00130

33 Pranto, T.H., Noman, A.A., Mahmud, A. et al. (2021). Blockchain and smart contract for IoT enabled smart agriculture. *Peer Journal of Computer Science* 7: e407. doi:10.7717/peerj-cs.407

34 Zhaofeng, M., Lingyun, W., Xiaochang, W. et al. (2020). Blockchain-9 enabled decentralized trust management and secure usage control of IoT big data. *IEEE Internet of Things Journal* 7 (5): 4000–4015. doi:10.1109/jiot.2019.2960526

35 Lu, R., Lin, X., Zhu, H. et al. (2008). ECPP: Efficient conditional privacy preservation protocol for secure vehicular communications. IEEE INFOCOM 2008 – The 27th Conference on Computer Communications. doi:10.1109/infocom.2008.179

36 Lin, X. Sun, X. Ho, P-H. et al. (2007). Gsis: a secure and privacy-preserving protocol for vehicular communications. *IEEE Transactions on Vehicular Technology* 56 (6): 3442–3456. doi:10.1109/tvt.2007.906878

37 Jesus, E.F., Chicarino, V R., De Albuquerque, C.V. et al. (2018). A survey of how to use blockchain to secure internet of things and the stalker attack. *Security and Communication Networks* 2018: 1–27. doi:10.1155/2018/9675050

38 Lu, Z., Liu, W., Wang, Q. et al. (2018). A privacy-preserving trust model based on blockchain for vanets. *IEEE Access* 6: 45655–45664. doi:10.1109/access.2018.2864189

39 Singh, M. and Kim, S. (2017). Introduce reward-based intelligent vehicles communication using Blockchain. 2017 International SoC Design Conference (ISOCC). doi:10.1109/isocc.2017.8368806.

40 Jeong, S., Dao, N., Lee, Y. et al. (2018). Blockchain based billing system for electric vehicle and charging station. 2018 Tenth International Conference on Ubiquitous and Future Networks (ICUFN). doi:10.1109/icufn.2018.8436987

41 Baldini, G., Hernandez-Ramos, J. L., Steri, G. et al. (2019). Zone keys trust management in vehicular networks based on Blockchain. 2019 Global IoT Summit (GIoTS). doi:10.1109/giots.2019.8766375

42 Baldini, G., Hernandez-Ramos, J. L., Steri, G. et al. (2020). A review on the application of distributed ledgers in the evolution of road transport. *IEEE Internet Computing* 24 (6): 27–36. doi:10.1109/mic.2020.3023295

43 Bhalerao, S., Agarwal, S., Borkar, S. et al. (2019). Supply chain management using Bockchain. 2019 International Conference on Intelligent Sustainable Systems (ICISS). doi:10.1109/iss1.2019.8908031

44 Butt, T. A., Iqbal, R., Salah, K. et al. (2019). Privacy management in social internet of vehicles: review, challenges and Blockchain based solutions. *IEEE Access* 7: 79694–79713. doi:10.1109/access.2019.2922236

45 Das, D., Banerjee, S., Mansoor, W. et al. (2020). Design of a secure Blockchain-based smart IOV architecture. 2020 3rd International Conference on Signal Processing and Information Security (ICSPIS). doi:10.1109/icspis51252.2020.9340142

46 Gao, J., Obour Agyekum, K.O., Sifah, E.B. et al. (2020). A Blockchain-SDN-enabled internet of vehicles environment for fog computing and 5g networks. *IEEE Internet of Things Journal* 7 (5): 4278–4291. doi:10.1109/jiot.2019.2956241

47 Hatim, S.M., Elias, S.J., Ali, R.M. et al. (2020). Blockchain-based internet of vehicles (BIOV): an approach towards Smart CITIES Development. 2020 5th IEEE International Conference on Recent Advances and Innovations in Engineering (ICRAIE). doi:10.1109/icraie51050.2020.9358355

48 Hu, W., Hu, Y., Yao, W. et al. (2019). A Blockchain-based byzantine consensus algorithm for information authentication of the internet of vehicles. *IEEE Access* 7: 139703–139711. doi:10.1109/access.2019.2941507

49 Jabbar, R., Fetais, N., Kharbeche, M. et al. (2021). Blockchain for the internet of vehicles: how to use Blockchain to secure vehicle-to-everything (v2x) communication and payment? *IEEE Sensors Journal*, 1–1. doi:10.1109/jsen.2021.3062219

50 Jabbar, R., Krichen, M., Shinoy, M. et al. (2020). A model-based and Resource-Aware testing framework for parking system payment using Blockchain. 2020 International Wireless Communications and Mobile Computing (IWCMC). doi:10.1109/iwcmc48107.2020.9148212

51 Kukharenko, E.G., Korkunov, I.A., Gorodnichev, M.G. et al. (2019). On the introduction of Digital economics in the transport industry. 2019 Systems of Signals Generating and Processing in the Field of on Board Communications. doi:10.1109/sosg.2019.8706797

52 Lakhan, A., Ahmad, M., Bilal, M. et al. (2021). Mobility aware Blockchain enabled offloading and scheduling in vehicular fog cloud computing. *IEEE Transactions on Intelligent Transportation Systems* 22: 4212–4223. doi:10.1109/tits.2021.3056461

53 Kumar, A., Abhishek, K., Nerurkar, P. et al. (2020). Secure smart contracts for cloud-based manufacturing using Ethereum Blockchain. *Transactions on Emerging Telecommunications Technologies* doi:10.1002/ett.4129

54 Lamssaggad, A., Benamar, N., Hafid, A S. et al. (2021). A survey on the current security landscape of intelligent transportation systems. *IEEE Access* 9: 9180–9208. doi:10.1109/access.2021.3050038

55 Lei, A., Cruickshank, H., Cao, Y. et al. (2017). Blockchain-based dynamic key management for heterogeneous intelligent transportation systems. *IEEE Internet of Things Journal* 4 (6): 1832–1843. doi:10.1109/jiot.2017.2740569

56 Mei, Q., Xiong, H., Zhao, Y. et al. (2021). Toward Blockchain-enabled IOV with Edge Computing: Efficient and privacy-preserving vehicular communication and dynamic updating. 2021 IEEE Conference on Dependable and Secure Computing (DSC). doi:10.1109/dsc49826.2021.9346240

57 Sangeerth, P.S. and Lakshmy, K.V. (2021). Blockchain based smart contracts in automation of shipping ports. 2021 6th International Conference on Inventive Computation Technologies (ICICT). doi:10.1109/icict50816.2021.9358634

58 Shwetha, A.N. and Prabodh, C.P. (2019). Blockchain – BRINGING accountability in the public distribution system. 2019 4th International Conference on Recent

Trends on Electronics, Information, Communication & Technology (RTEICT). doi:10.1109/rteict46194.2019.9016903

59 Zhou, Y., Liu, Q., Liu, M. et al. (2020). Research on blockchain-based identity verification between IOV entities. 2020 International Conference on High Performance Big Data and Intelligent Systems (HPBD&IS). doi:10.1109/hpbdis49115.2020.9130597

60 Knirsch, F., Unterweger, A., and Engel, D. (2017). Privacy-preserving blockchain-based electric vehicle charging with dynamic tariff decisions. *Computer Science – Research and Development* 33 (1–2): 71–79. doi:10.1007/s00450-017-0348-5.

61 Saxena, S., Bhushan, B., and Ahad, M.A. (2021). Blockchain based solutions to secure Iot: background, integration trends and a way forward. *Journal of Network and Computer Applications* 181. 103050. doi:10.1016/j.jnca.2021.103050

62 Goyal, S., Sharma, N., Kaushik, I. et al. (2021). Blockchain as a solution for security attacks in named data networking of things. *Security and Privacy Issues in IoT Devices and Sensor Networks* 211–243. doi:10.1016/b978-0-12-821255-4.00010-9

63 Bhushan, B., Sahoo, C., Sinha, P. et al. (2021). Unification of Blockchain and internet of things (BIoT): requirements, working model, challenges and future directions. *Wireless Networks* 27: 55–90. doi:10.1007/s11276-020-02445-6

64 Bhushan, B., Sinha, P., Sagayam, K.M. et al. (2020). Untangling Blockchain technology: A survey on state of the art, security threats, privacy services, applications and future research directions. *Computers and Electrical Engineering* 90: 106897. doi:10.1016/j.compeleceng.2020.106897

65 Dorri, A., Steger, M., Kanhere, S.S. et al. (2017). Blockchain: A distributed solution to automotive security and privacy. *IEEE Communications Magazine* 55 (12):119–125. doi:10.1109/mcom.2017.1700879

66 Ahluwalia, M.V. and Gangopadhyay, A. (2008). Privacy preserving data mining. *Computer Security, Privacy and Politics, IGI Global*, 70–93. doi:10.4018/978-1-59904-804-8.ch005

9

Utilization of Agro Waste for Energy Engineering Applications: Toward the Manufacturing of Batteries and Super Capacitors

S.N. Kumar[1], Akhil Sabuj[1], Nishitha R. P.[1], O Lijo Joseph[1], Aju Mathew George[2], and I. Christina Jane[3]

[1]*Department of EEE, Amal Jyothi College of Engineering, Kanjirapally, Kerala, India*
[2]*Department of Civil Engineering, Amal Jyothi College of Engineering, Kanjirapally, Kerala, India*
[3]*Department of ECE, Mar Ephraem College of Engineering, Elavuvilai, Tamil Nadu, India*

9.1   Introduction

Advances in science and technologies over the past few centuries have arguably become a prominent aspect for human living conditions. But, with a holistic approach, it can be concluded that even though the quality of life is apparently improving, our ecosystem is deteriorating in a critical manner; the major reason being the nonscientific approaches being made to meet our day-to-day energy demands. This section outlines why our energy sources should be made "clean" in the context of their ecologically friendly nature and how this can be achieved. Also is outlined the contributions that biomass can provide in this area.

The present era in which we live is termed the Anthropocene, which is primarily characterized by anthropogenic climatic changes. These changes are due to global warming occurring as a result of increased levels of greenhouse gases in the atmosphere, which are caused by human activities around the globe. The chart provided as Figure 9.1 depicts the contribution of different sources of energy toward Total Primary Energy Consumption (TPEC) of the world, as of 2020 [1]. It is evident from this chart that a major share of TPEC goes to conventional carbon-based fuels such as oil, coal, and natural gas. Combustion reactions of these fuels, which are carried out to extract desired amounts of energy, emit carbon dioxide (CO_2) as a by-product. Therefore, an increase in energy demand, as per the present scenario, implies an increase in carbon emissions. It is widely known that the increase in the concentration of CO_2 will lead to an increase in the Earth's global surface temperature. This is one of the core factors responsible for anthropogenic climate change.

The plot in Figure 9.2 has been formulated with the help of data from the National Aeronautics and Space Administration (NASA) and other similar agencies [2, 3]. It depicts an exponential increase in CO_2 emissions right from the seventeenth century when industrial activities started gaining momentum. Expert opinions are that a reduction in the energy demand will be extremely difficult to achieve for the upcoming generation. Therefore, the only viable and practical option for us, provided the present technological

Smart and Sustainable Approaches for Optimizing Performance of Wireless Networks: Real-time Applications, First Edition.
Edited by Sherin Zafar, Mohd Abdul Ahad, Syed Imran Ali, Deepa Mehta, and M. Afshar Alam.

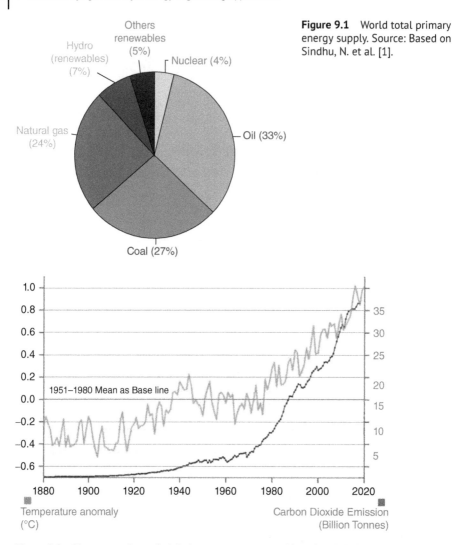

Figure 9.1 World total primary energy supply. Source: Based on Sindhu, N. et al. [1].

Figure 9.2 Plot comparison of global temperature anomalies with global carbon emissions which dates back to the Industrial Revolution.

barriers remain unchanged, is to carry out a transition from the carbon-based traditional sources of energy to novel carbon-free "clean energy."

9.2 Super Capacitors and Electrode Materials

The capacitors which come under the category of Electric Double Layer Capacitors (EDLC), which in turn are electrochemical devices, are most widely referred to as super capacitors, owing to their enhanced energy density when compared with traditional capacitors. The categorization of existing capacitor technologies is illustrated in Figure 9.3, which is prepared based on the work of Pandolfo et al. [4]. The concept of EDLC was first developed and

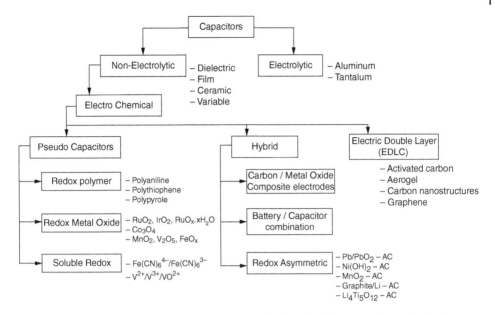

Figure 9.3 Block diagram representing the categorization of existing capacitor technologies.

modeled by von Helmholtz in the nineteenth century. However, it was not until 1957, when H.I. Becker of General Electric demonstrated the practical use of the double-layer capacitor, that it began to be used for the storage of electrical charge [4]. The patented design from Sohio (Standard Oil of Ohio) was licensed by Nippon Electric Company (NEC). By the end of the 1970s, they were marketed as super capacitors to be used as backup power devices for computer memory. The first generation of EDLCs had high internal resistance, which limited their current output. However, with further research and with the development of electrolytes with better conductivity, discharge currents increased.

9.2.1 Energy Density

For any given capacitor of capacitance value, say C, with a voltage rating of V, the maximum energy E that can be stored within it is governed by Equation 9.1:

$$E = \frac{1}{2}CV^2 \tag{9.1}$$

As the energy storage of a capacitor is directly proportional to the square of its voltage, researchers began work on increasing the breakdown voltage. Early electrochemical capacitors consisted of two aluminum foils coated with activated carbon, which were soaked in an electrolyte and separated by a thin porous insulator. The first EDLC made by Sohio consisted of carbon paste electrodes and was formed by soaking porous carbon in an electrolyte separated by an ion-permeable separator. EDLCs made from activated carbon have an energy storage density of 1–10 Wh kg^{-1} and a power density of 500–10 000 W kg^{-1}.

A Ragone plot with energy density (Wh kg^{-1}) and power density (W kg^{-1}) on logarithmic scale on both axes, for select electrical energy storage devices, is presented in Figure 9.4, partly relying on previous work by Pandolfo et al. [4].

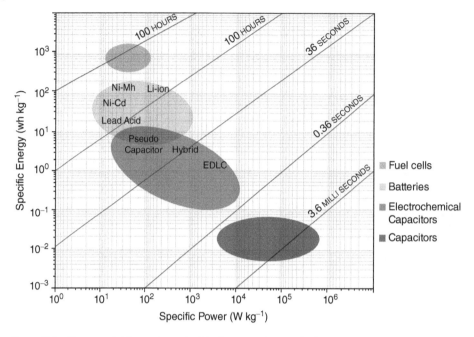

Figure 9.4 Ragone plot for various electrical energy storage devices.

EDLCs have charging and discharging time of only seconds to minutes, with efficiencies from 85 to 99%. They have a life span of more than 10 years. However, they only have a maximum voltage of less than 3 V. They have a gravimetric surface area of 1000–3000 $m^2\,g^{-1}$ and gravimetric capacitance values of 200–400 $F\,g^{-1}$ in aqueous electrolyte, and 100–150 $F\,g^{-1}$ in organic solvents and ionic liquids. These figures are promising in the energy storage market, especially in the case of home energy storage solutions.

9.3 Related Works in the Utilization of Agro-Waste for Energy Engineering Applications

As per the census of 2019, India has a rural population of 65.53%. The total area of Kerala is 38 852 km^2, including 31 253.20 km^2 rural area and 7598.80 km^2 urban area. In rural areas, agriculture is the main occupation of most people and the waste generated from farmland can be utilized as an ecofriendly compost material that will create an income for the people. Efficient utilization of agro-waste, thereby empowering the rural area people, will pave the way toward a sustainable green economy. The need for empowering the rural community is a vital one and the efficient utilization of green waste is also a needy one for a sustainable green circular economy. There is also a need to push green forms of energy in the rural sector. Part of realizing a green economy in the rural sector is battery technology. The activated carbon generated from the rural agricultural waste goes toward making eco-friendly battery electrodes for use in these batteries. Thus, the farmer not only gets an incentive from the sale of agricultural waste, but will also benefit from improved battery technology that can be used to power rural households.

Agro-waste is produced from the waste generated by agricultural crops, agro-industries, horticulture, aquaculture, animal husbandry, and hydroculture, etc. [1]. Rapid increase in population, increased urbanization, and growing living standards are the main causes for the many types of the solid agricultural wastes in the environment. The Ministry of New and Renewable Energy (MNRE) reports that every year about 500 million metric tons of agro-waste is generated in India (MNRE 2019) [5]. The agro-waste could be managed through proper bio-treatments to reduce and reuse the waste. The agro-waste is generated in large quantities during harvesting from crops like rice, wheat, sugarcane, and jute, etc. In India, about 10–20% of these agro-wastes are disposed of without proper waste management [6]. A large quantity of the agro-waste is decomposed into compost or manure, which have less impact on the environment. Consumption of energy by humans has rapidly increased in various applications like transportation, electricity for households, and industry. Commonly, regarding carbon-based electrode materials that are used in energy storage devices, most of the activated carbon is produced from fossil fuels, which are not environmentally-friendly. Therefore, biomass can be good alternative for the production of activated carbon [7]. For various sustainable industries, green composite materials have become a potential alternative solution. Green materials have a wide range of functional abilities for various applications [8]. The materials used in storage devices have great potential due low cost, easy processing techniques, stability, and versatility, when made from green sources [9].

Sharma and Singh (2019) highlight recycling and utilization of sugarcane waste and rice husk ash which find applications in the semiconductor industry [10]. To reduce the melting temperature of rice husk and sugarcane leaf ash, egg shell powder is used. The agro-waste materials like sheep and goat fecal matter are employed for the synthesis of zinc oxide nanoparticles (ZnONPs) for different applications, explained in Chikkanna et al. (2018) [11]. New ways of cost-effective synthesis and reducing the use of chemicals are developed with the help of this research [11]. The utilization of solid waste in the fabrication of super capacitors is discussed in Dutta et al. (2020) [12]. In Liu et al. (2019) [13], the activated carbon electrodes are prepared from rice husk. In Mehare (2019) [14], bio-waste from onion peelings were used for the preparation of porous agro-waste derived carbon for super capacitor applications. In general, the energy gap between conventional capacitors and batteries are filled by super capacitors. Nitrogen and sulfur-enriched porous carbon are used in high energy, high temperature Ultracapacitors, which are produced from water melon seeds [15]. In Haridas et al. (2019), the construction of iron sulfide nanoparticle loaded graphitic carbon capsules from waste biomass for sustainable lithium-ion storage was highlighted [16]. The electrode materials of a lithium-ion battery are composed of iron sulfide (FeS). A lithium-ion battery was synthesized using sawdust, which was carbonized and used as the conductive carbon matrix. For carbonization of sawdust, the obtained sample was dried and heated until the temperature increased to 800 °C at a heating rate of 5 °C in a nitrogen atmosphere.

The main aim of agro-waste management is to reduce the quantity and to reuse the organic waste material [17]. The waste management process includes collection, transport, processing, recycling, and disposal of waste. The agricultural waste is reused for composting, pyrolysis, energy production, fertilizers, animal feed, direct combustion, and other products. The agro-waste produced during agricultural activities are classified and categorized as cultivation activities, livestock wastes, aquaculture, agro-industrial waste, horticulture, etc. [18].

Pineapple waste contains various substances that are valuable for the development of composite materials and drugs, not just for biogas and bioethanol creation [19]. For various sustainable industries, green composite materials have become a potential alternative solution. Green materials have a wide range of functional abilities for high-tech applications [20], which usually come with improved mechanical and electrical properties. The performance of graphitic carbon Nano flakes derived from green tea waste was discussed by Sekar et al. (2019) [21]. By synthesizing green tea waste, high-quality carbon Nano flakes were obtained and used as an electrode source material for high-performance super capacitors [21]. Gandla et al. (2020) highlighted the design of high voltage, high energy, and activated carbon-based super capacitors using eco-friendly green tea waste [22]. Due to their high energy capacity and excellent cyclic durability, lithium-ion batteries (LIBs) are considered as the next generation of energy storage devices, among various electrical devices. Moro et al. (2019) synthesized waste materials for the fabrication of disposable electrodes for the electro-analytical applications [23]. The activated carbon was generated from pineapple leaf fiber by employing the simple hydrothermal technique and potassium hydroxide (KOH) chemical activation under heat treatment in an Ar atmosphere [24]. Na et al. (2018) produced a novel egg white gel polymer electrolyte and a green solid-state super capacitor from egg and rice waste [25]. The fabricated novel egg white gel polymer electrolyte has several advantages, such as high ionic conductivity, high electrolyte retention, adequate mechanical properties and thermal stability, and eco-friendliness, by combining egg white, egg shell, and optimization of electrolyte salt.

The Carbon Nano-Onions (CNOs) was synthesized from waste oil for energy storage devices [26]. Here, traditional pyrolysis techniques were used for synthesis using waste frying oils as free carbon sources. Fiu et al. (2019) used green self-assembly synthesis of porous lignin-derived carbo quasi-Nano sheets methods for high performance super capacitors [27]. Usually, porous carbons are obtained by calcinating carbonaceous precursors in high temperature with activation of strong corrosive agents such as KOH, $ZnCl_2$, and H_3PO_4. In this technique, non-toxic and low-corrosive zinc oxalates as pore-creating substances are adopted for this operation, which avoids equipment damage and the shrinkage of lignin during carbonization. The synthesis of bi-functional iron embedded nanostructure from collagen waste for Li-ion battery was discussed by Berhanu et al. (2018) for photo catalysis applications [28]. By structural and morphological analysis, core-shell types of nanostructures are formed by embedding different phases of Fe nanoparticles in a graphitized carbon matrix. The hydrothermal reaction is used for preparing an Fe_3O_4 core with a thin carbon layer of high electrical conductivity, excellent buffering effect, and mechanical strength. In this experiment, extraction of collagen is from the leather manufacturing industries and the nanoparticles are synthesized by simple heat treatment at 1000 °C for 2 hours. The obtained hybrid nano structure is used as an anode in Li-ion batteries. The aloe vera-based activated carbon finds its application in the fabrication of a high-energy super capacitor. The surface area of the activated carbon was approximately 1890 m^2/g with a high specific capacitance of 410 and 306 F/g. The activated carbon electrodes could power a red light emitting diode (LED) for a specific duration of 20 minutes upon charging for less than 20 seconds [29].

9.4 Inferences from Works Related with Utilization of Coconut, Rice Husk, and Pineapple Waste for Fabrication of Super Capacitor

The easily available agro-waste in the southern part of Kerala are coconut, rice husk, and pineapple waste, hence the following tables depict its usage in the fabrication of super capacitor. Tables 9.1, 9.2, and 9.3 represent the fabrication of super capacitors from coconut shell, rice husk, and pineapple waste.

9.5 Factors Contributing in the Fabrication of Super Capacitor from Agro-Waste

In the synthesis of activated carbon, two techniques are involved: carbonization and activation. The carbonization converts the waste into biochar by pyrolysis or gasification. The activation may be physical or chemical to increase the total pore areas. The pre-carbonization is involved in some applications; an example is ^{60}Co γ-irradiation to palm oil EFB (empty fruit brunch) fibers prior to carbonization. The pre-carbonization role is to increase the

Table 9.1 Fabrication of super capacitor from coconut shell waste.

Reference	Activation agent	Characteristics
[30]	Ferrocene	Prepared grapheme oxide has excellent properties as an electric double-layer capacitor (EDLC) with superior cyclic stability. rGO was fabricated under low temperature (300 °C, 15 min). The specific capacitance value of 111.1 F/g was observed in the sample.
[31]	Two-stage activation with 1 M H_2SO_4 and 1 M KOH	Cyclic voltammetry test was employed on generated activated carbon. The capacitor fabricated from activated carbon has a capacitance (C_s) of 600.89 F/g and energy density (E_d) of 46.94 Wh kg^{-1}.
[32]	Steam, 200° Celsius, 30–90 min	The capacitor fabricated from activated carbon has a capacitance of 228 F/g and energy density of 38.50 Wh kg^{-1}.
[33]	$ZnCl_2$ and $FeCl_3$ under N_2 gas conditions	The capacitor fabricated from activated carbon has a C_s of 159 F/g and E_d of 69.00 Wh kg^{-1}.
[34]	Pyrolysis gases	The capacitor fabricated from activated carbon has a C_s of 258 F/g.
[35]	KoH	The capacitor fabricated from activated carbon has a C_s of 356.20 F/g and E_d of 88.80 Wh kg^{-1}.

Table 9.2 Fabrication of super capacitor from rice husk waste.

Reference	Activation employed	Characteristics
[36]	NaOH	Rice husk-derived activated carbon (RHAC) has excellent discharge capacities. Electrochemical performance was superior; capacity was found to be consistent even after 100 cycles with a value of 448 mA hg^{-1} and As value was 2176 m^2 g^{-1}.
[37]	KOH	Pre-carbonization of rice husk and activation by KOH·C$_s$ value of 250 F g^{-1} and A$_s$ of 2523.4 m^2 g^{-1}was observed in the sample. Good cyclic stability over 10 000 cycles.
[38]	KOH	Graphene nanosheet is formulated. C$_s$ value was 115 Fg^{-1} with an A$_s$ of 1225 m^2 g^{-1}. Good cyclic stability over 2000 cycles.
[39]	NaOH	C$_s$ of the electrode obtained was 172.3 F g^{-1} with an A$_s$ of 2681 m^2 g^{-1}.
[40]	Chemical activation	Hybrid combination of rice husk and LiMn$_2$O$_4$. Energy density is 29.5 Wh kg^{-1}. Energy retention is 81.3% over 10 000 cycles.

Table 9.3 Fabrication of super capacitor from pineapple waste.

Reference	Activation employed	Characteristics
[41]	Chemical activation [0.9 M KOH activator] and physical activation	C$_s$ of the electrode obtained was 127 F g^{-1}; elemental analysis results reveal the percentage of carbon and oxygen of 86.03 and 9.49%. The energy density was observed to have the value of 4.41 Wh Kg^{-1}and power density of 10.49 W Kg^{-1}.
[42]	KOH activation and single stage pyrolysis comprising of carbonization and physical activation	C$_s$ of the electrode obtained was 150 F g^{-1}. Hybrid combination of leaf wastes are used in this work; nano fibers comprises of 90% of activated carbon.
[43]	KOH chemical activation and heat treatment in Ar atmosphere	C$_s$ of the electrode obtained was 202 F g^{-1}; pineapple leaf derived nanosheets are stable at 10 000 cycles, and an A$_s$ of derived nanosheet is 1681 m^2 g^{-1}.
[44]	KOH chemical activation by N$_2$ and CO$_2$	C$_s$ of the electrode obtained was 150 F g^{-1}. The energy density was observed to have a value of 5.2 Wh Kg^{-1} and power density of 42 W Kg^{-1}.
[45]	KOH chemical activation	C$_s$ of the electrode obtained was 175 F g^{-1} and an A$_s$ of 945 m^2 g^{-1}. Pineapple leaf and fiber hybrid combination is used in this work.

efficiency of the activation procedure by changing the structure of the molecules. Chemical activation is the process by which the carbon-based materials are combined with strong acid, base, or salt as an activation agent. The chemical activation agent increases the surface area of the material and intense research proves that KOH is an efficient one. The post activation treatment is also employed to increase the efficiency of the electrodes for super capacitor applications. The wet oxidation by HNO_3 introduces oxygen functional group into the biochar. The oxygen functional group increases the porosity nature of the electrode and thereby the ions in the electrolyte travel freely into the pores of the electrodes. The nature of the agro-waste plays a vital role; corn waste is rich in protein and mineral contents. The biochar produced from corn waste reveals that the ash content is higher when compared with other wastes. The microstructure analysis shows that nano sheets are formed from the high proteins and ash contents of the corn-based waste materials.

The capacitance effect was found to be good in the case of electrodes with macro pores. The diffusion rate of electrolyte ions is high from the bulk electrolyte and due to the macroscopic pores, diffusion resistance is minimized. The presence of the oxygen functional group increases the specific capacitance and power density of the electrodes. The surface functional group such as phenol and carboxyl tends to increase the formation of reversible redox reaction on the interface of the electrode.

9.6 Conclusion

This chapter focuses on the utilization of agro-waste for energy engineering applications. The disposal of agro-waste is a crucial problem in many rural areas and improper disposal leads to pollution, thereby affecting the ecological balance. Proper treatments of agro-waste by carbonization and activation produce biochar that can be employed in the fabrication of energy storage elements. The various steps involved in biochar generation are discussed in this chapter with respect to various sources of agro-waste. The factors contributed to the biochar generation from agro-waste is also discussed in this work. The outcome of this chapter will be beneficial to a researcher who aims to develop eco-friendly energy storage elements. The efficient utilization of agro-waste generates a sustainable circular green economy.

Acknowledgment

The authors would like to acknowledge the Schmitt Center for Biomedical Instrumentation of Amal Jyothi College of Engineering for support in this work.

References

1 Sindhu, N.P., Seharawat, S.P., and Malik, J.S. (2015). Strategies of agricultural waste management for better employment and environment. *International Journal of Current Research* 7 (12): 24604–24608.

2 GISTEMP Team (2021). GISS Surface Temperature Analysis (GISTEMP), version 4. NASA Goddard Institute for Space Studies. Dataset accessed 20YY-MM-DD at https://data.giss.nasa.gov/gistemp

3 Friedlingstein, P. et al. (2020). Global Carbon Budget 2020. *Earth Systym Scientific Data* 12: 3269–3340. https://doi.org/10.5194/essd-12-3269-2020.

4 Pandolfo, T., Ruiz, V., Sivakkumar, S. et al. (2013). General properties of electrochemical capacitors. In *Supercapacitors: Materials, Systems and Applications.* pp. 69–110.

5 Shehrawat, P.S. and Sindhu, N. (2012). Agricultural waste utilization for healthy environment and sustainable lifestyle. Proceedings of the Third International Scientific Symposium Agrosymposium, Jahorina. pp. 393–399.

6 Lakshmi, M.V., Goutami, N., and Kumari, A.H. (2017), Agricultural waste concept, generation, utilization and management. *International Journal of Multidisciplinary Advanced Research Trends IV* 1 (3): 1–4.

7 Chaudhary, R., Chaudhary, S.M.R., Maji, S. et al. (2020) Jackfruit seed-derived Nanoporous carbons as the electrode material for supercapacitors, *Journal of Carbon Research* 11: 1–13.

8 Gao, M., Shih, C.C., Pan, S.Y. et al. (2018). Advances and challenges of green materials for electronics and energy storage applications: from design to end-of-life recovery. *Journal of Materials Chemistry* A6 (42): 20546–20563.

9 Chaudhary, S., Mohan, R., Sinha, O.P. (2020). Green synthesis of ternary-doped layered graphene nanosheets (DGNS) synthesized from waste onion peel for supercapacitors. *Applied Physics A*, Springer 126 (10): 1–7.

10 Sharma, G. and Singh, K. (2019). Recycling and utilization of agro-food waste ashes: syntheses of the glasses for wide-band gap semiconductor applications: *Journal of Material Cycles and Waste Management* 21 (4): 801–809.

11 Chikkanna, M.M., Neelagund, S.E., and Rajashekarappa, K.K. (2018). Green synthesis of zinc oxide nanoparticles (ZnO NPs) and their biological activity, *SN Applied Sciences* 1 (1): 1–10.

12 Dutta, A., Mahanta, J., and Banerjee, T. (2020). Supercapacitors in the light of solid waste and energy management: a review. *Advanced Sustainable Systems* 4 (12): 2000182.

13 Liu, D., Zhang, W., Huang, W. (2019). Effect of removing silica in rice husk for the preparation of activated carbon for supercapacitor applications. *Chinese Chemical Letters* 1 30 (6): 1315–1319.

14 Mehare, M.D., Deshmukh, A.D., and Dhoble, S.J. (2019). Preparation of porous agro-waste-derived carbon from onion peel for supercapacitor application, *Energy Materials* 55 (10): 4213–4224.

15 Thangavel, R., Kannan, A.G., Ponraj, R. et al. (2018). Nitrogen- and sulfur-enriched porous carbon from waste watermelon seeds for high-energy, high-temperature green Ultracapacitors. *New Journal of Chemistry-Royal Society of Chemistry* 6 (36): 17751–17762.

16 Haridas, A.K., Jeon, J., Heo, J. et al. (2019) *In-situ* construction of iron sulfide nanoparticle loaded graphitic carbon capsules from waste biomass for sustainable lithium-ion storage. *ACS Sustainable Chemistry and Engineeing* 7: 6870–6879.

17 Scaglia, B. and Adani, F. (2008) An index for quantifying the aerobic reactivity of municipal solid wastes and derived waste products. *Science of the Total Environment* 394 (1): 183–191.

18 Obil, F.O., Ugwuishiwu, B.O., Nwakaire, J.N. (2016) Agricultural waste concept, generation, utilization and management. *Nigerian Journal of Technology* 35 (4): 957–964.

19 Rabiu, Z., Maigari, F.U., Lawan, U. et al. (2018). Pineapple waste utilization as a sustainable means of waste management. In *Sustainable Technologies for the Management of Agricultural Wastes*, pp. 143–154. Singapore: Springer.

20 Faresa, O., AL-Oqlab, F.M., and Hayajneh, M.T. (2019) Dielectric relaxation of mediterranean lignocellulosic fibers for sustainable functional biomaterials. *Materials Chemistry and Physics* 229: 174–182.

21 Sekar, S., Lee, Y., Kim, D.Y. et al. (2019) Substantial LIB anode performance of graphitic carbon nanoflakes derived from biomass green-tea waste, edn. *Nanomaterials, MDPI* 9: 1–10.

22 Gandla, D., Chen, H., and Tan, D.Q. (2020) Mesoporous structure favorable for high voltage and high energy supercapacitor. *Materials Research Express* 7: 1–13.

23 Moro, G., Bottari, F., Van Loon, J. et al. (2019) Disposable electrodes from waste materials and renewable sources for (bio) electroanalytical applications. *Biosensors and Bioelectronics* 146: 1–17.

24 Kingsakklang, S., Roddecha, S., and Sriariyanan, M. (2019) The interconnected open-channel highly porous carbon material derived from pineapple leaf fibers as a sustainable electrode material for electrochemical energy storage devices. *Trans Tech Publications, Switzerland: Materials Science and Technology X* 798: 97–104.

25 Na, R., Wang, X., Lu, N. et al. (2018) Novel egg white gel polymer electrolyte and a green solid-state supercapacitor derived from the egg and rice waste. *Journal of International Society of Electrochemistry* 274: 316–325.

26 Jung, S.H., Myung, Y.S., Das, G.S. et al. (2019) Carbon Nano-onions from waste oil for pplication in energy storage devices. *New Journal of Chemistry-Royal Society of Chemistry* 44 (18): 7369–7375.

27 Fu, F., Yang, D., Zhang, W. et al. (2019) Green self-assembly synthesis of porous lignin-derived carbon quasi-nanosheets for high-performance supercapacitors. *Chemical Engineering Journal* 392: 123721.

28 Berhanu, T.M., Ashokkumar, M., Murali, R. et al. (2018) Bi-functional iron embedded carbon nanostructures from collagen waste for photocatalysis and Li-ion battery applications: a waste to wealth approach. *Journal of Cleaner Production* 210: 190–199.

29 Karnan, M., Subramani, K., Sudhan, N. et al. (2016). Aloe vera derived activated high-surface-area carbon for flexible and high-energy supercapacitors. *ACS Applied Materials & Interfaces* 8 (51): 35191–35202.

30 Tamilselvi, R., Ramesh, M., Lekshmi, G.S. et al. (2020). Graphene oxide–based supercapacitors from agricultural wastes: a step to mass production of highly efficient electrodes for electrical transportation systems. *Renewable Energy* 151: 731–739.

31 Omokafe, S.M., Adeniyi, A.A., Igbafen, E.O. et al. (2020). Fabrication of activated carbon from coconut shells and its electrochemical properties for supercapacitors. *International Journal of Electrochemical Science* 15: 10854–10865.

32 Mi, J., Wang, X.R., Fan, R.J. et al. (2012). Coconut-shell-based porous carbons with a tunable micro/mesopore ratio for high-performance supercapacitors. *Energy & Fuels* 26: 5323.

33 Sun, L., Tian, C., Li, M, et al. (2013). From coconut shell to porous graphene-like nanosheets for high-power supercapacitors. *Journal of Materials Chemistry A* 1: 6463.

34 Jain, A., Aravindan, V., Jayaraman, S. et al. (2013). Activated carbons derived from coconut shells as high energy density cathode material for Li-ion capacitors. *Scientific Reports* 3: 4.

35 Sun, K., Leng, C.-Y., Jiang, J.-C. et al. (2018). Microporous activated carbons from coconut shells produced by self-activation using the pyrolysis gases produced from them, that have an excellent electric double layer performance. *Carbon* 130: 844.

36 Yu, K., Li, J., Qi, H. et al. (2018). High-capacity activated carbon anode material for lithium-ion batteries prepared from rice husk by a facile method. *Diamond and Related Materials* 86: 139–145.

37 Xu, H., Gao, B., Cao, H. et al. (2014). Nanoporous activated carbon derived from rice husk for high performance supercapacitor. *Journal of Nanomaterials* 1: 1–7.

38 Sankar, S., Lee, H., Jung, H. et al. (2017). Ultrathin graphene nanosheets derived from rice husks for sustainable supercapacitor electrodes. *New Journal of Chemistry* 41 (22): 13792–13797.

39 Le Van, K. and Thi, T.T. (2014). Activated carbon derived from rice husk by NaOH activation and its application in supercapacitor. *Progress in Natural Science: Materials International* 24 (3): 191–198.

40 Rong, C., Chen, S., Han, J. et al. (2015). Hybrid supercapacitors integrated rice husk based activated carbon with $LiMn_2O_4$. *Journal of Renewable and Sustainable Energy* 7 (2): 023104.

41 Amri, A., Taslim, R., and Taer, E. (2020). The physical and electrochemical properties of activated carbon electrode derived from pineapple leaf waste for super capacitor applications. *Journal of Physics: Conference Series* 1655 (1): 012008.

42 Taer, E. and Taslim, R. (2020). A high potential of biomass leaves waste for porous activated carbon nanofiber/Nanosheet as electrode material of supercapacitor. *Journal of Physics: Conference Series* 1655 (1): 012007.

43 Sodtipinta, J., Ieosakulrat, C., Poonyayant, N. et al. (2017). Interconnected open-channel carbon nanosheets derived from pineapple leaf fiber as a sustainable active material for supercapacitors. *Industrial Crops and Products* 104: 13–20.

44 Taer, E., Apriwandi, A., Ningsih, Y.S. et al. (2019). Preparation of activated carbon electrode from pineapple crown waste for supercapacitor application. *International Journal of Electrochemical Science* 14: 2462–2475.

45 Taer, E., Agustino, A., Awitdrus, A. et al. (2021). The synthesis of carbon nanofiber derived from pineapple leaf fibers as a carbon electrode for supercapacitor application. *Journal of Electrochemical Energy Conversion and Storage* 18 (3): 1–8.

10

Computational Intelligence Techniques for Optimization in Networks

Ashu Gautam and Rashima Mahajan

MRIIRS Faridabad, Haryana, India

10.1 Introduction Focussing on Pedagogy of Impending Approach

Intelligent devices. Smartphones. Intelligent cars. Smart households. Smart towns. A better world! These concepts have been promoted for years. To date, many different and frequently disjointed research networks have explored the achievement of these objectives. In computing, a communication between a group of two or more devices forms a network. In practice, a network forms a number of different computer systems connected through physical and/or wireless connections. The scale can range from a single PC sharing out basic peripherals to massive data centers located throughout the world, to the Internet itself. Irrespective of scope, all networks allow computers and/or entities to share information and resources.

MANET (Mobile Ad Hoc Network) is a wireless network decentralized type. Since it does not rely on a pre-existing infrastructure, such as wired network routers or wireless network access points, the network is ad hoc. Instead, each node contributes to the routing by transmitting data to other nodes, so that the data transmitted by the nodes is determined dynamically based on network connectivity and the routing algorithm used.

Wireless sensor networks (WSN) are a new category of networked systems with strong computer and energy resources and an ad hoc operating environment as shown in Figure 10.1. When wireless sensor networks are deployed in a hostile environment, safety is extremely important because they are prone to various kinds of malicious attacks [1]. The WSN comprises of hundreds or thousands of self-organizing, low-power, low-cost wireless nodes and is used in a variety of applications, including military sensing and tracking, environmental monitoring, and disaster management, etc.

A Cyber Physical System (CPS) consists of a collection of devices that interact and communicate with each other. It integrates computer and communication, and control and monitoring techniques. Various CPS applications can be found in almost all areas of human life, such as manufacturing systems, intelligent grids, robotics, transport systems, medical devices, the military, home networks, and intelligent buildings, etc.

Smart and Sustainable Approaches for Optimizing Performance of Wireless Networks: Real-time Applications, First Edition.
Edited by Sherin Zafar, Mohd Abdul Ahad, Syed Imran Ali, Deepa Mehta, and M. Afshar Alam.
© 2022 John Wiley & Sons Ltd. Published 2022 by John Wiley & Sons Ltd.

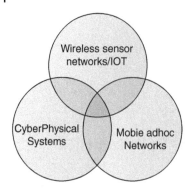

Figure 10.1 Various complex networks for smart world.

Securing CPS is non-trivial and requires techniques and understanding that encompass a variety of computer and physical components. This diversity is often exploited by attackers: a spoiling of expertise by researchers and/or developers can mean that vulnerabilities are overlooked, introduced, or interoperated in ways that are not anticipated.

By using CPSs, the synergy of computational and physical network components leading to the Internet of Things, Data, and Services was made possible. CPS engineering promises a wide range of fields from healthcare, manufacturing, and transport, to aerospace and warfare, to impact system condition monitoring. CPS for applications in environmental monitoring completely transforms human-to-human, human-to-machine, and machine-to-machine interactions with Internet Cloud usage [12]. A recent trend is to benefit from the fusion of virtual networking and physical performance in order to perform all conventional and complex sensing and communication tasks reliably.

10.1.1 Security Challenges in Networks

With the gradual combination of the Internet and social life, the Internet changes how people learn and work, but also exposes us to increasingly serious threats to security. The identification of various network attacks, especially unprecedented attacks, is a key issue that needs to be resolved urgently. Cybersecurity is responsible for technology and processes designed to protect computers, networks, programs, and data against attacks and unauthorized access, alteration, or destruction [4]. A system of network security consists of a network security system and a system of PC security. Firewalls, antivirus, code, and intrusion detection systems (IDS) are included in each system.

When developing a secure network, the following considerations must be taken into account:

- Accessibility: authorized users have access to and from a specific network.
- Privacy: network information remains private, disclosure should not be easily available.
- Authentication: makes sure that the users of the network are the person they claim to be.
- Integrity: Ensures that the message in transit has not been changed; the information must be the same as it is sent.
- Non-repudiation: makes sure that the user does not reject the use of the network.

Safety needs to be incorporated in order to design a completely safe WSN into every node of the system. This is due to the possibility that an implemented component can easily become an attack point without any security. This means that safety must cover every aspect of the design of a WSN application that requires a high level of security [5]. In recent years, wireless sensor systems have received a lot of attention. The severe energy constraints and the demanding deployment of WSNs make computer security more challenging than conventional networks for these systems. Components designed safely can easily become an attack point. It is therefore essential to integrate security into every component to pervade security and privacy into every aspect of the design.

A few general research challenges are discussed as follows:

a) Real-Time System Abstraction: given the massive amounts of actuators and sensors and computer-based information exchange of various information classes, new frameworks that allow abstracting salient system options in real time should be developed [14]. For example, network topologies for CPSs may be dynamically modified according to the physical environment.
b) Robustness, Security, and Safety: unlike logical computations in cyber systems, interactions with the physical world inevitably exhibit particular levels of uncertainty because of issues, such as environmental randomness, errors on physical devices, and possibility of security threats.
c) Hybrid System Control and Modeling: cyber and physical systems differ in that the former evolves continuously in real time, whereas the latter changes consistently with discreet logic. Hybrid systems and control techniques, which include cyber and physical elements, should be developed for CPSs.
d) Control over Networks: the implementation and design of the CPS network control pose many issues, such as time- and event-driven computations, transmission failures, time-varying delays, and system reconfigurations.
e) Sensor–Actuator Networks: WSNs have been extensively studied in the last decade. However, wireless sensor–actuator networks have been rarely explored, especially from the CPS perspective.
f) Validation and Verification: hardware and software, OS, and middleware components must undergo complete testing and integrative verification to ensure that the overall requirements of CPS are satisfied.
g) Scheduling Co-Design and Control: scheduling and control co-design is a well-studied field in embedded systems and real-time communities. Various aspects of co-design have been reconsidered with the emergence of CPSs.

10.1.2 Attacks Vulnerability in Complex Networks

When considering network security, it must be stressed that the whole network should remain secure. Network security does not only concern the security in the computers at each end of the communication chain. Network security is constantly evolving, due to traffic growth, usage trends, and the ever-changing threats to the landscape [15, 16]. For example, the widespread adoption of cloud computing, social networking, and bring-your-own-device (BYOD) programs are introducing new challenges and threats to an already complex network.

Due to the unavailability of the centralized system, MANETs are exposed to different network layer attacks. Worm Hole, Black Hole, Gray Hole, Byzantine, and Sybil Attacks are some of the examples of network layer attacks, which demolish network topology, resulting in data loss and network deprivation. In the Black Hole Attack, a node proclaims itself as having closest paths to all the destinations. This node absorbs all the data packets of networks by exploiting the routing protocol, thus degrading network performance.

Eavesdropping refers to the attack that adversely can intercept any information communicated by the system [7]. It is called a passive attack because the attacker does not interfere with the working of the system and simply observes its operation. CPS is particularly susceptible to eavesdropping through traffic analysis, such as intercepting the monitoring data transferred in sensor networks collected through monitoring. Eavesdropping also violates user's privacy. such as a patient's personal health status data transferred in the system. In Figure 10.2, attack 5 can represent the eavesdropping attacks on data aggregation processes; attack 8 can represent the tapping of controller demands.

A Compromised-Key Attack is a secret code which is necessary to interpret secure information. Once an attacker obtains a key, then the key is considered a compromised key [8]. An attacker can gain access to a secured communication without the perception of the sender or receiver by using the compromised key. The attacker can decrypt or modify data by the compromising key, and try to use the compromised key to compute additional keys, which could allow the attacker access to other secured communications or resources.

Man-in-the-Middle Attack is where false messages are sent to the operator, and can take the form of a false negative or a false positive. This may cause the operator to take an action, such as flipping a breaker, when it is not required, or it may cause the operator to think everything is fine and not take an action when an action is required.

Denial-of-Service Attack Denial of Service (DoS) attack [9] is one of the network attacks that prevent the legitimate traffics or requests for network resources from being processed

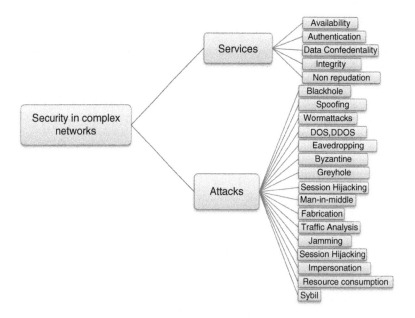

Figure 10.2 Security aspects in complex networks.

or responded to by the system. This type of attack typically transmits a large amount of data to the network to make busy handling of the information so that usual services cannot be provided. Thus, the denial-of-service attack prevents usual work or use of the system. After gaining access to the network of cyber physical systems, the invader can always do any of the following:

- Flood a controller or the complete sensor network with traffic until a shutdown occurs due to the overload.
- Send invalid information to controller or system networks, which causes abnormal termination or behavior of the services.
- Block traffic, which ends in a loss of access to network assets by authorized elements in the system.

This introduction has focussed on concepts of networks and attacks. In Section 10.1, the relevant analysis has been done on the various research carried out for tackling security in various types of networks, e.g., WSN, IoT (Internet of Things) [10], CPS, etc. Section 10.2 focusses on the various protocols for networks, advantages; and disadvantages of different types of protocols, and further different types of existing Hybrid Protocols are compared. Section 10.4 emphasizes the computational intelligence for optimizing the various hybrid protocols for securing different networks. Section 10.5 has provided the various objectives of the research to be conducted.

10.2 Relevant Analysis

Relevant Analysis attends to the various investigations carried out in the field of security for various types of attacks in different types of complex networks.

Zhou et al. (2018) presents a Blackhole Attack simulation based on AODV routing protocol and network performance analysis under attack, realized using an NS-3 network simulator [25]. They simulate network states with different moving speeds, different numbers of communication nodes, and malicious nodes. In order to better simulate the complications and randomness of real networks, their simulation scheme introduces random factors to achieve the randomness of communication objects. The experimental results show the impact of Blackhole Attacks on QoS parameters, such as packet loss rate (PLR), average throughput (ATP), and average end-to-end delay (EED), which promotes the research of MANET safety measure.

Xiao et al. (2018) investigates the attack model for IoT structures, and reviews the IoT security solutions based on machine learning methods including supervised learning, unsupervised learning, and reinforcement learning [13]. Machine learning-based IoT authentication, access control, secure offloading, and malware detection schemes to protect data privacy are focussed upon. It is emphasized that backup security mechanisms have to be provided to protect IoT systems from the exploration stage in the learning process.

Chelli et al. tend to outline the major aspects of WSNs security and discuss some security attacks and their classification mechanisms [1]. The sensitive nature of collected information collected makes security in these special networks of important concern. Owing to the hostile nature of their preparation environments, the wireless medium, and the constrained nature of resources on the tiny sensor devices used in such networks, security poses more severe challenges compared to the traditional networks. Such attacks to any a part of the hardware or software package might cause critical damage to those networks. Indeed, the growth of effective and efficient defense mechanisms against these attacks must be addressed at every stage of the system plan.

A new machine learning based strategy to detect spoofing attacks in WSNs has been proposed. Based on comprehensive analytical models for the mobile radio channel, the proposed algorithm conglomerates two classifiers to process and analyze the instant samples of received signal strength to detect attacks. The algorithm is enhanced for scenarios where the legitimate node and the attacking node are at the same distance or are very close to each other relative to the landmark, which is the worst-case scenario. The results show that the proposed strategy improves the performance of attack detection in about 10% of cases, regarding a similar approach in the literature.

Deng et al. (2019) [2] focusses on the current problems of IoT in network security, and points out the necessity of intrusion detection. Several kinds of intrusion detection tools are discussed, and its application on IoT architecture is investigated. The application of different intrusion detection technologies are analyzed, and made a prospect of the next phase of research. Using data-mining and machine-learning methods to study network intrusion technology has become a hot issue. A single class feature or a detection model make it very difficult to develop the detection rate of network intrusion detection. The performance of the proposed model is validated through the public databases.

Wu et al. (2017) emphasizes the effect of degree heterogeneity on structural vulnerability of interdependent networks when they suffer targeted attacks [4]. Firstly, they construct an interdependent system model composed of two network components. By adjusting a parameter, the extent of degree heterogeneity of each network can be controlled.

The role of the IoT devices in healthcare and the role of IT in handling the massive volume of high security patient's medical data have been further explored. All the physical objects will work seamlessly with machine-to-machine and human-to-machine interfaces. This level of interconnection is a bonus for healthcare, where health influencing factors, both internal and external to the human body, can be analyzed based on this model.

Software has been set-up as an innovative network architecture which provides network control through software logic. It decouples control and data plane to tailor the network according to the user's needs. OpenFlow, a standardized network protocol, acts as an interface between controllers and switches. The softwarized controllers are extremely susceptible to Distributed Denial of Service (DDos) attacks. The proposed detection system uses an unsupervised stochastic Restricted Boltzmann Machine algorithm to self-learn the reliable network metrics. This algorithm detects and classifies the type of DDoS attacks in a dynamic network environment by framing a new context. The results prove that an RBM-based DDoS detection system achieves higher accuracy than existing methods.

Focus has been on the network layer, which is the most significant part of realizing the IoT environment, and a recommended well-designed network architecture for the IoT, such as IoT network management, connection management, grouping, and privacy.

The most often-mentioned machine learning algorithms or techniques in the primary studies are: Support Vector Machine; Artificial Neural Network; Naïve Bayes; Decision Tree; kNN; k-Mean; Random Forest; and Deep Learning. All other algorithms and techniques were discussed only in one paper. These outcomes show that Support Vector Machine (SVM) is the most used machine learning method for IoT security. This method is used in the major studies, both for intrusion detection and for authentication. The SVM is a supervised learning technique that generates input output mapping functions (either a classification function or a regression function) from a set of labeled training data (Arel et al. 2010).

Mehdi et al. (2014) discuss some advanced machine learning techniques, namely Deep Learning (DL), to simplify the analytics and learning in the IoT domain [14]. The potential of using emerging DL techniques for IoT data analytics have been discussed, and its promises and challenges are introduced. Several foremost architectures of DL are used in the framework of IoT applications followed by several open-source outlines for development of DL architectures.

Sridevi et al. (2017) have proposed a new structure called the pigeon principle and hybrid encryption to sense and overcome the watchdog weakness [23]. Partial dropping, Collusion, and Ambiguous collisions are the three main problems in MANETs, which have not been overcome by other schemes. A combination of hybrid encryption schemes as well as the pigeon principal overcomes all the six weakness of watchdog, even where a Receiver collision have been identified and detected by EAACK. The pigeon principle and hybrid encryption overcome all the six weakness of watchdog and give good results. The results still have to complete energy harvesting, and to improve the rate of data transmission further they be should considered.

Govindasamy and Punniakody (2017) analyze the performance of the Ad hoc On-Demand Distance Vector (AODV) routing protocol, Optimized Link State Routing (OLSR), and Zone Routing Protocol (ZRP) in IEEE 802.15.4-based WSN in the presence of wormhole attacks [24]. The metrics used to study the performance of the reactive routing protocol (AODV), proactive (OLSR), and hybrid routing protocol (ZRP) in WSNs are throughput, average end-to-end delay, Packet Delivery Ratio (PDR), and total energy consumption of sensor network. The simulation is done by using qualnet simulator 5.0. ZRP has highest throughput and OLSR has the lowest average end-to-end delay, due to trade-off between the throughput and energy consumption of ZRP and OLSR. AODV has better overall performance in terms of throughput and average end-to-end delay than the other two routing protocols comparatively. Finally, it is concluded that the routing protocols AODV, OLSR, and ZRP are vulnerable to wormhole attacks. Hence, it is required to design a secure routing protocol or energy-efficient intrusion detection system to prevent the security threats in resource constrained WSNs in the future.

10.3 Broad Area of Research

The enormous growth of the Internet has led to greater awareness and attention in network security and routing. Four top-level Internet security needs have been recognized: end-system security, end-to-end security, secure service quality (QoS), [6] and secure network infrastructure. End system security is typically achieved by setting up a single host security wall, but fresh firewall technology has been used to enlarge "defense perimeter" for complete organizations by shielding the intranet with a small number of firewall

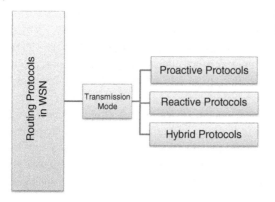

Figure 10.3 Routing protocols in wireless sensor networks.

systems. End-to-end security is well appropriate to provide confidentiality, authentication, and integrity. Secure QoS presents a number of new security challenges: authentication and authorization of users who require expensive network resources, both to prevent resource holdup and to prevent denial of service due to unauthorized traffic, etc. The three overhead areas are being studied actively and many of them are used on the Internet today. There is another important security class that has difficulties that alarm the network infrastructure. Network operation depends upon management and control protocols to organize and operate the network infrastructure, plus routers, DNS servers, etc. An attack on the system infrastructure can cause denial of service from the point of view of the user, but for a network manager or operator, the attacker is taking benefit of the lack of authenticity, integrity, and possibly privacy.

Figure 10.3 shows the latest routing protocols. These are intended to deal with simple network letdowns (e.g., links going up and down, nodes crashing, and restarting, etc.), and thus can show several vulnerabilities when facing a strategically located intruder.

10.3.1 Routing Protocols

A. Proactive or Table-Driven Protocols

Proactive or table-driven protocols are routes that are well-defined in Table 10.1. As the network topology changes, routes are updated [9, 11]. This is achieved by sending a broadcast message to the entire network. In these routings, overhead is more due to maintaining up-to-date information and delay is less due to already defined routes in the table. Because they maintain routing information, even before it is required, they are called proactive. They are not suitable for large networks because they preserve node entry for each and every node in the routing table of every node. Examples of Proactive routing protocols are: DSDV, OLSR, and WRP. OLSR [14] is enhanced for mobile ad hoc networks, which can also be used for WSN applications. This protocol enables efficient flooding of control messages throughout the network by using selected nodes called Multipoint Relays (MRPs). The problem of flooding the network with control messages is overcome by the MPR nodes. MRPs are selected by each node and are used to forward control messages resulting in a distributed operation of the protocol. Additionally, a node constantly maintains routes to all destinations in the network, thus making the protocol suitable for traffic pattern that is random and sporadic. Furthermore, its

Table 10.1 Hybrid/secured routing protocols.

Protocols	Parameters				
	Multiple routes	Routes information stored in	Route metric	Advantages	Disadvantages
ZRP [20]	No	Intrazone and interzone tables	Shortest path	Reduced transmissions	Overlapping zones
ZHLS	Yes	Intrazone and interzone tables	Shortest path	Low control overhead	Static zonemap required
DST	Yes	Route tables	Forwarding using tree table	Reduced transmission	Root node
DDR	Yes	Intrazone and interzone tables	Stable routing	No zone coordinator	Neighbors may become bottleneck.
Anthocnet [30]	Yes	Pheromone and neighbor table	Using pheromone	High delivery ration, robustness, adaptable, scalable.	No centralized processor to guide the system toward good solution
HCR [31]	Yes	Host clustered list, member node list, neighbor cluster list, global cluster table	ELECT_CH, INTRA_CB, INTER_CB	Head set approach, less overhead delay	Cluster head changes are expensive. Cluster head CH, selection. Imbalance of energy consumption

proactive nature makes OLSR suitable for networks where communicating pairs change over time. The protocol is an optimization of a classical link state routing algorithm and uses the concept of MultiPoint relays (MPRs). This protocol is more favorable for traffic patterns where a large subset of nodes is communicating with another large subset of nodes, and where the source and destination pairs are changing over time. This protocol is predominantly suited for large and dense networks.

B. Reactive or On-Demand Routing Protocols

To send the packet reactive protocols from source to destination, search the route on demand. Reactive protocols have fewer overheads and delays [2]. The route is discovered by flooding the route request packets (RREQ) over the entire network [5]. AODV, DSR, and TORA are the protocols used to discover the route. AODV uses an on-demand approach to find routes, i.e., a route is only set when a source node is required to transmit data packets. AODV has two basic routes, discovery and maintenance operations. Route REQuest (RREQ), Route REPly (RREP), and Route ERRor (RERR) messages are used by AODV to locate and maintain routes. When a source node requires a route to the destination node for which it does not have a route, it transmits an RREQ packet to the network. An RREQ packet includes IP source, IP address, source sequence number, destination sequence number, request ID, and hop count. If a node receives a route request with the same source address and requests ID fields as in previous route request packets, the packet will be rejected. Otherwise, it will examine whether there is an entry for the destination address in its routing table. If that address is present, the destination sequence number in the table is compared to the destination in its routing table and if it cannot reach the destination via that route, it increases the destination sequence number and sends a route request. The destination sequence number therefore indicates the freshness of the route. In route maintenance, when a link breakage in an active route is detected, the node notifies this link breakage by sending an RERR message to the source node. The source node will reinitiate the route discovery process if it still has data to send.

C. Hybrid Routing Protocol

Hybrid routing protocols combine some of the features of proactive protocols and some of the features of reactive protocols to achieve better results. These protocols solve the routing overhead of proactive routing protocols and the delay of reactive routing protocols. Networks are separated into zones and suitable for large networks [17]. Hybrid routing protocols use reactive routing protocols for route discovery. It uses proactive routing protocols for table maintenance. Examples of Hybrid routing protocols which are used for route discovery are: ZRP, HCR, and ANTHOCNET: ZRP is a wireless hybrid routing protocol. which consists of proactive and reactive routing protocols for data communication over the network.

10.3.2 Hybrid Protocols

Hybrid protocols are a new generation of protocols, a combination of both proactive and reactive protocols and Table 10.1 shows various categories. They have the potential to provide higher scalability than pure reactive or proactive protocols. Another novelty of hybrid routing protocols is that they attempt to eliminate single point of failures and creating bottleneck nodes in the network [21–23, 29].

How to organize the network according to network parameters is the challenge for all hybrid routing protocols [24–26]. These protocols can offer a greater scalability than the other two classes. They try to reduce the number of re-broadcast nodes by defining an area that allows the nodes to function together. The best or most appropriate nodes can then be used to discover the route.

10.4 Problem Identification

The various concepts studied in the literature and with the increasingly in-depth integration of the Internet and social life, the Internet is not only changing how people learn and work, but it also exposes us to increasingly serious security threats. How to identify various network attacks, particularly not previously seen attacks, is a key issue to be solved urgently. Computational intelligence characteristics (CI) [27] systems, such as adaptation, fault tolerance, high computational speed, and error resilience in the face of noisy information, meet the requirements of building a good intrusion detection model. Computational intelligence characteristics (CI) [27] systems, such as adaptation, fault tolerance, high computational speed, and error resistance to noisy information, meet the requirements of a good intrusion detection model. Here we provide an overview of the research progress in applying CI methods to the problem of handling security. The scope of this review will encompass core methods of CI, including artificial neural networks, fuzzy systems, evolutionary computation, artificial immune systems, swarm intelligence, and soft computing. The research contributions in each field are systematically summarized and compared, so that we can clearly define existing research challenges and highlight promising new directions for research. The results of this review should provide valuable insights into the current safety literature and be a good source for anyone interested in applying CI security approaches in hybrid routing protocols [30].

10.5 Objectives of the Study

The objectives of study are as follows:

(i) To focus on various types of attacks and the security vulnerabilities of various networks and also to study limitations of various traditional security approaches in networks.
(ii) To provide in-depth analysis of various attacks that tend to hamper security of networks like WSNs and IoT Networks.
(iii) Utilize computational Intelligence-based algorithms to improve security and performance of networks.
(iv) To compare traditional security approaches and their performances of the proposed algorithm.
(v) To validate the proposed approach by using a specified simulator.
(vi) To deliver better performance of proposed secured algorithm optimized through various computational strategies.

10.6 Methodology to be Adopted

The broad of area of research and objectives of study has focussed upon various networks, their security vulnerabilities attacks, and computational approaches that are implied in the network. The proposed algorithm in this synopsis will focus on application of the Computational Intelligence algorithm to optimal network performance. Figure 10.4 shows a flowchart explaining the methodology.

With rising interest toward security owing to the increase in Internet usage in a smart world, there arises a need for efficient and secure optimization of security in hybrid networks. CI is an increasingly popular sub-discipline of AI and focuses on big data. CI differs from the well-known artificial intelligence field of AI.

10.7 Proposed/Expected Outcome of the Research

It is planned to implement in the proposed approach, an efficient hybrid technique for networks algorithms with simulation and analysis with the help of existing frameworks for computational intelligence. We will review the core computational intelligence approaches that have been proposed to solve intrusion detection problems. The scope of the research

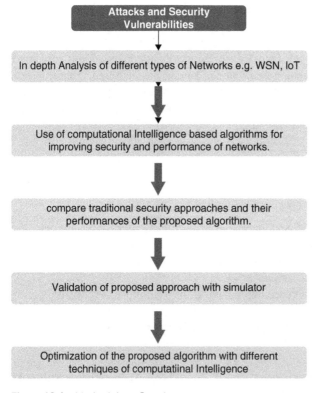

Figure 10.4 Methodology flowchart.

the core methods of CI, unlike the rest of the methods, has the synergistic power to intertwine the pros and cons of these methods in such a way that their cons will be compensated for. Future improvement of the algorithm will be contingent on the results after simulation and analysis.

References

1 Chelli, K. (2015). Security issues in wireless sensor networks: Attacks and countermeasures. Proceedings of the World Congress on Engineering 1: 1–3. http://www.iaeng.org/publication/WCE2015/WCE2015_pp519-524.pdf

2 Azzawi, M.A., Hassan, R., and Bakar, K.A.A. (2016). A review on internet of things (IoT) in healthcare. *International Journal of Applied Engineering Research* 11 (20): 10216–10221. https://www.researchgate.net/profile/Mustafa_Azzawi/publication/309718253_A_Review_on_Internet_of_Things_IoT_in_Healthcare/links/581e46e008aea429b295c6eb/A-Review-on-Internet-of-Things-IoT-in-Healthcare.pdf

3 Alippi, C., D'Alto, V., Falchetto, M. et al. (2017). Detecting changes at the sensor level in cyber-physical systems: methodology and technological implementation. In 2017 International Joint Conference on Neural Networks (IJCNN), 1780–1786. IEEE. https://ieeexplore.ieee.org/abstract/document/7966066

4 Wu, Y., Sun, S., Wang, L. et al. (2017). Attack vulnerability of interdependent local-world networks: The effect of degree heterogeneity. In 43rd Annual Conference of the IEEE Industrial Electronics Society, IECON, 8763–8767. IEEE. https://ieeexplore.ieee.org/abstract/document/8217540

5 Wang, F., Vetter, B., and Wu, S.F. (1997). *Secure Routing Protocols: Theory and Practice.* Technical report, North Carolina State University. https://pdfs.semanticscholar.org/6fb9/80832510e40cae925df94e9a796c62816b83.pdf

6 Prabha, R., Krishnaveni, M., Manjula, S.H. et al. (2015). QoS aware trust metric based framework for wireless sensor networks. *Procedia Computer Science* 48: 373–380. https://www.sciencedirect.com/science/article/pii/S187705091500705X

7 Deng, L., Li, D., Yao, X. et al. (2019). Mobile network intrusion detection for IoT system based on transfer learning algorithm. *Cluster Computing* 22 (4): 9889–9904. https://link.springer.com/article/10.1007/s10586-018-1847-2

8 Arel, I., Rose, D.C., and Karnowski, T.P. (2010). Deep machine learning-a new frontier in artificial intelligence research. *IEEE Computational Intelligence Magazine*, 5 (4): 13–18. https://www.researchgate.net/profile/Thomas_Karnowski/publication/224183837_Deep_Machine_Learning_-_A_New_Frontier_in_Artificial_Intelligence_Research_Research_Frontier/links/570bb11b08aee06603519bf8/Deep-Machine-Learning-A-New-Frontier-in-Artificial-Intelligence-Research-Research-Frontier.pdf

9 Li, G., Yan, Z., and Fu, Y. (2018). A study and simulation research of blackhole attack on mobile ad hoc network. 2018 IEEE Conference on Communications and Network Security (CNS), 1–6. IEEE https://ieeexplore.ieee.org/abstract/document/8433148.

10 Yeh, K.H. (2016). A secure IoT-based healthcare system with body sensor networks. *IEEE Access* 4: 10288–10299. https://ieeexplore.ieee.org/abstract/document/7779108

11 Brill, C. and Nash, T. (2017). A comparative analysis of MANET routing protocols through simulation. In 12th International Conference for Internet Technology and Secured Transactions (ICITST), 244–247. IEEE. https://ieeexplore.ieee.org/document/8356392

12 Zeng, D., Guo, S., and Cheng, Z. (2011). The web of things: a survey. *JCM*, 6 (6): 424–438. http://citeseerx.ist.psu.edu/viewdoc/download?doi=10.1.1.348.4644&rep=rep1&type=pdf#page=6

13 Xiao, L., Wan, X., Lu, X. et al. (2018). IoT Security Techniques Based on Machine Learning. arXiv preprint arXiv 1801: 06275. https://arxiv.org/abs/1801.06275

14 Mehdi, M., Al-Fuqaha, A., Sorour, S. et al. (2018). Deep learning for IoT big data and streaming analytics: A survey. *IEEE Communication Surveys and Tutorials* 20 (4): 2923–2960. https://ieeexplore.ieee.org/abstract/document/8373692

15 Mahdavinejad, M.S., Rezvan, M., Barekatain, M. et al. (2018). Machine learning for Internet of Things data analysis: a survey. *Digital Communications and Networks* 4 (3): 161–175. https://www.sciencedirect.com/science/article/pii/S235286481730247X

16 Zhao, Z., Ding, W., Wang, J. et al. (2015). A hybrid processing system for large-scale traffic sensor data. *IEEE Access* 3: 2341–2351. https://ieeexplore.ieee.org/abstract/document/7328238

17 Whitmore, A., Agarwal, A., and Da Xu, L. (2015). The Internet of Things – a survey of topics and trends. *Information Systems Frontiers* 17 (2): 261–274. https://link.springer.com/article/10.1007/s10796-014-9489-2

18 Baljinder, S. and Tejpreet, S. (2018). Modified EGSR for network security. *Pure and Applied Mathematics* 118 (16): 121–138. ISSN: 1311-8080 (printed version) Singh, B. and Singh, T. (2017) Modified EGSR for network security. In 51st Annual Conference on Information Sciences and Systems (CISS), 1–7. IEEE. https://ieeexplore.ieee.org/abstract/document/7926098

19 Kaur, H. and Singh J. (2017). Comparative analysis of hybrid routing protocols ZRP, HCR and ANTHOCNET. *International Journal of Advanced Research in Computer Science*, 8 (7). https://search.proquest.com/openview/61cdec56b2101f80a262f526277ce90b/1?pq-origsite=gscholar&cbl=1606379

20 Sengupta, A., Sengupta, D., and Das, A. (2017). Designing an enhanced ZRP algorithm for MANET and simulation using OPNET. In Third International Conference on Research in Computational Intelligence and Communication Networks (ICRCICN), 153–156. IEEE. https://ieeexplore.ieee.org/abstract/document/8234498

21 Alvarez, J., Maag, S., and Zaidi, F. (2017). Monitoring dynamic mobile ad-hoc networks: A fully Distributed Hybrid Architecture. In IEEE 31st International Conference on Advanced Information Networking and Applications (AINA), 407–414. IEEE. https://ieeexplore.ieee.org/abstract/document/7920938

22 Kaur, J. and Singh, G. (2017). Hybrid AODV algorithm for path establishment in MANET using bio inspired techniques. In 2017 IEEE International Conference on Power, Control, Signals and Instrumentation Engineering (ICPCSI), 823–829. IEEE. https://ieeexplore.ieee.org/abstract/document/8391828

23 Sridevi, N. and Nagarajan, V. (2017). Enhanced secure wireless communication in MANETs. In International Conference on Communication and Signal Processing (ICCSP), 2256–2261. IEEE. https://ieeexplore.ieee.org/abstract/document/8286818

24 Govindasamy, J. and Punniakody, S. (2017). A comparative study of reactive, proactive and hybrid routing protocol in wireless sensor network under wormhole attack. *Journal of Electrical Systems and Information Technology* 5 (3): 735–744. https://www.sciencedirect.com/science/article/pii/S2314717217300181

25 Zhou, W., Jia, Y., Peng, A. et al. (2018). The effect of IoT new features on security and privacy: new threats, existing solutions, and challenges yet to be solved. *Journal of the IEEE Internet of Things* 6 (2): 1606–1616. https://ieeexplore.ieee.org/abstract/document/8386824

26 Sanzgiri, K., Dahill, B., Levine, B.N. et al. (2002). A secure routing protocol for ad hoc networks. In Proceedings of the 10th IEEE International Conference on Network Protocols, 78–87. IEEE. https://ieeexplore.ieee.org/abstract/document/1181388

27 Fu, Y. and Ding, Z. (2017). Hybrid channel access with CSMA/CA and SOTDMA to improve the performance of MANET. In 17th International Conference on Communication Technology (ICCT), 793–799. IEEE. https://ieeexplore.ieee.org/abstract/document/8359746

28 Vazhayil, A., Vinayakumar, R., and Soman, K.P. (2018). Comparative study of the detection of malicious URLs using shallow and deep networks. In 9th International Conference on Computing, Communication and Networking Technologies (ICCCNT), 1–6. IEEE. https://ieeexplore.ieee.org/abstract/document/8494159

29 Pratomo, B.A., Burnap, P., and Theodorakopoulos, G. (2018). Unsupervised approach for detecting low rate attacks on network traffic with autoencoder. In International Conference on Cyber Security and Protection of Digital Services (Cyber Security), 1–8. IEEE.https://ieeexplore.ieee.org/abstract/document/8560678

30 Meng, F., Fu, Y., Lou, F. et al. (2017). An effective network attack detection method based on Kernel PCA and LSTM-RNN. In International Conference on Computer Systems, Electronics and Control (ICCSEC), 568–572. IEEE. https://ieeexplore.ieee.org/abstract/document/8447022

31 Sokolov, A.N., Pyatnitsky, I.A., and Alabugin, S.K. (2018). Research of classical machine learning methods and deep learning models effectiveness in detecting anomalies of industrial control system. In Global Smart Industry Conference (GloSIC). 1–6. IEEE. https://ieeexplore.ieee.org/abstract/document/8570073

32 Spaulding, J. and Mohaisen, A. (2018). Defending Internet of Things against malicious domain names using D-FENS. In IEEE/ACM Symposium on Edge Computing (SEC), 387–392. IEEE. https://ieeexplore.ieee.org/abstract/document/8567696

11

R&D Export and ICT Regimes in India

Sana Zeba

Department of Computer Engineering, Jamia Hamdard, New Delhi, India

11.1   Introduction

Policy inerventions towards research and Development regime entrusts its varied technology culture in economic scenarios of every country, where exports in Research and Development (R&D) plays a considerable role in the country's growth profile and its various characteristic models of development. The international monetary fund defines the status of science, and its ratio within various interdisciplinary areas of technology and implementation.

It also includes biomedical and pharmaceutical sciences, as its commercial research criteria, along with patenting and copyright criteria that entails basics of research and innovative development in science, as accorded by central product classification. This ranges from tangible and intangible patenting and copyright system that are one of the key reported areas of the R&D export regime. Tangible pantenting lies under intellectual property and intangible takes leap ahead of software and product designing and development. The same plays a swift role in Foreign Direct Investment (FDI) exchange system of technology across international borders by Multinational Enterprises as well as national think tanks and their R&D investment regimes.

Ministry of Electronics and Information technology-MEITy, along with its various nationally located export units, plays an advantageous role in tapping the resource analysis for R&D export, with world R&D exports having grown from US$14 billion in 2000 to US$ 87 billion in 2010 and stood at US$ 173 billion in 2019.

As per various recorded information, USA is world leader in R&D exports and Switzerland, the smallest exporter, and according to the International Trade Centre, India is world's 8th largest country in R&D exports. R&D export depends on the FDI and R&D import services, where Indian FDI was only 0.8% until 2018, as compared to South Korea and China. In case of R&D imports Ireland and USA are the largest importers in today's world.

R&D in exports will be considered as an asset in forthcoming times, where there may be varied in sectors which India can put forward for export regimes including profound changes and development in science technology and innovation [1, 2].

In case of R&D driven software exports policy standouts, regimes have to be revisited in the context of the RBI compliances and policies of Export & Import Policy of Government

Smart and Sustainable Approaches for Optimizing Performance of Wireless Networks: Real-time Applications, First Edition.
Edited by Sherin Zafar, Mohd Abdul Ahad, Syed Imran Ali, Deepa Mehta, and M. Afshar Alam.

of India and export promotion regimes need to be revisited at MeiTy policy desk. There is a need to have a holistic view on effective policies in export of software-based technology systems and intellectual property, to attain insights into stakeholders and specialized departments in technology translation. Direction towards the end of knowledge driven arena of technology, can thus be blurred at the boundaries and hurdles coming ahead in technology development and translation.

Nowadays, disruptive technology is the key role player in each sector of research and its further development scenario, where its further integrative implementation is far more important than any advised or specifically devised hypothesis.

Sustainable and smart approach in information technology and its allied domains is a key research highlight in this ongoing tech-driven landscape. Disruptive technology in IT and IT-enabled areas highlights the areas where novel technologies play prominent roles in establishing superior pathways in competing with disruptive technologies. As technology is the non-linear tool available to humanity, this can always affect the fundamental changes on grounds of competitiveness and economic index opportunities. Technology is linked to economy and environmental development with all its best possible prospects, to promote the knowledge linked internalization and translation of knowledge products. Until 1970, technological innovation had its smart outcomes from limited resource of knowledge parks and throughput clusters, established at national and international level, occurring from national and international laboratories. In the early 1990s, the growth in Information and Communication Technology (ICT) was non-linear, following stagnation in technology breakdowns, the growth simulation up to 25–27% only, delivering up to $60 billion technology exports by 2010. This increase in technology exports is based on:

- Technology clusters development in the outreach of scientists and innovators
- Data assimilation and increase in quality of consolidated information
- Adoption of the innovative- and strategic-based startup opportunities and cost-effective manufacturing in IT systems.

The invention of the telegraph in 1837, the invention of the telephone in 1876, the development of shortwave radio by 1926, and then the more consistent high-frequency microwave radio in 1946, has permitted instantaneous long-distance communication, emancipating the channels for visual telecast, that has lead the way forward for satellite and space communication since 1957. The same has advanced the way forward for development of the 1970s mobile communication handset, that led to the discovery of the Internet and the worldwide web, growing at a rapid pace since 1980.

This emanative stride has proven its presence in far more developed and developing nations, proving its outreach in retail, banking, education, agriculture, and many more areas of human intelligence, including AI and machine-learning power systems, a new learning challenge.

The advent of web 3.0, the semantic web system is characterized by connective intelligentia, data concepts, and applications for various human services with (artificial) intelligence, including reading, execution, information personalization, machine commanding, cloud computing, and immersion. Here ICT becomes a chief intelligent prospect where various prospects of semantic web invention has put forth the quality learning and efficient program management via websites, blogs, forums, and other applications. Similarly, web 4.0 has created another major impact in machine learning throughout its human machine interface in symbiotic interaction relations, creating faster and much transparent quality web content. Cluster of web 5.0 is below the development platform, which will understand the emotional challenge of the web user providing the user-friendly interface, along with emotional uptakes.

ICT is the most primitive and most challenging technology that will bring advancement to every area of technology, wherever advancement and innovation will pave its advented way for user-machine interaction. Artificial intelligence has become an expert training tool in training various methodologies of decision-making and its allied networking science. Networking through technology integration and its stake implementation toward its large-scale translation initiated the way forward of smart system formation, to provide user-friendly solutions and thus socio-economic development. As Information and communication technology go hand-in-hand in every respect, to transform the imaginary scientific pool of expertise at the globalized platform to face up to infrastructure capabilities, is to revolutionize the support in decision-making.

11.2 Artificial Intelligence the Uptake of Infrastructure Development

One such development regime has been followed in the case of formation of smart cities, taking up one of the few robust and citizen centric technological outputs. The paradigm of artificial intelligence (AI) is revolutionizing this role by supporting intelligent and wireless decision-making by means of coordination, data delivery, analyzing data trends, and forecasting, quantifying the uncertainty and thus suggesting its course of action. Horizons of AI networking open its avenues using symbolic logics, artificial neural networking (which mimics biological systems), fuzzy systems, evolutionary computing, intelligent agents, and models of probabilistic theories. Such systems provide access to collect, transmit, and analyze the data for monitoring and operation. The methodology used to designing computer decision systems is based on application of AI techniques that combines fuzzy inference system with machine learning, including decision trees as major support tools (Figure 11.1).

One of the most sincere uses of AI in resource utilization, while making the informed decisions, is, making the smart city more relevant, efficient, and benchmarking the data processing techniques, by efficient connectivity and performance fabrication (Figures 11.2 and 11.3).

Regulations in data intelligence have created machine-learning impacts to be more fruitful, creating more reflective impacts in decision-making process of machine learning, AI, and Internet of Things, creating an impactful policy-making process. Development of software modules for decision support and data analytics have become important tools in enterprise information systems for decision-making and creating enterprise competition modules. Nowadays, AI is providing an efficient way for heterogenous device connections by enabling generation of a multitude of novel AI-based semantic IoT or AI-SI-IoT architectures. (Figure 11.4). One of the key enablers of SI-IoT stands large in developing smart

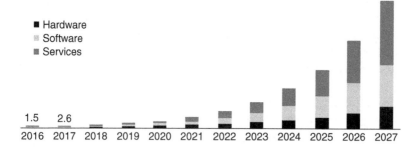

Figure 11.1 Market survey report for AI and ML use worldwide until 2027. Source: With permission of Grand View Research, Inc.

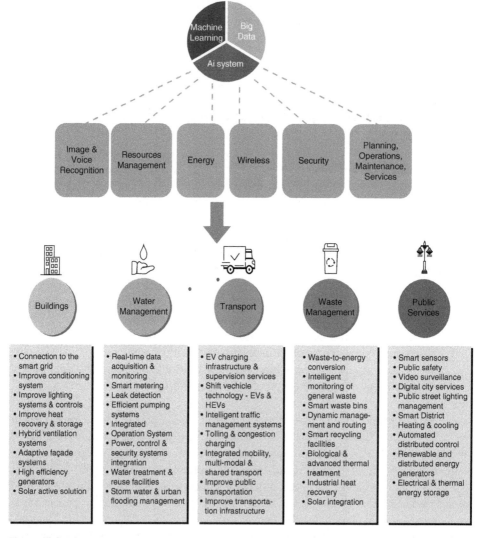

Figure 11.2 Smart energy management.

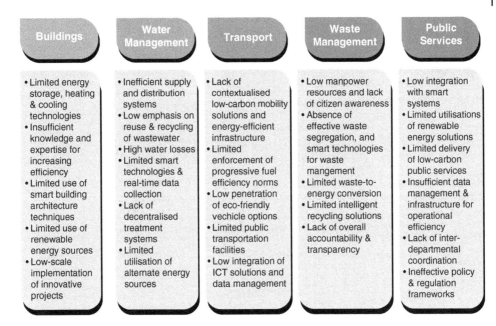

Figure 11.3 Challenges of smart energy management.

city infrastructure, where multi-meta data devices are being hybridized to generate large amounts of processed and unprocessed data to be used further for sustainable services.

In particular, efficient data analysis and management is powered and achieved by AI-enabled machine learning. Smart city in a broader concept covers many domains that has conglomerated many new technologically throughput solutions, in creating the urban infrastructure. This can be more user-friendly, and embeded with all basic information systems available on citizen centric dashboards, of virtual connectivity. Integrated sensors and local positioning systems in devices, that are being utilized for global connectivity modules, have provided catalytic solutions anticipating consumer's need.

Connective AI is key role player in making decision support systems viz. Group Decision Support System (DSS), Collaborative DSS, Adaptive DSS, Executive DSS, and augmented DSS, etc., assisted by a broad range of interactive data, models, and knowledge. Main aim of these decision support systems is to solve semi structured, ill structured, and unstructured problems (Figure 11.5). Features of AI used to support intelligent decision-making, nowadays has brought forward a scenario of well-developed decision support in developing global platforms in machine learning and artificial intelligence (Table 11.1).

In augmented DSS, there is an active and ongoing interaction between human and machine, with mutual information, involved in automated decision-making by machine with ML continuously. It is also becoming an important tool in biomedicine and bio-diagnostic tools, as in the case of diabetic retinopathy using queries and root cause analysis.

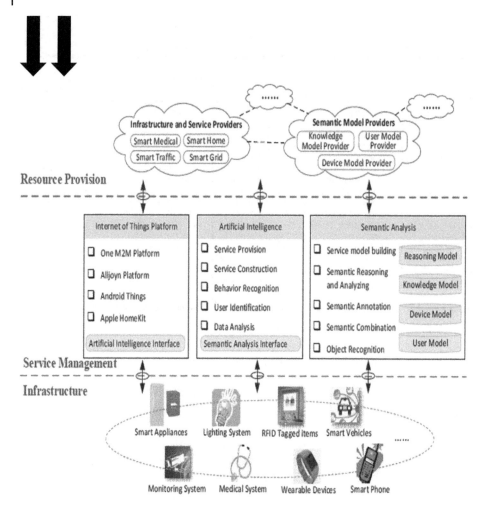

Figure 11.4 AI-SI-IoT architecture [1].

11.3 Future Analysis and Conclusion

Embedding AI within forthcoming technological scenarios has created a widened strive to be fill, in and to address many issues and challenges ahead of human centric technology domain. As per the data available, use of IoT technology has increased dramatically during the last few years, in amount of instrumentation installed worldwide, which is at 15.41 billion devices installed in 2015, which further is going to increase subsequently up to 50.11 billion in 2023, and 75.44 billion by 2025 (Figure 11.6). The infrastructural supply side intervention in IoT, AI, and Quantum technology will help to develop the connectivity-based R&D focussing on common supercomputing facilities at academic centers as well as inter-university-academia intelligence formation, as a more committed infrastructure, a future workforce (Figure 11.7).

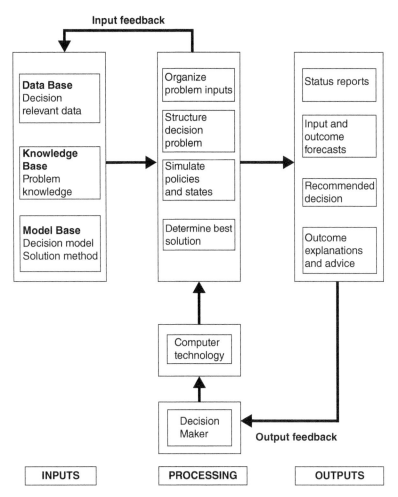

Figure 11.5 Structure of decision support system.

Table 11.1 Features of artificial intelligence.

Type of intelligence	Definition	Level of human involvement	Example	Example in health care
Assisted	System providing and automating repetitive tasks	Little or none	Industrial robots	UR robots for blood work (Copenhagen Hospital)
Augmented	Humans and machines collaboratively make decisions	Some or high	Business analytics	Watson for Oncology (Memorial Sloan Kettering)
Autonomous	Decisions made by adaptive intelligent systems autonomously	Little or none	Autonomous vehicle	IDx-DR for retinal images (University of Iowa)

Source: Universal Robots

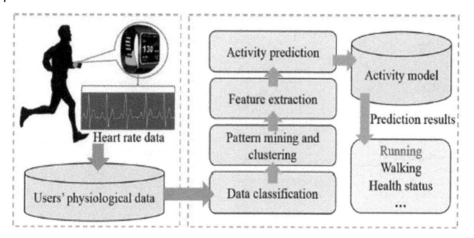

Figure 11.6 Running prediction based on heart rate [1].

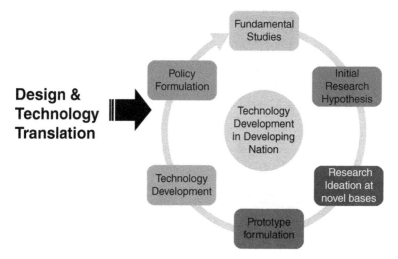

Figure 11.7 Emerging technology system.

This emerging technological system will in other ways augment total factor production at industrial as well as capital formation levels from an economic perspective, of cyber physical and IT sector growth (Figure 11.8).

This sharp but subsequent rise in number of IoT devices installed has shown a large landscape to be developed with IoT embed technology in every area of technology and its implementation. The digital space offered and shown by digitization of industry, has been taken mostly by IoT towards telecom, software, and electronic hardware, offering tremendous opportunities for various other industrial as well as public sector skill development. Thus the advent of IoT has been raised in billions to trillions of dollars toward intelligent system and its applications, which further has increased industrial and human resource solutions.

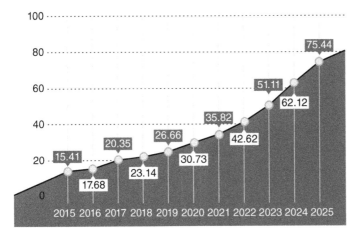

Figure 11.8 Number of IoT devices predicted to be installed by 2025. Source: Deep Learn Strategies.

This advent of IoT can, and will thrive further to trillions of dollars in IT industry and a new and fruitful IoT industrial sector, providing automated solutions like: data collection hotspots, health, energy, agriculture, education, security, smart city, disaster management, etc., through remotely connected devices.

In present-day scenario, there is a need to develop better national strategies, intergovernmental support systems and international share of platform, for putting IT steps into wider spheres and to put in place a more systematic approach, up to the bottom world economy. This in turn will boost up the best of its professional efficiency, management, proactive monitoring and strategic advisory systems.

In applying this paradigm to national strategy, it is important to consider the challenges faced by individual sectors, and its varied manifestations under the aegis of AI, IoT, or the IT sector can provide. Cementing of collaborative sectorial stakeholders in technology applications, with a strong and sustainable framework, analyzes the best possible policy interventions and technology enablers at the globalized and unified platform levels [3–9].

References

1 Guo, K., Lu, Y., Gao, H. et al. (2018). Artificial intelligence-based semantic Internet of Things in a user-centric smart city. *Sensors*, 18 (5): 1341. doi:10.3390/s18051341. Licensed under CC BY.

2 Niti Ayog National strategy for AI. (2018). #AI for all, Discussion Paper.

3 India's Hi-Tech Exports (2014) Potential Markets and Key Policy Interventions Export-Import Bank of India, Occasional Paper No. 164.

4 Parthasarathy, B. (2004) Globalizing Information Technology: The Domestic Policy Context for India's Software Production and Exports Iterations Journal of Software History.

5 Sorin-Iulian, C., Cristache, S-F., Vuţă, M. et al. (2020) Assessing the Impact of ICT Sector on Sustainable Development in the European Union: An Empirical Analysis Using Panel Data doi:10.3390/su12020592

6 Greening Emerging IT Technologies: Techniques and Practices (2017) *Journal of Internet Services and Applications* 8 (1): 9. doi:10.1186/s13174-017-0060-5

7 Zhang W. (2019) Constitutional governance in India and China and its impact on National Innovation. In: Liu, K.C. and Racherla, U. (eds) *Innovation, Economic Development, and Intellectual Property in India and China*. ARCIALA Series on Intellectual Assets and Law in Asia. 39–67. Singapore: Springer. https://doi.org/10.1007/978-981-13-8102-7_3

8 Woolley, T. and Raasch, B. (2005). Predictors of sunburn in North Queensland recreational boat users. *Health Promotion Journal of Australia* 16 (1): 26–31.

9 Draft 5th National Science, Technology, and Innovation Policy for public consultation.

12

Metaheuristics to Aid Energy-Efficient Path Selection in Route Aggregated Mobile Ad Hoc Networks

Deepa Mehta[1], Sherin Zafar[2], Siddhartha Sankar Biswas[2], Nida Iftekhar[2], and Samia Khan[2]

[1]*Department of EEE, GD Goenka University, Sohna, Haryana, New Delhi, India*
[2]*Department of Computer Science and Engineering, School of Engineering Sciences and Technology, Jamia Hamdard, New Delhi, India*

12.1 Introduction

Ad hoc networks [13] have endured several issues, [15, 25] arising consequentially due to their unique features. Some of the main issues among others are restricted battery life [52], scarce memory, and limited processing capabilities [17]. However, the advantages associated with MANETs make them uniquely popular for certain crucial applications, [49] such as in military battlegrounds for the soldiers to remain connected, provincial applications including conferences, etc., and commercial applications including disaster management [4], etc. With increasing interest toward the use of MANETs as new applications evolve, there arises a need for efficient [3] routing protocols that are capable of reducing the constraints [5] characterizing MANETs. Energy efficiency [50] is one of the crucial requirements of the protocol, as it will eventually help to increase the lifetime of the network, reducing the communication overheads, [25] as well as other limited resources such as memory, which also becomes one of the critical requirements of the routing protocol. It is a well-known that the ad hoc network faces a restricted energy capacity [26] owing to its infrastructureless operation.

Thus, a mechanism should be introduced into the ad hoc network [32] operation, which renders the path selection for forwarding the packets incorporating the energy element of the path along with the number of nodes in the path. It is a well proven fact that the lifetime of the network depends on the battery power available with the nodes, to sustain the network operations. Battery research observes a dearth of significant development and therefore it becomes crucial to place emphasis on energy management, as in the absence of any significant research for improving the battery life of the devices, the entire consequence is borne by the routing protocol, as it possesses the central role in the functioning of MANET. Energy-efficient routing in ad hoc networks includes discovering paths with minimal energy consumption, reducing the dissipation of processor power [29] of the computing devices. Mobile devices in ad hoc networks are dependent on restricted power sources (which cannot be generated by the devices on their own), which determines the longevity of

Smart and Sustainable Approaches for Optimizing Performance of Wireless Networks: Real-time Applications, First Edition.
Edited by Sherin Zafar, Mohd Abdul Ahad, Syed Imran Ali, Deepa Mehta, and M. Afshar Alam.
© 2022 John Wiley & Sons Ltd. Published 2022 by John Wiley & Sons Ltd.

the network. It becomes extremely important for the protocol to be energy-efficient so that the network can be sustained for a longer amount of time. The highly mobile nodes forming an ad-hoc network are powered with the help of exhaustive resource known as the battery. The lifetime of a network largely depends on the approaches incorporated within the protocols, in order to reduce the energy consumption in the absence of any advancements in the field of battery performance.

To ensure a longer battery life, there is a need to incorporate efficient use of battery energy, which can be achieved either by reducing the energy dissipation [6–8] during the active state in which the node actively takes part in the communication by either transmitting or receiving packets, or during the inactive state where the node is idle and listens to the wireless channel anticipating any routing request from the neighboring nodes. Since one of the main apprehensions in ad hoc networks is how to work toward maximizing the lifetime of the network, which can only be possible through maximum saving of energy where a need arises for a routing algorithm capable of incorporating energy constraints. This problem of efficient usage of energy can be formulated as an optimization problem and metaheuristics can be employed to choose an optimal route considering the energy level of the route.

Over the years, Swarm Intelligence [16] has been the driving force behind the evolution of multiple algorithms. Nature has provided a major stimulus to the development of swarm intelligence. This branch of Artificial Intelligence [47] models and designs computational methods based on complex but insightful group behavior of social insects [14] such as ants, bees [48], termites, etc. One of the major features related to the collective behavior of swarms is self-organization. These unique features of swarms and their study has led to the development of swarm intelligence-based artificial systems constituting several artificially generated agents following a structured behavior without the central administrator. This behavior is modeled on the basis of real social insect behaviors, with themselves and with the environment. Thus, such systems are able to find solutions to problems pertaining to practical relevance.

The use of the swarm intelligence methodology offers multiple advantages when compared to the classical methodologies. Some of them are:

- Adaptability: the group of created swarms are flexible to the changes in the environment.
- Robust nature: in case of failure of one or more individuals, the entire group is still capable of performing the task.
- Self-organizing: the overall group require little administration and is majorly capable of self-organization.

These attractive properties of swarm intelligence render the approach to be entitled as an eminent design paradigm.

Natural group activities based on behavioral patterns of several swarms without the controlling of any individual swarm in the group have been investigated. The various swarms included in these studies are ants, bees, termites, wasps, etc. The researchers [10] have concluded that the ants possess the most captivating swarm-level behaviors. The pattern of ants while looking for food is quite fascinating. They not only follow the shortest route leading to the food source but also communicate in a special way (called stigmergy) in order to share the information with the other ants. Stigmergy [23] is the mode of social networking approach of indirect coordination used by ants in order to solve foraging problems. The realization of stigmergetic coordination is capable when one ant leaves a trail in the environment by emitting a chemical known as pheromone that acts as a stimulus for the other ants. The observation and in-depth study of real colonies of ants has led to the development of ant algorithms [18] and to their application to multiple diverse optimization problems.

12.2   Framework

12.2.1   Route Aggregation

Route aggregation (RA) works by superseding a group of routes and replacing it with a lone route for advertisement. This proves effective in curtailing the size of routing tables, as the space for a group is now taken by only a single route. The various advantages of RA [30, 39] are:

a) Rendering the routing process significantly efficient
b) Lessening the load on the processor, as well as memory requirements are reduced
c) The topology changes are isolated

12.3   Clustering

The clustering [24, 31, 51] approach is a method which involves splitting the entire network into interconnected groups of substructures known as clusters. Every cluster [46] elects a coordinator within itself, responsible for communication in the cluster and is known as the Cluster Head [20]. The advantages of clustering are:

1. Attempts to strive toward attaining communication scalability [12, 41] in case of an increase in the number of nodes and high node mobility
2. It also helps in reducing the number of retransmissions and collision.
3. It also minimizes the amount of information to be stored and propagated within the network

12.4   Ant Colony Optimization

The ant colony behavior, [2, 40] while discovering a food source, has attracted a huge amount of interest when dealing with combinatorial optimization issues. Ant Colony Optimization (ACO) [42] metaheuristic bases itself on the reverse engineering of the natural ant colony behavior. A metaheuristic [19] is formed of two words "meta" meaning of high level and "heuristic" meaning to search. Metaheuristic thus refers to the high-level algorithmic abstraction providing a blueprint to evolve heuristic approaches to find solutions to optimization problems. A metaheuristic framework requires lesser modifications and possesses applicability to all general purpose and diverse optimization problems. Among the various popular metaheuristic algorithms, ACO metaheuristics [21] has proven to be one of the most promising. ACO [44] works by incorporating a colony of artificial ants in order to develop solutions to a particular optimization problem. It also incorporates a mechanism of providing feedback pertaining to the quality of these solutions by reminiscing the same communication procedure adopted by ants in the real world. Artificial ants work as agents and are allocated with computational resources and are instilled with capabilities of indirect communication with the help of stigmergy.

Stigmergetic communication is based on the emission of a pheromone laid by the ant while returning, thus attracting other members to follow the same path. This reinforces the route by adding more pheromones along the same path. Over a period of time, the pheromone evaporates and loses its attractive potentiality, hence leading to a solution that the shorter path gets visited more frequently and thus less pheromone evaporates, thus providing an optimal solution. The pseudo code of ACO is detailed as:

Algorithm Ant Colony Optimization Metaheuristic

Set parameters, initialize pheromone trails
While (not_termination)
Generate solutions
Pheromone update
End while

12.4.1 Setting Parameters and Initializing Pheromone Trails

ACO operation is based on the process of simulating a large number of artificial ants in order to model the natural behavior of real ants. However, artificial ants are different from real ants in many ways. Artificial ants possess memory capabilities which help them to remember the routes followed by them, which makes it easier for them to retrace their return path. Artificial ants deposit the pheromone only while returning and the memory stored path decides the amount of pheromone to be deposited. The evaporation of the pheromone follows an exponential function.

12.4.2 Generating Solutions

The artificial ants thus created on each iteration explore the network to build solutions by visiting a unique node each time and the selection of the next node is based on the pheromone value associated with that path. The probabilistic rule drives the ant while choosing the solution element based on the intensity of deposited pheromone as well as heuristic data. During or after the construction of a solution, it is the ant's duty to evaluate the solution and determine the amount of pheromone to be deposited along the path. The probability rule guiding the movement of the ant using a stochastic decision-making policy based on pheromone and heuristic information is depicted in Equation (12.1).

$$P_{ij}^k(t) = \begin{cases} \frac{[\tau_{ij}(t)]^\alpha * [\eta_{ij}]^\beta}{\sum_{l \in \mathcal{N}_i^k} [\tau_{il}(t)]^\alpha * [\eta_{il}]^\beta}, & j \in \mathcal{N}_i^k \\ 0, & j \notin \mathcal{N}_i^k \end{cases}. \tag{12.1}$$

The term $P_{ij}^k(t)$ refers to the probability with which ant k will drift from node i toward node j during the tth iteration. N_i^k refers to the group of neighborhood nodes corresponding to the kth ant at the ith node. The condition when $P_{ij}^k(t)$ is equal to zero refers to the condition where the ants are forbidden to move to any node outside the neighborhood. $[\tau_{ij}(t)]^\alpha$ refers to the amount of pheromone deposited on the path between node i and j with α as the weight. $[\eta_{ij}]^\beta$ refers to the heuristic value on the path between i and j with β as weight. η_{ij} remains constant while the algorithm executes. The weights α and β control the level of pheromone and heuristic weights. If α is high then it indicates the ants follow the pheromone strictly. If low, then algorithm works heuristically.

12.4.3 Pheromone Update

Initially, the amount of pheromone trails pertaining to all the routes are initialized to a minute constant value. Then after constructing a solution path, the pheromone trails undergo updating in two ways, depicted by Equation. (12.2). This equation includes the evaporation phenomenon determining the amount of the pheromone trails pertaining to the paths decreased following an evaporation rate (ρ) that is important for the ants so that they can ignore the suboptimal routes previously converged. The rate of pheromone evaporation is usually fixed to be appropriately fast so as to support the investigation of new parts of the search area, and evade the algorithm from prematurely converge toward local optimum.

$$\tau_{ij}(t+1) \leftarrow (1-\rho) * \tau_{ij}(t) + \sum_{k=1}^{m} \Delta\tau_{ij}^k(t), \quad \forall i,j \; \varepsilon A, 0 \leq \rho < 1. \tag{12.2}$$

Equation 12.3 involves the increase of the values of pheromone trails deposited at the visited paths which in directly proportional with their tour quality. This increase in the level of pheromone deposited helps implement a useful procedure of exploitation of the better-quality paths by increasing their probability of being utilized yet again by future ants. The quality paths selected will include the solution elements that have been used by a number of ants previously, or would have been used by at most one ant and which resulted in a high-quality solution.

$$\Delta\tau_{ij}^k(t) = \begin{cases} Q/C^k(t), & \text{if arc } (i,j) \in T^k(t) \\ 0, & \text{otherwise} \end{cases}. \tag{12.3}$$

where Q is application specific constant and $C^k(t)$ is the overall cost function of the tour $T^k(t)$

The several iterations performed in the algorithm render each and every node to be informed of the best neighbors to be forwarded packets to when looking for a specific destination node. However, this algorithm can be modified in order to optimize the memory utilization and also incorporate the energy element of the path into the algorithm, and one such modification is explained in the next section.

12.5 Methodology

The elemental objective of this chapter is to proffer a highly reliable and dominantly efficient MANET routing protocol which is capable of performing energy-efficient route aggregation with the help of ACO metaheuristics. The proposed and validated approach Zone Routing Protocol-Route Aggregated (ZRP-RA) of this chapter has optimized all the Quality of Service (QoS) issues of ZRP [1] including energy efficiency [45, 53], but these results need to prove their worth in a real-time scenario. All the performance improvement percentages through the ZRP-RA [34–38] approach provided huge optimization but energy issues need to be optimized further. Keeping this in view, the proposed ACO-ZRP-RA approach utilizes the unique property of bio inspired techniques, that of never degrading the previously achieved performance parameters and allowing an algorithm to come out of global or local minima issue. So, further ACO approach is applied on efficient ZRP-RA methodology and when compared with traditional ZRP [9, 22, 27, 28] the performance percentages are enhanced further, to be adapted well for real-time applications.

The algorithm is initiated with the cluster head determination in which the node with the highest residual energy is chosen to work as cluster head. This aids in giving a unified element to the activity of ZRP. The incorporated head presently ends up in charge of the sending of information and directing packets inside the group. Algorithm commences its operation, as shown in Figure 12.1, by initializing the zone list and adding the nodes within the selected ZRP [33, 43] zone of a particular node. Then a cluster head is selected from the list using an ACO algorithm based on the highest residual energy saved. The algorithm forms zone clusters by checking the border nodes for the selected zone and adding more nodes into the same cluster with the help of the ACO algorithm, thus creating zone cluster

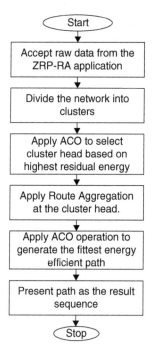

Figure 12.1 Proposed algorithm.

with the cluster head as the administrator. ACO algorithm for cluster head selection and formation are described further in Figure 12.1 which provides a flowchart for the same.

With a view to achieving the desired aim, the research reaches the following objectives:

- To design and develop an efficient scheme for cluster design, as well as an election mechanism for head node with maximum residual energy
- To aggregate all routes at the Head of the Cluster
- To apply ACO algorithm for energy-efficient route generation
- To implement ACO algorithm for updating the head node based on the maximum residual energy and amalgamating the route aggregation approach to ensure reduced overheads and longer network life.

Figure 12.1 shows the steps followed in the proposed algorithm. Each step is elaborated on in the following sections.

12.5.1 Energy Efficient ACO Algorithm

Consider the memory record in order to search for the ant identification. In the case of not finding the record, save the required data. The node restarts a timeout timer for the ant. To avoid any loop, the node eliminates the ant if a similar record is found in the memory. On receiving a backward ant, the node explores its memory and locates for the next node toward which the ant must be forwarded. The timeout timer helps to eliminate the record identifying the backward ant if the timer times out.

The Energy-Efficient Ant-Based Routing (EEABR) incorporated few improvements to the conventional ant colony optimization algorithm, with a view to reducing the memory utilization in the network nodes and also incorporated a consideration for the energy quality related to the paths explored by the ants. The conventional algorithm includes sending off the forward ants irrespective of the destination node, thus requiring communication between the nodes which leads to the storage of identification pertaining to the neighboring nodes as well as their corresponding levels of pheromone trail into the routing tables of the nodes. However, it becomes a problem for the larger networks, owing to the amount of memory needed to store such information.

To save memory utilization in Wireless Sensor Networks (WSNs), the forwarded ants can be sent straight toward the sink node, thus reducing the memory to be used only for the neighbors in the same direction as the sink node. This considerably reduces the size of the routing tables and, consequently, the memory required. The improved algorithm in an attempt to reduce memory reduces each ant's memory Mi to just a couple of records, which is only the two latest visited nodes. The path taken by the ants in this algorithm is stored in the memory at each node, which now records the details of each ant. A memory record is created at each node, which stores the previous node visited by the ant, the next node to which the ant was forwarded, identification number of the ant, and an associated timeout value. The process followed by the node after receiving a forwarded ant is as follows:

Consider the memory record in order to identify the ant. In the case of not finding the record, save the required data. The node restarts a timeout timer for the ant.

To avoid any loop, the node eliminates the ant if a similar record is found in the memory. On receiving a backward ant, the node explores its memory and locates the next node toward which the ant must be forwarded. The timeout timer helps to eliminate the record identifying the backward ant if the timer times out.

The proposed and validated approach ZRP-RA of this chapter has optimized all the QoS issues of ZRP [1] including energy efficiency [45, 53], but these results need to prove their worth in a real-time scenario. All the performance improvement percentages through the RP-RA [34–38] approach provided huge optimization but energy issues need to be optimized further. Keeping this in view, the proposed ACO-ZRP-RA approach utilizes the unique property of bio inspired techniques of never degrading the previously achieved performance parameters and allowing an algorithm to come out of global or local minima issue. So, a further ACO approach is applied on efficient ZRP-RA methodology, and when compared with traditional ZRP [9, 22, 27, 28] the performance percentages are enhanced further, to be adapted well for real-time applications.

The algorithm initiates with the cluster head determination in which the node with the highest residual energy is chosen to work as the cluster head. This aids in giving a unified element to the activity of ZRP. The incorporated head presently winds up in charge of sending of information and directing packets inside the group. The algorithm commences its operation, as shown in Figure 12.2, by initializing the zone list and adding the nodes within the selected ZRP [33, 43] zone of a particular node. Then a cluster head is selected from the list using an ACO algorithm based on the highest residual energy saved. The algorithm forms zone clusters by checking the border nodes for the selected zone and adding more nodes into the same cluster with the help of ACO algorithm, thus creating zone cluster with the cluster head as the administrator. ACO algorithm for cluster head selection and formation are described further in Figure 12.3, which provides a flowchart for the same.

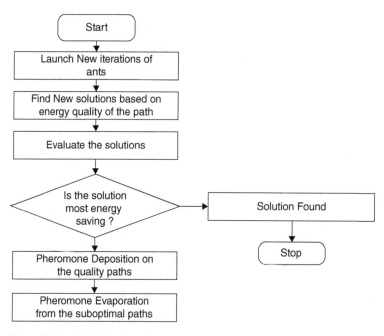

Figure 12.2 Algorithm.

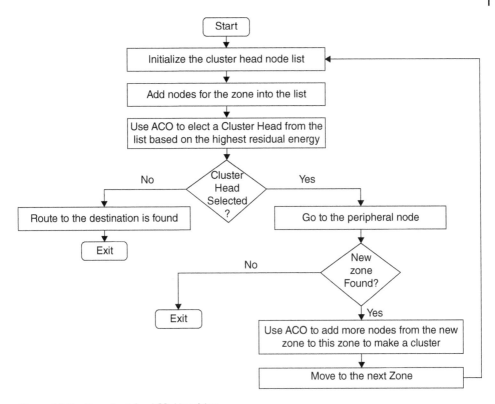

Figure 12.3 Flowchart for ACO Algorithm.

12.5.2 ACO-Aided Cluster Design and Head Selection

ACO is a developmental calculation that employments a meta heuristic approach. It is motivated by the foraging behaviour of ants. The ants discharge chemical called pheromones on the way whereas moving along the path. As a greater number of ants moves along the way, the pheromone concentration increments. The more the pheromone concentration, more is the chance for an unused ant to select that way to reach the nourishment from the colony. The way chosen by the ants will be the most limited path from settle to nourishment. The method could be a kind of distributed optimization component, in which they discover the most limited remove from the food to colony. Each single insect contributes to the arrangement, collaborating within the work. Counterfeit ants are utilized to find the arrangement of difficult optimization issues. Fake ants utilize an incremental valuable approach to explore for a feasible solution.

A cluster head is chosen based on two viewpoints, the pheromone esteem related with each hub and its visibility. Perceivability alludes to the number of hubs that will be secured in the event that the hub is included into the cluster headset. Perceivability keeps changing as topology changes. The pheromone esteem related with a hub is overhauled for each emphasis of the calculation. For each cycle, a node is chosen as the cluster head and another cluster head is chosen based on the pheromone and perceivability of its neighbour hubs. This handle proceeds until all the hubs within the organize are secured. A hub is said to be covered if it may be a cluster head or falls within the extend of an as of now chosen cluster head.

12.5.3 ACO-Aided Route Aggregation

Algorithm for ACO-Aided Route Aggregation:

1. For each Cluster Head in the cluster
2. Apply Route Aggregation (using ACO)
3. Update Routing Tables within the Cluster
4. Maintain and update the cluster Head with highest residual energy
5. Move to next Cluster Head

12.5.4 ACO-Aided Energy: Efficient Path Selection

The ants help us to determine routes between the nodes with maximum residual energy. Among the multiple routes discovered by the ants, the routes with maximum residual energy of nodes are chosen for data transmission. The algorithm used for route determination for data transmission is exhibited in Figure 12.2.

12.6 Results

The proposed mechanism enables the perfect exploitation of residual energy of the node and prevents the premature dying of the node led by energy starvation and thus improving the network lifetime. The energy-efficient ZRP-RA achieves enhancement by further applying Route Aggregation [11] at the Cluster Head. Table 12.1 details the simulation parameters while simulating the proposed approach using NS2.

	Packet Delivery Ratio			
Number of nodes	25	50	75	100
ZRP	42.21	98.38	51.41	35
ACO-ZRP-RA	71.78	99.82	76.28	63.71

Table 12.1 Simulation parameters while simulating NS2.

Simulation Parameters	Value
Network area	$1000 \times 1000 \text{ m}^2$
Velocity	$1 \sim 2 \text{ m/s}$
Number of nodes	255 075 100
Packet size	512 bytes
Traffic type	CBR
Number of connection	20
Packet rate	2 P/s
Pause time	0
Simulation time	100 sec

Packet Delivery Ratio Vs Number of Nodes

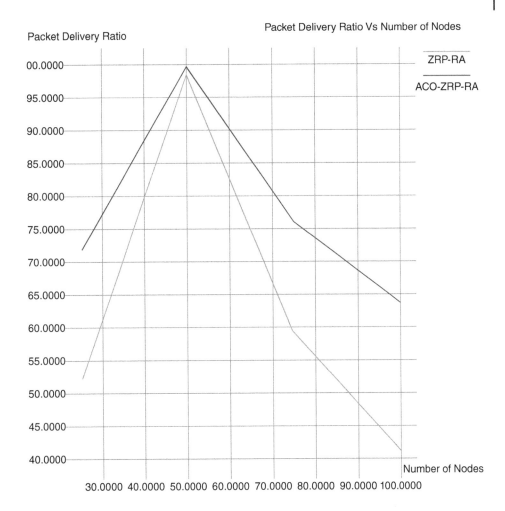

	Average energy consumption			
Number of nodes	25	50	75	100
ZRP	0.69	0.62	0.67	0.72
ACO-ZRP-RA	0.61	0.53	0.57	0.63

Table 12.2 reveals that ant colony optimization ensures a marked decrease in the energy consumption as compared to ZRP. Thus, accomplishing performance enhancement when compared to the conventional hybrid protocol.

Table 12.2 Ant colony optimization

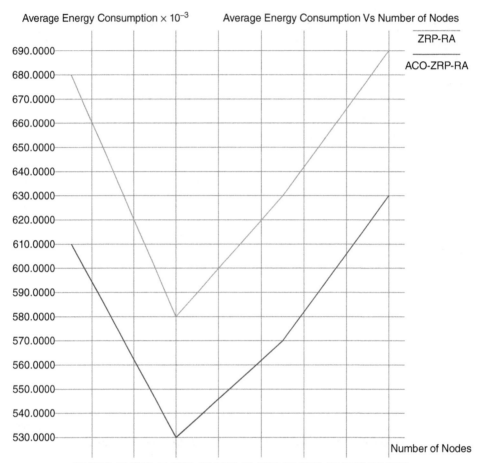

	Throughput			
Number of nodes	25	50	75	100
ZRP	46.3	106.15	56.23	38.02
ACO-ZRP-RA	97.37	126.28	103.72	81.67

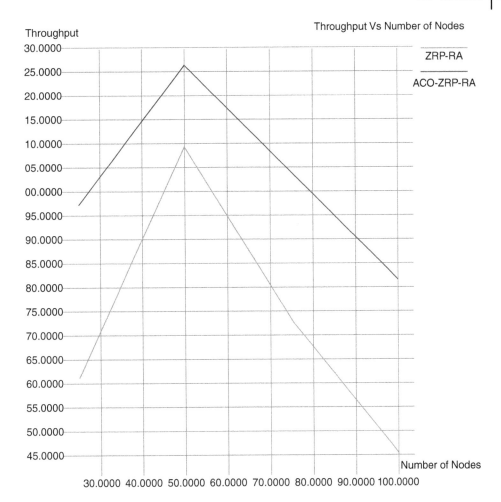

Throughput Vs Number of Nodes

ZRP-RA

ACO-ZRP-RA

Throughput

30.0000
25.0000
20.0000
15.0000
10.0000
05.0000
00.0000
95.0000
90.0000
85.0000
80.0000
75.0000
70.0000
65.0000
60.0000
55.0000
50.0000
45.0000

Number of Nodes

30.0000 40.0000 50.0000 60.0000 70.0000 80.0000 90.0000 100.0000

12.7   Discussion

The ACO technique provides energy-efficient optimal paths for data transmission and the Route Aggregation mechanism efficiently reduces overheads by aggregating the information at the head, hence reducing the energy consumed, even in high-density conditions. It also reduces the queuing of packets by reducing congestion in the network as a result of efficient aggregation of information at the head, along with the circulation of the cluster head as the energy level decreases, even under high-density conditions, thus ensuring an improvement in the amount of packets delivered, which consequently results in an improved throughput and thus achieves further enhancement in the performance to produce better results.

12.8 Conclusion

The Meta Heuristic Algorithm-based ACO approach is utilized for further energy-efficient enhancement of the proposed ZRP-RA protocol. The importance of the ACO applied ZRP-RA protocol for MANET is further specified by simulation and comparing it with the ZRP-RA approach; 30.62% increase in packet delivery ratio, reduction in delay value by 23.96%, a 14.989% reduction in average jitter value. Throughput improvement has increased by 49.22%, the average energy consumption has been reduced by 9.2835%, and the Packet loss ratio is reduced by an impressive 51.61%, hence when scalability is increased, average delay and jitter of the proposed approach does not rise too high, delivery ratio is also maintained, and throughput also does not fall too low. Therefore, it can be specified that the proposed methodology is effective with respect to the proposed ZRP-RA approach of mobile ad-hoc networks, as various QoS parameters like delivery ratio, average delay, jitter, and average throughput are maintained to provide an enhanced energy-efficient optimized solution for MANET.

References

1 Ahmad, I., Barukab, O.M., and Khan, S.A. (2016). Reducing flooding of zone routing protocol in Mobile Adhoc networks. *VAWKUM Transactions on Computer Science* 9 (2): 8–20.

2 Ahmed, M., Honsy, M.I., Marghny, H.M. et al. (2012). Ant colony and load balancing optimizations for AODV routing protocol. *International Journal of Sensor Networks and Data Communications* 1: 1–14.

3 Alkalbani, A.S. and Mantoro, T. (2016). Residual Energy Effects on Wireless Sensor Networks (REE-WSN). International Conference on Informatics and Computing (ICIC). IEEE, Mataram, Indonesia, 288–291.

4 Alslaim, M.N., Alaqel, H.A., and Zaghloul S.S. (2014). A comparative study of MANET routing protocols. In: The Third International Conference on e-Technologies and Networks for Development (ICeND2014). IEEE, Beirut, Lebanon, 178–182.

5 Asher, Y., Feldman, S., Feldman, F. et al. (2010). Scalability issues in ad-hoc networks: metrical routing versus table driven routing. *Wireless Personal Communications* 52 (3): 423–447.

6 Awad, O.A. and Mariam, R. (2016). An efficient energy aware ZRP-fuzzy clustering protocol for WSN. *International Journal of Scientific and Engineering Research* 7 (3): 1060–1065.

7 B. Chen., Jamieson, K., Balakrishnan, H. et al. (2002). SPAN: an energy-efficient coordination algorithm for topology maintenance in ad hoc wireless networks. *Wireless Networks* 8 (5): 481–494.

8 Bashir, S.A. and Sharma, N. (2016). Energy and memory efficient routing protocol in Manet. *International Journal of Engineering Development and Research* 4 (4): 2321–9939.

9 Beijar, N. (2002), Zone routing protocol (ZRP), Networking Laboratory, Helsinki University of Technology, Finland. April 1–2.

10 Blum, C. (2005), Ant colony optimization: introduction and recent trends. *Physics of Life Reviews* 2: 353–373.

11 Bohdanowicz, F. and Henke, C. (2014). Loop detection and automated route aggregation in distance vector routing. In: 2014 IEEE Symposium on Computers and Communications (ISCC). IEEE, 1–6.

12 Byung-Jae, K., Song, N., and Miller, L.E. (2004). On the scalability of Ad hoc networks. *IEEE Communications Letters* 8 (8): 503–505.

13 Cheng, X. and Huang, X. (2000). *Ad Hoc Wireless Networking*, Kluwer Academic Publishers.

14 Constantinou, C., Kwiatkowska, M.Z., and Liu, Z. (2005). A Biologically Inspired QoS Routing Algorithm for Mobile Ad Hoc Networks. In: 19th International Conference on Advanced Information Networking and Applications (AINA'05). IEEE, 1: 426–431.

15 Corson, S. and Macker, J. (1999). Mobile Ad hoc Networking (MANET): Routing Protocol Performance Issues and Evaluation Considerations. Network Working Group, RFC2501, Available: https://tools.ietf.org/html/rfc2501 (Accessed: March 2017)

16 Deepa, O. and Senthilkumar, A. (2016). Swarm intelligence from natural to artificial Systems: Ant Colony Optimization. *International Journal on Applications of Graph Theory in Wireless Ad hoc Networks and Sensor Networks (GRAPH-HOC)*, 8 (1): 9–17.

17 Dharam, V., Agarwal, S.K., and Imam, S.A. (2012). A simulation study on node energy constraints of routing protocols of Mobile Ad hoc networks use of QualNet simulator, international journal of advanced research in electrical. *Electronics and Instrumentation Engineering* 1 (5): 401–410.

18 Dorigo, M. and Caro, G.D. (1999). Ant colony optimization: a new meta-heuristic. In: Proceedings of the 1999 Congress on Evolutionary Computation-CEC99, Washington, DC, 2: 1477.

19 Floriano, D.R. and Socievole, A. (2011). Meta-Heuristics Techniques and Swarm Intelligence in Mobile Ad-hoc Networks, In: Wang, X. *Mobile Ad-hoc Network and Applications*, Europe: InTech.

20 Ghode, S. and Bhoyar, K.K. (2016). NEMA: Node Energy Monitoring Algorithm for Zone Head Selection in mobile ad-hoc network using residual battery power of node. In: International Conference on Wireless Communications, Signal Processing and Networking (WiSPNET), Chennai, India, 1999–2004.

21 Gupta, A.K., Sadawarti, H., and Verma, A.K. (2012). MANET routing protocols based on ant colony optimization, *International Journal of Modelling and Optimization* 2 (1): 42–49.

22 Haas, Z.J., Pearlman, M.R., and Samar, P. (2002). The Zone Routing Protocol (ZRP) Ad Hoc Networks, IETF Internet Draft, Available: https://tools.ietf.org/html/draftietf-manet-zone-zrp-04 (Accessed: March 2015)

23 Heylighen, F. (2016). Stigmergy as a universal coordination mechanism. I: Definition and components. *Cognitive Systems Research* 38: 4–13.

24 Huang, J. (2016). A clustering routing protocol for Mobile Ad hoc networks. *Mathematical Problems in Engineering* 2016: 1–10. https://doi.org/10.1155/2016/5395894

25 Jacquet, P. and Viennot, L. (2000). Overhead in Mobile Ad-hoc Network Protocols. [Research Report] RR-3965, INRIA. 2000. <inria-00072683> Available: https://hal.inria.fr/inria-00072683 (Accessed: March 2015).

26 Kanakala, S., Ananthula, V.R., and Vempaty, P. (2014). Energy-efficient cluster based routing protocol in Mobile Ad hoc networks using network coding. *Journal of Computer Networks and Communications* 2014: 1. https://doi.org/10.1155/2014/351020

27 Kaur, S. and Kaur, S. (2013). Analysis of zone routing protocol in MANET, *International Journal of Research in Engineering and Technology* 2 (9): 520–524.

28 Kaushik, G. and Goyal, S. (2013). A Clustering based AODV approach for MANET. In: Proceedings of the International Conference on Emerging Trends in Engineering and Technology 901–904.

29 Kim K.J. and Koo, H.W. (2007). Optimizing power-aware routing using zone routing protocol in MANET. In: IFIP International Conference on Network and Parallel Computing Workshops. Liaoning, 670–675.

30 Le, F., Xie, G.G., and Zhang, H. (2011). On route aggregation. In: The Proceedings of the 7th Conference on Emerging Networking Experiments and Technologies, coNEXT'11 ACM, New York, 1–12.

31 Lim, M.L. and Yu, C. (2004). Does cluster architecture enhance performance scalability of clustered mobile Ad Hoc networks. In: International Conference on Wireless 2004: 71–80.

32 Loo, J., Mauri, J.L., and Ortiz, J.H. (2012). *Mobile Ad Hoc Networks: Current Status and Future Trends*. CRC Press.

33 Mann, S., Sharma, A.. and Gupta, A. (2015). Parametric analysis of zone routing protocol on the basis of mobility. *International Journal of Innovations in Engineering and Technology* 5 (2): 263–269.

34 Mehta D., Kashyap I., and Zafar S. (2018). Random cluster head selection based routing approach for energy enrichment in MANET. In: International Conference on Recent Innovations in Signal processing and Embedded systems (RISE-2017). IEEE, Bhopal, India, 119–123.

35 Mehta, D., Kashyap, I., and Zafar, S. (2017). Neoteric RA approach for optimization in ZRP. *Studies in Computational Intelligence* 713 : 1–13. ISSN: 1860-949X.

36 Mehta, D., Kashyap, I., and Zafar, S. (2017). Routing optimization in cloud networks, *International Journal of Advanced Research in Computer Science* 8 (2): ISSN 0976-5697.

37 Mehta, D., Kashyap, I., and Zafar, S. (2017). Consummate scalability through clustered approach in ZRP. *International Journal of Sensors Wireless Communication and Control (IJSWCC)* 7 (3): 178–187. ISSN: 2210-3279.

38 Mehta, D., Kashyap, I., and Zafar, S. (2017). Synthesized hybrid ZRP through aggregated routes. *International Journal of Information Technology* 10 (1): 83–89. ISSN:2511-2112.

39 Fuji, M., Naito, K., Mori, K. et al. (2011), *Multicast Routing Protocol for Ad-Hoc Networks with Route Aggregation and Transmission Power Control, Department of Electrical and Electronic Engineering*, Japan: Mie University.

40 Nancharaiah, B. and ChandraMohan, B. (2014). Modified ant colony optimization to enhance MANET routing in Adhoc on demand distance vector. In: 2nd International Conference on Business and Infonnation Management (lCBIM). IEEE, Durgapur, India, 81–85.

41 Natarajan, K. and Mahadevan, G. (2017). Evaluation of seven MANET routing protocols using scalability scenario. *International Journal of Computer Science and Network.* 6 (2): 131–141.

42 Okazaki, A.M. and Fröhlich, A.A. (2011). AD-ZRP: ant-based routing algorithm for dynamic wireless sensor networks. In: 18th International Conference on Telecommunications, IEEE, Ayia Napa, Cyprus, 15–20.

43 Perkins, C.E. (2001). *ZRP: a hybrid framework for routing in Adhoc networks*. In: *Ad Hoc Networking*, 221–253. South Asia: Pearson.

44 Rafsanjani, M.K., Asadinia, S., and Pakzad, Z. (2010). A hybrid routing algorithm based on Ant Colony and ZHLS routing protocol for MANET. In: FGCN 2010, Part II, CCIS 120, 112–122.

45 Ravi, G. and Kashwan, K.R. (2013). A new routing protocol for energy efficient mobile applications for ad hoc networks, *International Review on Computers and Software (http://I.R.E.CO.S.)* 48 (10): 77–85.

46 Razaee, M. and Yaghmaee, M. (2009). Cluster based routing protocol for Mobile Adhoc networks. *Infocom Journal of Computer Science* 8: 30–36.

47 Sadollah, A., Choi, Y., and Kim, J.H. (2015). Metaheuristic optimization algorithms For approximate solutions to ordinary differential equations. In: 2015 IEEE Congress On Evolutionary Computation (CEC), Sendai, Japan, 792–798.

48 Saleem, M., Ullah, I., and Farooq, M. (2012). BeeSensor: a bee-inspired, energy efficient and scalable routing protocol for wireless sensor networks. *Information Sciences* 200: 38–56.

49 Sivakami, R. and Nawaz, G.M. (2011), Secured communication for MANETS in Military, In: 2011 International Conference on Computer, Communication and Electrical Technology (ICCCET). Tamilnadu, 146–151.

50 Wang, Y. (2010). Study on energy conservation in MANET, *Journal of Networks* 5 (6): 708–715.

51 Wei, D. and Chan, H.A. (2006). Clustering Ad Hoc networks: schemes and classifications. In: *3rd Annual IEEE Communications Society on Sensor and Ad Hoc Communications and Networks*. Reston, VA: IEEE. 920–926.

52 Yoshimachi, M. and Manabe, Y. (2016). Battery power management routing considering participation duration for Mobile Ad hoc networks. *Journal of Advances in Computer Networks* 4 (1): 13–18.

53 Zungeru, A.M., Seng, K.P., Ang, L.M. et al. (2002). Energy efficiency performance improvements for Ant-based routing algorithm in wireless sensor networks. *Journal of Sensors* 2013: 1–17. https://doi.org/10.1155/2013/759654

13

Knowledge Analytics in IOMT-MANET Through QoS Optimization for Sustainability

Neha Sharma[1], Nida Iftekhar[2], and Samia Khan[2]

[1]*Department of DEEE, GD Goenka University, Sohna, Haryana, India*
[2]*Department of Computer Science and Engineering, Jamia Hamdard, New Delhi, India*

13.1 Introduction

Internet of Things (IoT) is a well-known term nowadays and the reason behind its popularity is its vast range of applications to target real-time problems. IoT is basically a cluster of inter-related and inter-connected devices that work together in order to achieve common goals. All the devices in the cluster work collaboratively to attain efficiency and quality in the task. IoT is considered efficient and promising for future aspects only because of its functionality and unbeatable efficacy in terms of supporting numerous real-time applications in the fields of healthcare, agriculture, education, the military, and many more. It has much to contribute to the cyberspace world and the list of its advantages is endless. One of its major contribution is in the healthcare sector [5].

Internet of Medical Things (IoMT) is a real-time application of IoT only. Similar to IoT, IoMT is a cluster of distinct medical equipment that works digitally with various healthcare applications connecting together via a computer network. All the devices in the cluster are equipped with basic mechanisms that support connection between all the devices like Wi-Fi, availability of network, and other basic essential elements of networking. IoMT is the future of the healthcare domain. Basic features of IoMT that makes it promising from future aspects are monitoring of remote patients, keeping a strict check on a patient's medication, monitoring wearable mobile healthcare devices, reducing the gap between a doctor and his patient, sharing reports remotely with the doctors and patients, etc. [5, 6].

IoT has made Internet use smart and much more reliable and IoMT is a living example of this. IoMT has made it possible to collaborate distinct well-efficient and well-established mechanisms of the healthcare domain to achieve this level of smartness and intelligence in the network. Figure 13.1 depicts an IoMT environment.

The functionality and efficiency of IoMT is entirely dependent on the working of all the participating devices. As mentioned above, IoMT is a cluster of intercommunicating devices, so the connection of all the devices for data transmission is the primary aspect of an IoMT environment. In order to maintain interoperability among the participating devices, here comes into the picture a remote, self-organized, and infrastructureless system called MANET (Mobile Ad-hoc Network). MANET has evolved into one of the most

Smart and Sustainable Approaches for Optimizing Performance of Wireless Networks: Real-time Applications, First Edition.
Edited by Sherin Zafar, Mohd Abdul Ahad, Syed Imran Ali, Deepa Mehta, and M. Afshar Alam.
© 2022 John Wiley & Sons Ltd. Published 2022 by John Wiley & Sons Ltd.

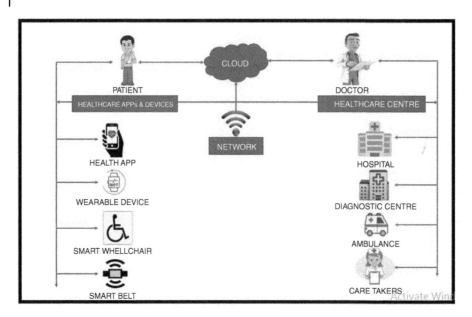

Figure 13.1 IoMT environment.

vigorous domains of communication in cord-free technology. The consolidated endeavor of MANET and IoMT has opened up novel ventures in the field of communication. The contribution of MANET has uplifted the performance of the IoMT environment [2, 7].

MANET has supported IoMT functionality in terms of efficient routing by applying one of its best hybrid routing protocols, i.e., ZRP (Zone Routing Protocol) which, being a hybrid routing protocol of MANET joins the benefits of proactive and reactive routing protocols in order to maintain a harmony between control overhear and inactivity in the network. ZRP divides the entire communication network into distinct zones and then further carries out the transmission. ZRP is best known for its swift delivery and negligible transmission overheads. By applying this functionality of ZRP to IoMT, the performance of communication in IoMT can be uplifted [13].

Another aspect of IoMT that requires more attention is optimized communication. The communication network must be optimized in terms of various parameters, such as PDR (Packet Delivery Ratio), of a considered network must be high, and End-to-End Delay of the network must be as low as possible, Throughput must always be high, NRL (Normalized Routing Load) in a network must be high, and PLR (Packet Loss Ratio) of the network must be low. Optimized QoS of any network is of utmost importance for an efficient network. In this chapter, it has been proposed to achieve optimization in communication of IoMT by applying the functionality of a nature-inspired optimization approach IWD (Intelligent Water Drop).

The IWD approach is well adapted for optimization and is one of the mechanisms that has been very little explored to date. The primary concept of IWD is based upon the behavior water drops in rivers. It has been observed that these water drops pursue a course of action that leads them to their destination by covering the shortest path. The two most substantial aspects of the IWD approach are the water drops in a river and amount of soil particles present on river's bed. These aspects form the foundation of this approach. IWD was proposed by Shah Hosseini first in 2007 [20–23].

The rivers contain many moving water drops, which is one of the substantial aspects of this approach. When a water drop moves toward its destination, it disturbs other substantial aspects of this approach, such as the soil particles present on the river's bed. A water drop carries soil particles while moving and so, in this way, at some places the amount of soil particles becomes less, and in other places becomes more. Also, the soil particles present on the river's bed also obstruct the path for water drop and hence effects its speed.

In order to evaluate the efficiency of the proposed approach, a group of standard performance evaluation parameters have been adopted that consist of parameters such as PDR, End-to-End Delay, Throughput, NRL, and PLR. Efficient simulation results depicted through these parameters have enhanced the performance of the proposed approach as expected in comparison with ZRP and a route aggregated ZRP, i.e., RA-ZRP [8–11].

Sections of this chapter are as follows. Section 13.2 elaborates on related work for a deep understanding of the various topics like IWD, IOT, Optimization, and many more. Section 13.3 describes the proposed neoteric nature inspired IWD algorithm for ZRP. Furthermore, Section 13.4 contains the various simulation results and the last section concludes the entire topic and exhibits future scope.

13.2 Related Work

This section comprises a review of distinct relevant papers and different investigations that have been executed in the domain of IoT, IoMT, and optimization through nature-inspired algorithms.

In [7], the authors describe the use of IoT and its impact on the healthcare domain. IoMT has been referred to as the future of the healthcare sector. IoMT has immensely enhanced the patient care system by making it easy and flexible to share reports and other data remotely. Even if this system has emerged so well, security is one aspect that still needs a lot of attention. The authors discuss various vulnerabilities of sharing data in between the various participating devices. Furthermore, the researchers discuss potential uses of IoT and security-related risks of IoMT.

In [6], researchers elaborate on the efficiency of IoMT in the healthcare domain. IoMT has certainly improved reliability, accuracy, and outcome of the electronic devices that support various healthcare applications. Furthermore, they highlight useful aspects and challenges encountered by IoMT. IoMT has certainly lowered the gap between a doctor and their patients, as doctors can treat their patients remotely, discuss their reports, and prescribe medication as well. However, IoMT has experienced distinct challenges while sharing data in terms of security and optimization.

In [32], the authors elucidate on smart healthcare in terms of IoMT, which has been referred to as the future of the healthcare industry. IoMT is one of the real-time applications of IoT. Various challenges of IoMT in terms of security and privacy, attacks and malware, confidentiality and authenticity, have been discussed. A comparative study of existing mechanisms to target all the above-mentioned security-related threats has been carried out by these researchers.

In [13], the authors focus on the rapid utilization of IoMT in the healthcare industry. IoMT has been adopted by numerous patients and other healthcare experts worldwide and has been proved to be an effective way of patient monitoring over the Internet. Distinct cost-effective in-home and work-related modules have been proposed, implemented, and adopted by different experts. No doubt, IoMT is termed as the future of the healthcare industry, as it has so much to contribute.

In [5], researchers discuss how valid is it to call IoMT the "Future of healthcare industry." IoMT joins various healthcare devices and applications into one group in order to provide more functionality on a single platform. One of the most promising uses of IoMT is remote patient monitoring. The authors cross-verify this particular aspect of IoMT with an intention to validate its use. They then focus on various challenges of IoMT based on date collected from various healthcare databases.

In [1], researchers focus on challenges related to routing between various nodes of IoT. Intercommunication of the participating nodes in an IoT may consist of multimedia content and in such scenarios the quality and security of data becomes of utmost priority. Similarly, the content shared between the nodes in an IoMT is extremely critical and confidential, so the security and quality of data is very crucial.

13.3 Proposed Neoteric Nature Inspired IWD Algorithm for ZRP

The neoteric nature-inspired IWD approach basically imitates the course of actions of natural water flow, particularly regarding rivers that start their journey from their source to a larger water body, i.e., an ocean or sea. During their movement from their source, rivers have to conquer many hurdles in terms of uneven ground, highs, and lows, etc. Despite these hurdles, it is observed that they take the shortest path from their origin to their destination and this formation of an optimum route is termed as 'intelligence of water drops'. This intelligence is implemented in this neoteric IWD optimization approach. Also, the intelligence factor of water drops is implemented in this approach with the creation of man-made water drops which are referred as intelligent water drops. These drops are designed while retaining two basic properties that are:

- amount of soil particles a water drop holds
- velocity at which a water drop moves along its path

These properties effect the overall movement pattern of an intelligent water drop and as this water drop moves along its path, the value of these properties keeps on changing, which builds up an environment for this neoteric approach [19].

From the technical outlook, an ever-changing environment represents a real-time problem. The target to achieve is the desired objective and the objective can be different in distinct scenarios. For instance, if the task is to reach a destination from a source, then there can be two different aspects. If the objective is to locate a destination in terms of quality, it clearly signifies that the destination is unknown. On the other hand, if the objective is to work out the shortest path among all available routes to reach a destination, then this indicates that the destination is already known. So, the intelligent water drops can work beautifully in non-static scenarios.

Research study analyses the pseudo code of IWD specified through the following steps:

- Input: Problem informational collection (formulated as completely associated chart)
- Output: An ideal arrangement
- Main advances:
  1. Introduce the static parameters which remain constants amid the pursuit procedure
  2. Introduce the dynamic parameters (i.e., parameters which are changed after every emphasis)
  3. Spread IWDs (IWD number) haphazardly on the built chart
  4. Restore the rundown of the already visited hubs of individual IWD one-by-one, to incorporate the hub just visited
  5. Rehash the accompanying strides for each IWD
     r∈ [1, iwd number] with halfway arrangement:
     (a) ai = the present hub for drop r
     (b) j = chose next hub
     (c) Move drop r from hub i to hub j
     (d) Update the accompanying parameters:
        – Velocity of the drop r
        – Soil particles incentive inside the drop r
        – Soil particles incentive inside the edge eij
     (e) End for:
  6. Apply a disentangling procedure utilizing expansiveness; first hunt to uncover the network structure for all arrangements developed by IWDs
  7. Select the Emphasis Best Arrangement (EBA) from all arrangements
  8. Update the dirt estimations of all edges incorporated into the Cycle Best Arrangement (CBA)
  9. Update the Absolute Best Arrangement (ABA). On the off chance that (quality [ABA] ≤ quality [EBA]) then

     ABA = EBA

  10. Increment iteration count by one
  11. Check the stop standard:
      While (the most extreme number of cycles has not arrived):
      – Do
      – Repeat stage 2 to stage 10
  12. Return to the complete best arrangement.

The objective function of IWDRA used is given in Figure 13.2.

13.4 Simulation Results

An arbitrary network scenario has been established by utilizing NS2 for implementation of the proposed IWDRA approach over ZRP and RA-ZRP and to exhibit comparison analysis.

Figure 13.2 Objective function IWDRA.

```
/* Zone Head Selection*/

void
ZoneHeadList::zad(nsaddr_t addr, int coveredFlag) {
    ZoneHead *newZH = iwd(fcount.ntable.nsaddr_t) ;
    if(newZH == NULL) {
        printf("### No Zone Head Seected ###");
        exit(0);
    }

    if(numHops == agent_->radius_) {
        if(zLst_.findPerNode(newFoundZone)
==max(fcount.ntable.nsaddr_t)) {
            newFoundZone->        newFoundZone=neighbour_table-
>iwd(nsaddr_t)
            zLst_.addPerNode(newFoundZone, TRUE);
        }
        newZH->next_ = zlist_;
        head_ = newZH;
        numNodes_++; }
}
}

/* Route Aggregation */
void
IERPAGENT::routeAggregation(head_, ztable_ent){
defaults {
        addRouteInPacket(nsaddr_t dest, Packet *p, ztable_ent){
        table_->AddEntry (*newZH, ztable_ent);
        if(foundIERPRoute == TRUE) {
        routeLen = handleToIERPRoute->numHops_+1;
        (agent_->pktUtil_).pkt_add_ROUTE_space(p,
routeLen);
            iwd->iwdpath[--iwd->currenthop];
            ZoneRoute *curNode = handleToIERPRoute;
}
}
}
```

Distinct parameters considered are simulation area, number of nodes, velocity, traffic load, data packet size, data packet rate, pause time, simulation time, and number of maximum connections, as listed in Table 13.1.

The simulation results have been depicted in distinct forms in this chapter, such as values in tabular format, bar-graphs, and X-graphs for the same values. Graphs are well-known for giving a visual representation of the results, whereas values give actual and accurate variation of the results. The performance evaluation parameters adopted in order to depict the simulation results are PDR, E2E Delay, overall Throughput, NRL, and PLR. The simulation results depicted in this chapter detail the comparison analysis of ZRP, RA-ZRP, and IWDRA-ZRP for varying number of nodes in the simulated network.

1. Packet Delivery Ratio:
   Packet Delivery Ratio, i.e., PDR for a network can also be termed $PDR = R_p/S_p$, where, R_p is the number of data packets received by the destination end and S_p is the number of data packets originally sent by the source end. Table 13.2 show that the performance of IWDRA-ZRP is better than both ZRP and RA-ZRP for varying number of nodes. Figure 13.3 exhibits the performance variations of the three protocols through a bar graph, and its X-graph is depicted in Figure 13.4, where the x-axis portrays number of nodes and the y-axis portrays PDR.

Table 13.1 Simulation structure.

Simulation parameters	Values
Area	1000×1000
Number of nodes	50, 100, 150, 200, 250
Velocity	$0 \sim 20$ m/s
Traffic	CBR
Packet size	512 bytes
Packet Rate	4 P/s
Pause time	0 s
Simulation time	600 s
Maximum connections	20

Table 13.2 Packet delivery ratio.

Protocols	PDR (varying nodes)				
	50	100	150	200	250
ZRP	65	69.5	73.2	74.8	77.1
RA-ZRP	72	75.9	77.2	77.8	82.5
IWDRA-ZRP	79	81.2	84.8	85.9	86.1

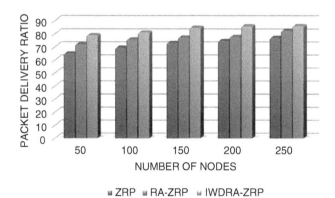

Figure 13.3 Varying PDR with dynamic nodes.

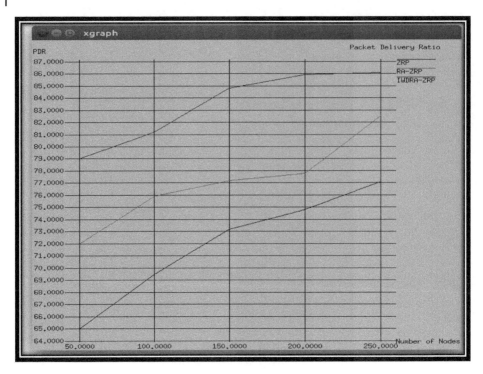

Figure 13.4 X-Graph for PDR with dynamic nodes.

Table 13.3 End-to-End Delay.

Protocols	End-to-End Delay (varying nodes)				
	50	100	150	200	250
ZRP	1.2	1.5	1.7	1.75	1.82
RA-ZRP	0.9	1.2	1.3	1.5	1.51
IWDRA-ZRP	0.51	0.7	0.8	0.9	0.9

2. End-to-End Delay:
   E2E Delay for a network can be processed in terms of values using function as: $D_e = (Rt - St)$, where R_t represents receiving time of the data packet and S_t represents sending time of the data packet. Table 13.3 show that the performance of IWDRA-ZRP is better than both ZRP and RA-ZRP for varying number of nodes. Figure 13.5 exhibits the performance variations of the three protocols through a bar graph, and its X-graph is depicted in Figure 13.6, where the *x*-axis portrays the number of nodes and the *y*-axis portrays E2E Delay.

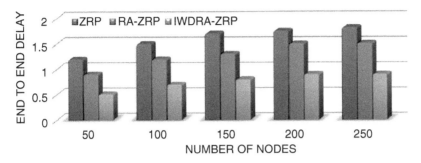

Figure 13.5 Varying End-to-End Delay with dynamic nodes.

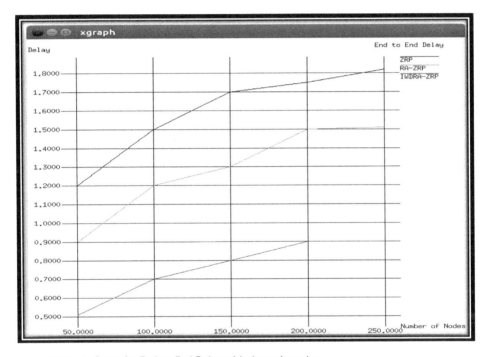

Figure 13.6 X-Graph for End-to-End Delay with dynamic nodes.

3. Throughput:

   Throughput for a network performance can be calculated by utilizing the function as: *Throughput* $= (D_p * P_s)/T_d$, where D_p symbolizes the total number of delivered packets, P_s gives the size of a packet, and T_d stands for total duration of time. It is measured in *bps*, i.e., bits per second. Table 13.4 show that the performance of IWDRA-ZRP is better than both ZRP and RA-ZRP for varying number of nodes. Figure 13.7 exhibits the performance variations of the three protocols through bar graph and its X-graph is depicted in Figure 13.8, where the *x*-axis portrays number of nodes and the *y*-axis portrays Throughput.

Table 13.4 Throughput.

Protocols	Throughput (varying nodes)				
	50	**100**	**150**	**200**	**250**
ZRP	78.4	85.21	91.46	94.73	99.39
RA-ZRP	84.67	89.12	94.83	100.43	108.32
IWDRA-ZRP	87.7	94.9	101.1	107.83	115.28

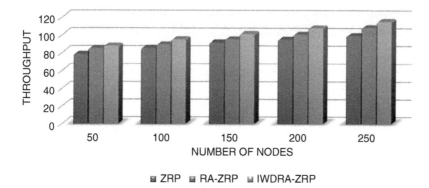

■ ZRP ■ RA-ZRP ■ IWDRA-ZRP

Figure 13.7 Varying throughput with dynamic nodes.

Table 13.5 Normalized routing load.

Protocols	NRL (varying nodes)				
	50	**100**	**150**	**200**	**250**
ZRP	7.382	6.791	8.26	8.628	9.158
RA-ZRP	6.842	5.826	6.982	6.891	7.735
IWDRA-ZRP	4.923	4.726	5.109	5.372	6.012

4. **Normalized Routing Load:**

Normalized Routing Load can be measured as:

Normalized routing load = total number of routing packets/total number of delivery packets. Also, each of the forwarded packets is considered as one transmission. Table 13.5 shows that the performance of IWDRA-ZRP is better than both ZRP and RA-ZRP for varying number of nodes. Figure 13.9 exhibits the performance variations of the three protocols through a bar graph and its X-graph is depicted in Figure 13.10, where the *x*-axis portrays the number of nodes and the *y*-axis portrays NRL.

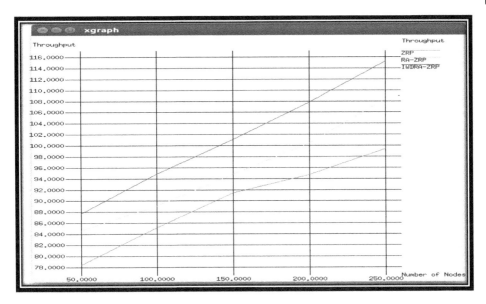

Figure 13.8   X-Graph for throughput with dynamic nodes.

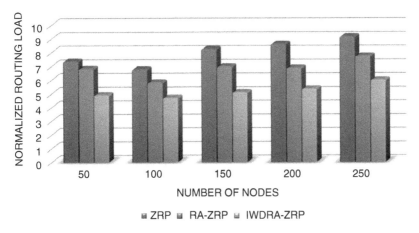

Figure 13.9   Varying NRL with dynamic nodes.

5. **Packet Loss Ratio:**

   Packet loss Ratio can be calculated by utilizing the function as: *Packet Loss Ratio* = (NP_s − NP_l) ∗ $100/NP_s$, where, NP_s is the number of sent packets and NP_l is number of lost packets. Table 13.6 show that the performance of IWDRA-ZRP is better than both ZRP and RA-ZRP for varying number of nodes. Figure 13.11 exhibits the performance variations of the three protocols through a bar graph and its X-graph is depicted in Figure 13.12, where the *x*-axis portrays the number of nodes and the *y*-axis portrays PLR.

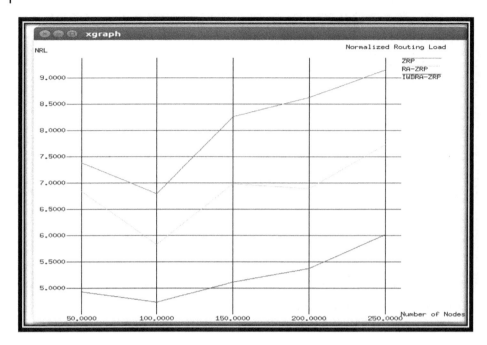

Figure 13.10   X-Graph for NRL with dynamic nodes.

Table 13.6   Packet loss ratio.

Protocols	PackPLRet loss ratio (varying nodes)				
	50	100	150	200	250
ZRP	35	30.5	26.8	25.2	22.9
RA-ZRP	28	24.1	22.8	22.2	17.5
IWDRA-ZRP	21	18.8	15.2	14.1	13.9

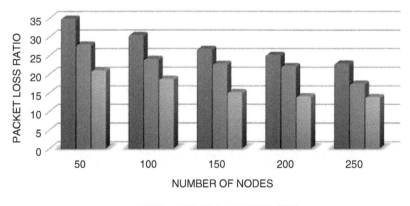

Figure 13.11   Varying PLR with dynamic nodes.

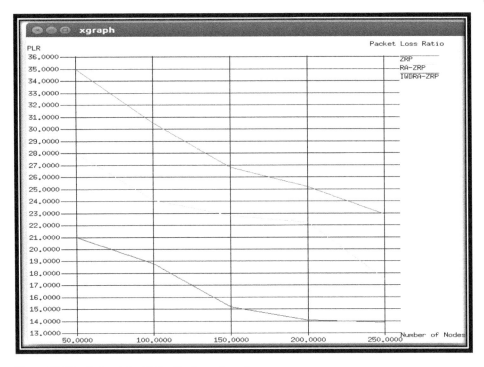

Figure 13.12 X-Graph for PLR with dynamic nodes.

13.5 Conclusion and Future Work

The current situation where the entire world is struggling to deal with the COVID-19 pandemic, the IoMT-MANET can contribute substantially. Regular patients can be monitored by their doctors digitally without having to visit the hospitals where they would be at the high risk of becoming infected by the virus. So, in this scenario, IoMT-MANET can enhance the efficiency of sharing data online and made it convenient for both doctors and patients. In IoMT-MANET, routing that is too normalized is the first and foremost requirement for quality-of-service improvement in MANET, in order to support the IoMT environment. The ZRP routing protocol of MANET is considered to be the most popularly utilized hybrid routing protocol. Although its performance is good, it still suffers from various vulnerabilities which need to be prioritized. So, the proposed nature-inspired approach intelligent water drop algorithm is utilized in this research study for performance optimization. The results analysis presented in this chapter validates the proposed approach through various performance factors such as PDR, delay, throughput, PLR, and NRL. The X-graph results show effectiveness and optimization of the proposed swarm intelligence-based nature-inspired IWD algorithm. The QoS optimization through the IWD approach validated in this chapter will serve as a basis for various IOMT services which is a need of the hour in a pandemic situation prevailing worldwide. Also, the proposed approach can be utilized for the betterment of the "Aarogya Setu" app launched by the Indian government.

References

1 Agnihotri, S. and Ramkumar, K.R. (2019). Content based routing algorithm to improve QoS in IoMT networks. In International Conference on Computational Intelligence, Security and Internet of Things, 195–206. Singapore: Springer.

2 Ahmed, T. and Rahman, R. (2016). Survey of anomaly detection algorithms: toward self-learning networks. In *Security of Self-Organizing Networks*, (Al-Sakib, K.P. ed.). 83–108. Auerbach Publications.

3 Beijar, N. (2002). *Zone routing protocol (ZRP)*. Finland: Netwoking Laboratory, Helsinki University of Technology 9: 1–12.

4 Bhatia, D. and Sharma, D.P. (2016). A comparative analysis of proactive, reactive and hybrid routing protocols over open source network simulator in mobile ad hoc network. *International Journal of Applied Engineering Research* 11 (6): 3885–3896.

5 Challoner, A. and Popescu, G.H. (2019). Intelligent sensing technology, smart healthcare services, and internet of medical things-based diagnosis. *American Journal of Medical Research* 6 (1): 13–18.

6 Joyia, G.J., Liaqat, R.M., Farooq, A. et al. (2017). Internet of Medical Things (IoMT): applications, benefits and future challenges in healthcare domain. *Journal of Communication* 12 (4): 240–247.

7 McFarland, R J. and Olatunbosun, S.B. (2019). An exploratory study on the use of Internet_of_Medical_Things (IoMT) in the healthcare industry and their associated cybersecurity risks. In Proceedings on the International Conference on Internet Computing (ICOMP) 1:15–121. The Steering Committee of The World Congress in Computer Science, Computer Engineering and Applied Computing (WorldComp).

8 Mehta, D., Kashyap, I., and Zafar, S. (2017). Consummate scalability through clustered approach in ZRP. *International Journal of Sensors Wireless Communications and Control* 7 (3): 178–187.

9 Mehta, D., Kashyap, I., and Zafar, S. (2017). Routing optimization in cloud networks. *International Journal of Advanced Research in Computer Science* 8 (2): 16–18.

10 Mehta, D., Kashyap, I., and Zafar, S. (2018). Neoteric RA approach for optimization in ZRP. *Innovations in Computational Intelligence* 103–114). Singapore: Springer.

11 Mehta, D., Kashyap, I., and Zafar, S. (2018). Protract route optimization in ZRP through novel RA approach. *International Journal of Sensors Wireless Communications and Control* 8 (1): 19–25.

12 Mehta, D., Kashyap, I., and Zafar, S. (2018). Synthesized hybrid ZRP through aggregated routes. *International Journal of Information Technology* 10 (1): 83–89.

13 Ning, Z., Dong, P., Wang, X. et al. (2020). Mobile edge computing enabled 5G health monitoring for internet of medical things: a decentralized game theoretic approach. *IEEE Journal on Selected Areas in Communications* 39: 468–478.

14 Paul, H. and Sarkar, P. (2013). A study and comparison of OLSR, AODV and ZRP routing protocols in ad hoc networks. *International Journal of Research in Engineering and Technology* 2 (8): 370–374.

15 Pearlman, M.R. and Haas, Z.J. (1999). Determining the optimal configuration for the zone routing protocol. *IEEE Journal on Selected Areas in Communications* 17 (8): 1395–1414.

16 Perkins, C.E. amd Bhagwat, P. (1994). Highly dynamic destination-sequenced distance-vector routing (DSDV) for mobile computers. *ACM SIGCOMM Computer Communication Review* 24 (4): 234–244).

17 Ravilla, D., Sumalatha, V., and Reddy, P.C.S. (2011). Performance comparisons of zrp and izrp routing protocols for ad hoc wireless networks. In 2011 International Conference on Energy, Automation and Signal, 1–8. IEEE.

18 Samad, A., Sinhab, G., Khalidc, M. et al. (2010). Registration and aggregate cache routing for ad hoc network. *International Journal of Advanced Networking and Applications* 2 (3): 699–706.

19 Sayad, L., Bouallouche-Medjkoune, L., and Aissani, D. (2017). IWDRP: an intelligent water drops inspired routing protocol for mobile ad hoc networks. *Wireless Personal Communications* 94 (4): 2561–2581.

20 Seyyedtaj, M. and Jamali, M.A.J. Security improvements zone routing protocol in Mobile ad hoc network. *International Journal of Computer Applications Technology and Research* 3 (9): 536–540.

21 Shah-Hosseini, H. (2007). Problem solving by intelligent water drops. In 2007 IEEE congress on evolutionary computation. 3226–3231. IEEE.

22 Shah-Hosseini, H. (2008). Intelligent water drops algorithm: a new optimization method for solving the multiple knapsack problem. *International Journal of Intelligent Computing and Cybernetics* 1 (2): 193–212.

23 Shah-Hosseini, H. (2009). The intelligent water drops algorithm: a nature-inspired swarm-based optimization algorithm. *International Journal of Bio-inspired computation* 1 (1–2): 71–79.

24 Sharma, N., Batra, U., and Zafar, S. (2019). A neoteric swarm intelligence stationed IOT–IWD algorithm for revolutionizing pharmaceutical industry leading to digital health. In *Emergence of Pharmaceutical Industry Growth with Industrial IoT Approach*, (Balas, V.E., Solanki, V.K., and Kumar, R. eds.). (1–19). Academic Press (Science Direct).

25 Sharma, N., Batra, U., and Zafar, S. (2019). Confine ingenious firefly algorithm correlating water drop algorithm for amassment in networks. *EAI Endorsed Transactions on Scalable Information Systems* 7: 1–11. doi: 10.4108/eai.13-7-2018.159625.

26 Sharma, N., Batra, U., and Zafar, S. (2019). Incalculating global optimization in ZRP through newfangled firefly algorithm. *International Journal of Computer Network and Information Security* 11 (2): 43.

27 Sharma, N., Batra, U., and Zafar, S. (2019). QoS optimization in networks through Swarm intelligence algorithm for sustainable big data management. International Conference on Emerging Trends in Information Technology 2019 for Lecture Notes in Electrical Engineering 605: 575–585.

28 Sharma, N., Batra, U., and Zafar, S. (2019). Remit accretion in IOT Networks encircling ingenious firefly algorithm correlating water drop algorithm. In International Conference on Computational Intelligence and Data Science (ICCIDS 2019), *Procedia*.

29 Sharma, N., Batra, U., and Zafar, S. (2019). Trust based hybrid routing approach for securing MANET. In 5th International Conference on Computing for Sustainable Global Development, IEEE Conference.

30 Sharma, N., Zafar, S., and Batra, U. (2019). Catechize global optimization through leading edge firefly based zone routing protocol. *Recent Patents on Computer Science* 12: 1–11. doi 10.2174/2213275912666181128121811.

31 Sivaparthipan, C.B., Rathinaraja, J., and Kumar, S.S. (2015). Intelligent water drop algorithm for workflow scheduling in cloud computing. *International Journal of Advance Information and Engineering Technology* 9 (9): 43–49.

32 Wazid, M., Das, A.K., Rodrigues, J.J. et al. (2019). IoMT malware detection approaches: analysis and research challenges. *IEEE Access* 7: 48576–48634.

33 Zafar, S. and Mehta, D. (2018). Neoteric iris acclamation subtlety. *Innovations in Computational Intelligence*, (1–15). Singapore: Springer.

34 Zafar, S. (2017). Cyber secure corroboration through CIB approach. *International Journal of Information Technology* 9 (2): 167–175.

35 Zafar, S. and Soni, M.K. (2014). Sustaining security in MANET: Biometric stationed authentication rotocol (BSAP) inculcating meta-heuristic genetic algorithm. *International Journal of Modern Education and Computer Science* 6 (9): 28.

14

Appraise Assortment of IoT Security Optimization

Ayesha Hena Afzal

Department of Computer Science and Engineering, FET, MRIIRS, Haryana, India

14.1 Introduction

With an intensive literature survey, it can be concluded that the Internet of Things (IoT) is a system of interrelated computing devices, digital and mechanical machines, objects, animals, or people. They all are provided with a unique identifier and can transfer data over a network without any involvement of human-to-human or human-to-computer interaction. An article published in an RFID journal in (1999) said: "Suppose if we have a computer that knows everything regarding to know about things with the help of all the relevant data gathered without human involvement." Rise in devices connected to the Internet is represented in Figure. 14.1.

IoT is gaining attention, both in the workplace and the environment. Its main objective is not only to impact on how to live life, but also how to work. Now we will throw some light on what are the impacts of IoT on society? This will lead to incredible reduce of loss, cost, and waste. For example, it will help us to know when appliances need repair, recall, and replacement, and whether they are still considered as new or past their expiry dates. The popular example is a smart fridge; for example, what if your fridge could alert you that the milk is out-of-date or there is no milk. With the help of its internal camera, it can see there is no milk present or the carton has passed its expiry date.

This chapter discusses the security of IoT and applications of IoT in the real world.

A. *Why IoT security so critical?*

IoT security is important because the growth of the Internet, connected data and devices, its application, and users have increased exponentially. IoT consists of such a wide array of networks, data, and services which are more vulnerable to hacking.

The main problem is because networking devices and appliances are very new, so robust security is still missing in product designs. The products are still sold with old and traditional operating security systems, and this increases the risk of attack to devices and applications.

One of the dangers of these attacks is breaches to private and confidential data. In 2013, hackers stole 70 million personal records of individuals by gaining access to the

Smart and Sustainable Approaches for Optimizing Performance of Wireless Networks: Real-time Applications, First Edition.
Edited by Sherin Zafar, Mohd Abdul Ahad, Syed Imran Ali, Deepa Mehta, and M. Afshar Alam.
© 2022 John Wiley & Sons Ltd. Published 2022 by John Wiley & Sons Ltd.

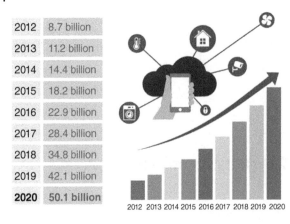

2012	8.7 billion
2013	11.2 billion
2014	14.4 billion
2015	18.2 billion
2016	22.9 billion
2017	28.4 billion
2018	34.8 billion
2019	42.1 billion
2020	**50.1 billion**

2012 2013 2014 2015 2016 2017 2018 2019 2020

Figure 14.1 Rise in Internet usage from 2002–2020.

connected heating, ventilation, and air conditioning (HVAC) system. Hacking of baby monitors on the market or in the home enables a third party to watch and record your personal life on video, and the video clips which are stored online. Also, hacking with a video camera can let a third party know your movements, whether you are at home or not, and make it easier for them to conduct crime, which poses a massive security risk. Therefore, security plays a crucial role in IoT.

Figure 14.2 represent the various components of the Internet of Things.

B. *Real-world application of IoT.*

As the previous section discusses why it is important to maintain security of IoT, now this part shows how IoT plays an important role in our daily lives. IoT is slowly changing into a vital part of our life. It extends from smart homes to healthcare to the clothes we wear. It not only enhance our comfort, but also simplifies and controls our daily work, personal tasks, work, lifestyle, etc. As is known, the IoT market is extremely large; however, some domain can mature more quickly than others.

We will discuss the application areas of IoT that have the potential for exponential growth.

Smart Home: The smart home is the one in which devices can communicate with each other, as well as with the environment and the Internet. With the assistance of house owners, IoT will manage and customize their home surroundings so as to extend security and energy management. For example, the Nest Learning Thermostat saves on average up to 12% on heating bills and 15% on air-conditioning.

Wearable devices: Wearable devices collect the data and information about the users with the help of software and sensors installed in the devices. Some examples include entertainment systems which can play music, play games, or watch movies according to their mood. Tracker belt is also an example of wearable devices.

Agriculture: This project is to assist the farmer in monitoring necessary information concerning their environment, such as humidity, air temperature, and soil quality with the assistance of remote devices that help to plan, improve yields, and create a harvest forecast.

Farming: A farm animal can be fitted with a biochip transponder. This comprises of an injectable chip, an electronic device that is inserted under the skin of the animal to provide

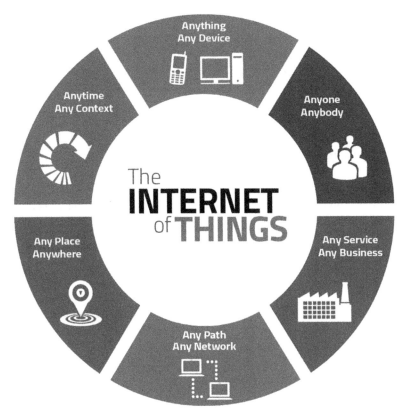

Figure 14.2 Components of Internet of Things.

a unique identification number of that animal. It helps to maintain the animal safely and efficiently, and it is less painful and more cost-effective than existing methods.

Until now we have seen the importance of IoT, how its security is so crucial, and the various applications based on IoT in everyday life. The upside of IoT is that it is now possible to do the things which we never dreamed of before. But like with all advantages, there are always disadvantages and this is also true of IoT. It is turning into an increasingly enticing target for cyber criminals. More connected devices result in an additional attack vector and additional risk for hackers to target us. Unless it is quickly addressed, this rising security concern means that we will be facing inevitable disaster.

"IoT security previously ignored has now become an issue of high concern."

Among various mechanisms, one of the strongest mechanisms to handle IoT security issues is Biometrics. This has been described as the identification of physical characteristics. These features help eliminate error in authentication.

Biometric features are categorized as Physiological which include (fingerprints, hand geometry, eye [iris and retina] face and ear), behavioral biometrics (signature, keystroke, voice, etc.), and another is the esoteric biometric (DNA, odor, palm vein, etc.). Among these, iris recognition is one of the most promising approaches due to stability, singularity, and non-invasiveness. As an introduction to IoT, its security, various application of IOT in the real world, and various strong mechanisms to secure IoT, were discussed in the previous section. The remainder of this chapter is organized as follows. Section 14.2 lists the literature that includes the related work and papers regarding IoT and its security attractions. Various security breaches are also discussed in this section. Analysis of various traditional security mechanisms are described in Sections 14.3 and 14.4, followed by conclusions and future scope.

14.2 Literature Review

An extensive literature survey about IoT and its security is provided in Table 14.1.

After performing the intensive literature surveys shown in Table 14.1, the following gaps have been highlighted in the IoT mechanism, which aims to develop an efficient, secure mechanism by using specific cryptographic/biometric techniques.

Various security breaches in IoT are now discussed:

There are various types of cyber attack. Here are some examples:

1. *BOTNETS*
   Botnet is a network of systems which are combined for the purpose of taking over the control of a system and distributing malware throughout that system. It is controlled by the operator via a command and control server. Criminals use them on a grand scale for various activities such as stealing of personal data and information, phishing emails, etc. [16].
2. *Man-in-Middle Concepts*
   Man-in-Middle is where primarily the attackers or the hacker interrupts and breaches communication between two systems. Attackers interrupt and transmit messages between two parties who believe that they are in communication with one another directly. The attackers trick the recipient into thinking that they are obtaining a legitimate message. The hackers can be extremely dangerous in IoT because of the nature of the "things" being hacked [16].
3. *Data and Identity Theft*
   Data and Identity Theft includes unpredictable hackers who perform unauthorized access to data and money with different types of impressive hacks. Internet users are often their own worst enemy, such as careless safekeeping of Internet connection devices, e.g., mobile phone, laptop, iPad, etc., which plays into the hand of hackers and malicious thieves [16].
4. *Social Engineering*
   Social Engineering is the act of manipulating people, such that they reveal their own confidential data and information. The attackers are mainly seeking the type of information that varies, but when they target the individual, they are trying to obtain their basic details like bank details, passwords, etc. They install the malicious software into the system of the user, which will provide them with access to personal information and gain control over that computer. An example of social engineering are phishing emails [16].

Table 14.1 Literature survey of IoT security.

Reference	Author name	Paper title	Year and location	Outcome of research
1	Suo et al. [1]	Security in the IoT: A review	March 2012 at Hangzhou, China	By analyzing the security, architecture, and features, various security requirements are provided. Various research status of key technologies is discussed, which include encryption mechanisms, protecting sensor data, communication techniques, and various cryptographic algorithms
2	Abomhara and Koien [2]	Security and privacy in the IoT: Current status and open issues	May 2014 at an international conference in Aalborg, Denmark	The authors, mainly point out the major security issues, security threats, and. major open challenges in the IoT domain As the IoT system is available everywhere, so issues related to its security will be of major concern
3	Kumar et al. [3]	Security in IoT: Challenges, solutions, and future directions	January 2016 at 49th Hawaii International Conference	Their main focus was on the challenges, solutions, and future directions related to security of IoT. The security of IoT became a major challenge as IoT applications are used in various fields, i.e., hospitals, cities, grids, buildings, and organizations. They discussed security attacks by the different layers which compromise IoT; the various methods which provide solutions to these problems with their limitations and what future works are recommended to overcome these limitations
4	Zhao and Ge [4]	A survey on the IoT security	December 2013, Leshan	They introduce the architecture and features of IoT security. Issues arise due to 3-layer structure and various solutions to these problems. Security can be provided by various algorithms, security protocols, etc.
5	Jie Lin et al. [5]	A survey on IoT: Architecture, enabling technologies, security and privacy, and applications	Mar 2017	They present a survey of IoT, architecture, technologies used, privacy, NS security, and application of IoT in the real world. They proposed Fog/Edge computing, integrated with IoT that enables various computing services. These help to improve the user's experience and recovery services in case of failure. To develop a Fog/Edge-based computing IoT infrastructure, first architecture-enabling technique issues related to IoT are investigated and then Fog/Edge computing and IoT is established

Table 14.1 (Continued)

Reference	Author name	Paper title	Year and location	Outcome of research
6	Granjal et al. [6]	Security for the IoT: a survey of existing protocols and open research issues	January 2015 at Portugal	They discuss how IoT has introduced the future vision of the Internet where users, computing devices, and objects lead to increase in their accuracy capabilities, processing, etc., which lead to economic benefits. Current architecture and protocols will play a key role in connectivity of devices in various IoT-based applications. Security will be the main concern: a different mechanism is developed to protect communication. Here, the survey recognizes the existing protocol challenges related to these protocols, as well as all open issues. Also, they cover various security fundamentals and secure communication in the IoT.
7	Zafar and Mehta [8]	Bi-orthogonal Wavelet Based Corner Template Encoding Methodology	June 2016, MR International Journal of Engineering and Technology	They focus on extracting features of the iris, which is one of the strong biometric features. Encoding is the most important operation. They discuss that template formation or encoding is performed through normalizing iris patterns with bi-orthogonal wavelength 3.5. They conclude that the major advantage of this scheme over classical construction methods is that it does not rely on the Fourier transform and the result is faster implementation of wavelength transform.
8	Zafar et al. [7]	An optimized genetic stowed biometric approach to potent QoS In MANET.	2015 International Conference on Soft Computing And Software engineering.	The main focus is on security breaches that occur in MANET (mobile ad-hoc network) due to self-configuration and less infrastructure. So, to secure MANET, an algorithm is proposed to cogitate biometric perception as the most innovative solution to incorporate iris features. With the help of genetic algorithm QoS-based issues, MANET is elucidated, resulting in optimized performance of the network. [7]

5. *Denial of Service*

Denial of Service (DoS) attacks occur when the service that would usually work suddenly becomes unavailable. This may be due to various reasons, but it usually refers to an infrastructure that cannot cope due to overloading of its capacity. A large number of systems are maliciously attacked in the distributed DoS attack. It is mostly done through BOTNET.

DoS, unlike other attacks, does not try to steal information or lead to security loss [16].

14.3   Analysis of Traditional Security Mechanisms in IOT

Table 14.2 shows an analysis of traditional security mechanisms in IoT and discusses their attacks and drawbacks.

After studying the various traditional cryptographic mechanisms presented in Table 14.2, it can be highlighted that all of these mechanisms face various attacks which could be destructive to IoT security. Therefore, this research analysis focusses on introducing

Table 14.2   Analysis of traditional security mechanisms in IoT and their attacks and drawbacks.

Reference	Name of mechanism	Description of mechanism	Different attacks or drawbacks
1	Advanced Encryption Standard (AES)	AES is a symmetrical block cipher standardized by NIST, which is used by the US Government to encrypt sensitive data worldwide. It is easy to implement on various hardware and software, as well as in a very restricted environment. Various features of AES are secure, cost-efficient, and capable of handling a 128-bit block while using key sizes of 128, 192, and 256 bits. It uses a substitution permutation network and works on a 4 × 4 matrix.	AES is vulnerable to Man-in-Middle attacks. [12]
2	High security and lightweight (HIGHT)	HIGHT is used for basic operations like XoR operation on a fiestel network. Key for HIGHT is generated while in the encryption and decryption phase. It has a block size of 64 bits, with a 128-bit key of 32 rounds. Main features of HIGHT are that it requires less power, fewer lines of code, and it improves the speed of RFID [11]	HIGHT is vulnerable to saturation attacks.
3	Bootstrap Security	Security is the most imperative factor in the success of IoT. Secure transmission of data will always be a challenge in this growing area. Today, Generic bootstrap architecture technology supports data integrity and authentication in IoT. [13]	To share the initial key with the smart phone or home gateway, a QR code is required. This QR code is in a Package of Things and an employee of a company can easily see the QR code and key. The attacker can perform a dictionary attack to get the key and plain text. Also, a long and random key cannot be stored in the device. [14]

Table 14.2 (Continued)

Reference	Name of mechanism	Description of mechanism	Different attacks or drawbacks
4	Elliptical Curve Cryptography (ECC)	ECC is a type of public key encryption technique which is based on elliptic curve theory and which is used for creating faster, smaller, and efficient cryptographic keys. It uses 164-bit keys to provide levels of security rather than 1024-bit keys used by other systems. They provide equivalent security with lower computing power and battery resource usage. It is a looping line intersecting two axes based on an equation created by a mathematical group.	An HP researcher, Nigel Smart, discovered a flaw in which some curves are extremely vulnerable. [9]
6	PRESENT	PRESENT is based on SPN and is used as an ultralight weight algorithm for security. It requires 4-bit input and output s-boxes, working mainly on a substitution layer. It works on 64-bit size block and key of 80- or 128-bit. [10]	PRESENT is vulnerable to Integral attacks. It is a powerful technique to recover secret keys. [15]
7	RC5	RC5 is proposed by Rivest for the rotations that are data independent. It is used mainly in wireless scenarios and works well as a lightweight algorithm. It possesses a fiestel structure and works on 32-bit size, which can also vary between 16, 32, and 64. It can work for 0,1, ...255 with 0,1, ...255 key bytes. But standard for RC5 is 16 bytes on 20 rounds.	RC5 is vulnerable to differential attacks.
8	RSA	RSA was invented by Ron Rivest, Adi Shamir, and Leonard Adleman in 1978. It works by selecting two large prime numbers and then generating public and private key pairs of them, after finding their modulus and choosing encryption key at random and then calculating the decryption key. The public key is published to everyone and the private key is made secure	RSA is vulnerable to various attacks, i.e., cycle attacks, searching message space, guessing, etc.

Biometric as one of the most reliable and secure mechanisms for maintaining security in an IoT network.

Biometric is a very efficient mechanism to provide convenient and secure authentication to a complete transaction. It is a method of recognition of a person based on their physiological or behavioral characteristics. The current Biometric-based solution is able to maintain confidentiality and privacy of important data efficiently.

14.4 Conclusion and Future Scope

In this chapter, an extensive literature survey focusses on IoT, various attacks that affect IoT security, and various traditional mechanisms that exist for securing IoT services, through Tables 14.1 and 14.2. The research analysis performed in this chapter specifies a security mechanism for IoT that will guarantee confidentiality, data integrity, and authentication. A future recommendation for this research study aims to build an extensive secure mechanism for IoT through various biometric mechanisms, e.g., DNA, iris, etc. Biometric systems are more user-friendly and surely will play a key role in developing a secure and advanced IoT system in the future.

References

1 Suo, H. Wan, J., and Zou, C. (2012).Security in the Internet of Things: a review. In 2012 International Conference on Computer Science and Electronics Engineering (ICCSEE).

2 Abomhara, M. and Koien, G.M. (2014). Security and privacy in the Internet of things: current status and open issues Privacy and Security. In 2014 International Conference on Mobile Systems (Prisms).

3 Kumar, S.A., Vealey, T., and Srivastav, H. Security in Internet of Things: challenges, solutions and future directions.

4 Zhao, K. and Ge, L. (2013).A survey on the Internet of Things security. In 2013 9th International Conference on Computational Intelligence and Security.

5 Lin, J., Yu, W., and Zhang, N. (2017). A survey on Internet of Things: architecture, enabling technologies, security and privacy, and applications. *IEEE Internet of Things Journal* 4 (5): 1125–1142.

6 Granjal, J., Monterio E., and Silva, J.S. (2015). Security for the Internet of Things: a survey of existing protocols and open research issues *IEEE Communications Surveys & Tutorials* 17 (3): 1294–1312.

7 Zafar, S., Soni, M.K,. and Beg, M.M.S. (2015). An optimized genetic stowed biometric approach to potent QOS in MANET. In 2015 International Conference on Soft computing and Software Engineering (SCSE).

8 Zafar, S. and Mehta, D. (2016). Bi-orthogonal wavelet based comer template encoding methodology *MR international Journal of Engineering and Technology* 8 (1): 7–15.

9 http://searchsecurity.techtarget.com/definition/elliptical-curve-cryptography

10 http://searchsecurity.techtarget.com/definition/block-cipher

11 http://searchsecurity.techtarget.com/definition/stream-cipher

12 http://searchsecurity.techtarget.com/definition/Advanced-Encryption-Standard

13 https://www.ericsson.com/en/publications/white-papers/bootstrapping-security–the-key-to-internet-of-things-authentication-and-data-integrity

14 https://link.springer.com/chapter/10.1007/978-3-319-02726-5_24

15 https://www.globalsign.com/en/blog/five-common-cyber-attacks-in-the-iot

16 https://www.globalsign.com/en/blog/five-common-cyber-attacks-in-the-iot/

15

Trust-Based Hybrid Routing Approach for Securing MANET

Neha Sharma[1] and Satrupa Biswas[2]

[1]*Department of EEE, G D Goenka University, Sohna, Haryana, India*
[2]*Department of Information Communication and Technology, Manipal University, MAHE*

15.1 Introduction

An ad-hoc network refers to a network that is generally created for a particular purpose. Such networks are mostly created for one-time use only, that is why they are commonly known as temporary networks. MANET (Mobile Ad-hoc Network) is an ad-hoc network only, a self-organizing gathering of wireless portable nodes that form a momentary and dynamic wireless network deprived of any infrastructure [1]. The rudimentary notion of using MANET is that the exchange of information in between the portable nodes count on the swift arrangement of a momentary network. Also, each node in a MANET can travel spontaneously in any direction and can change its links to the other nodes repeatedly. The rudiments in establishing a MANET is the constant preservation of routing information at each node. Figure 15.1 shows a basic structure of nodes in MANET. The routing conventions of such networks are isolated into three gatherings: Proactive Routing Protocols (Table Driven), Reactive Routing Protocols (Request Driven), and Hybrid Routing Protocols (Combination of Proactive and Reactive) [1–3].

The Hybrid Protocols attempt to fuse different parts of proactive and reactive routing conventions. They are habitually used to provide progressive routing. They use distance vectors to establish the best possible paths and report routing information only when there is a transformation in the topology of the network [4, 5].

Hybrid Routing Protocols joins the benefits of both protocols and attempts to achieve high-performance parameters [2, 6]. The hybrid protocols are proficient enough to render higher scalability than its corresponding protocols. The creativity of hybrid protocols is that they attempt to be liberated from the single point failures in the network and bottleneck problems by allowing all the nodes to route and forward data in case of unattainability of an ideal path, e.g., ZRP, TORA [7, 8].

After elaborating on the efficient Hybrid protocol, the next section of this chapter contains the various literature reviewed to learn and draw conclusions on this topic.

Smart and Sustainable Approaches for Optimizing Performance of Wireless Networks: Real-time Applications, First Edition.
Edited by Sherin Zafar, Mohd Abdul Ahad, Syed Imran Ali, Deepa Mehta, and M. Afshar Alam.
© 2022 John Wiley & Sons Ltd. Published 2022 by John Wiley & Sons Ltd.

Figure 15.1 Example of ad-hoc network.

15.2 Literature Review

The Literature Review shows the numerous studies carried out in the field of Hybrid Routing Protocols of MANET and also contrasts the hybrid routing protocols with reactive and proactive routing protocols.

The various challenges of MANET has been discussed in [9], which includes the infrastructureless scenario, limited memory, resource constraints, and the dynamic topology of MANET along with the limited computing capacity. All the above-discussed issues in MANET demand new or modified network strategies that can be used for better and efficient communication. Along with this, the routing serves a decisive role in enumerating the path to the destination from the source. Additionally, another major challenge in MANET is mobility modeling and control that bids for a noticeable human focus as it makes the topology of MANET more assertive.

Attention has been brought to the various network simulators in [4], which are found nowadays in the market, such as Ns2, Ns3, OMNET++, OPNET, NetSim, J-Sim, GloMoSim, REAL, QualNet, jist, and SWAN. The purpose of these open-source network simulators is to analyze the network performance and its metrics of measurement. The performance parameters for these simulators are generally Packet Delivery Ratio, Routing Overhead, Throughput, Average end-to-end delay, and Path Optimality [5]. Also, the major factors that affect the efficient routing in MANET during communication are Network Size and Ever-Changing Topology. During their research. it has been found that the OMNET++ and Ns3 are simulators that have been categorized as the most mature simulators. The OMNET++ provides better GUI support, whereas Ns3 performs tremendously in the case of larger models. Furthermore, the researchers have carried out a comparison between various MANET routing protocols with the help of more than two open-source network simulators.

The work presented in [7] had considered SZRP that has a major objective to achieve efficient secure neighbor discovery, secure routing packets, identification of noxious nodes, and preventing these nodes from damaging the network. This will certainly enhance the working of the basic ZRP.

In [10], the researchers have examined the MANET and concluded that in MANET all the nodes work together to transmit the data from a source to a destination. The above-mentioned situation can make the wireless channel prone to active and passive attacks by noxious nodes. The security threats in MANET can be Denial of Services, Eavesdropping, and Spoofing, etc. So, primarily it needs the implementation of security, and solutions of prevention, detection, and reaction mechanisms have been considered [11]. Furthermore, the evolution of a technique that involves the functionality of these mechanisms in a GUI environment can be carried out in order to overcome the security breach.

The study carried out in [8] has emphasized that in MANET, because of the swift displacement of mobile nodes, link damage takes place, which results in continual path failures and route discoveries. Also the mobility issue arises due to the arrival and departure of nodes at any point of time in the network. The path failures lead to data interruption, degrading the performance of the overall system. So performance and efficiency need to be improved.

In [5], the path discovery in MANET has been scrutinized. The mobile communication service is identical to an absolute communication approach attaining communication anywhere, anytime, and with anyone. The number of links and the fidelity of each link that establish the path are two major aspects on which the availability of the path depend [11]. A shortest path routing opts for a path that has minimum cost to forward the data to the destination node, and the selection of a routing algorithm that yields the shortest path depends upon direct traffic from source to destination, lessening the cost and amplifying the network performance. The elementary objective of such a routing algorithm is the correct and efficient route establishment in between two nodes [12]. This also ensures the timely delivery of messages in between the links. The bandwidth consumption, overheads, and cost must be minimum with route establishment.

The study of [13] has focused on CRAHNs, i.e., Cognitive Radio mobile Ad-Hoc Network. The social behavior of ants, bees, and termites has inspired these new problem-solving methods. Swarm intelligence provides self-organization, decentralization, adaptivity, robustness, and scalability that makes it ideal for CRAHNs. The performance parameters for evaluating swarm intelligence are mainly the packet delivery ratio, end-to-end delay or latency, and primary users' activity.

The intention of [14] has compared various routing protocols for MANETs. The routing in MANET is a cynical function and the reason behind this is the highly dynamic environment. Basically, the wireless links in MANET are eminently error prone and due to the mobility of nodes these links can go down frequently. Also, the interference and limited infrastructure makes it more vulnerable to bugs. Over the years, various remarkable protocols have been introduced for MANET, which have their own supports and restrictions. The discussed protocols can be further improved to serve MANET in much better ways.

The researchers in [1] has concluded that the extremely dynamic environment makes the routing a difficult task. Another crucial and challenging aspect of MANET that requires massive human focus is security issues along with their goals and attacks. Mainly two types of attacks have been considered, which are passive and active.

The researchers in [2] has elaborated on the various applications and challenges of WSN (Wireless Sensor Network). It is not possible to carry out very long-term operations with WSN, for which the limited battery life is responsible.

The hybrid WSN been discussed is the collaboration of the conventional static sensor network with the latest mobility technology. The hybrid WSN ensures the balanced consumption of energy by the nodes along with the extension of the network of the network. Furthermore, the hybrid WSN can be charged with renewable energy and the cost of sensor nodes needs to be lower, for which more research needs to be carried out.

The purpose of the researchers in [15] is to analyze the different swarm intelligence optimization techniques used for MANET. The limitations of WANET, such as increased energy consumption, requirement of high bandwidths, insecure routing, packet losses, etc., have been removed by MANET. Though the MANETs are very efficient in providing solutions for heuristic problems, they are not compatible to resolve the metaheuristic problems. Various solutions to these metaheuristic approaches have been proposed over time, some of which are PSO, ACO, ABC, BFOA, CSO, GCO, etc.

The researchers in [16] explain that the re-establishment of the already determined routes arises rapidly in MANET, as the mobile nodes frequently undergo link altercations. Thus, this route maintenance is a major responsibility of a routing algorithm. To overcome such issues, a routing algorithm needs to be adopted that can achieve an efficient, robust, and scalable routing in MANET. The routing algorithm needs to be aware of the already existing network topology and all the possible resources. A multi-objective MANET path optimization issue has been addressed by these researchers. The performance parameters for such problems are end-to-end delay, hop distance, load, cost, and reliability. A few of the multi-objective issues in MANET that have been considered are the problems of connectivity, QoS, multicast routing, energy-efficient clustering, etc. Section 15.3 of this chapter focusses on the gaps and objectives, which has been drawn from the literature review.

15.3 Gaps and Objectives from the Literature Review

Some gaps and objectives from the literature review are:

1. To overcome the security breach, the evolution of a technique that involves the functionality of prevention, detection, and reaction mechanism in a GUI environment can be carried out [10].
2. MANET experiences rapid path failures due to which data interruption occurs. It results in the degradation of the performance of the overall system. This issue needs to be addressed [5, 7].
3. A latest strategy under the swarm intelligence optimization is the Artificial Immune System, which can be explored [15].
4. Some parameters such as delay and congestion can be considered to check the accuracy of the routing algorithm [2].
5. The comprehensive performance of some metaheuristics implemented with multi-objective optimization model can be explored [12, 17].
6. After discussing these gaps and objectives, Section 15.4 focuses on the methodology to be adopted.

15.4   Methodology to be Adopted

Besides emerging diversion toward MANET, owing to the problems with infrastructure-less implementation of the network to reinforce communication, large-scale networks are required to be addressed. This can be facilitated with the help of Hybrid Routing Protocols. Mainly, the focus of this research will be inclined toward the betterment of routing and security [18, 19].

After focussing on relevant analyses and the objectives presented in the previous sections, the methodology of the proposed algorithm (Figure 15.2) will focus on utilization of one

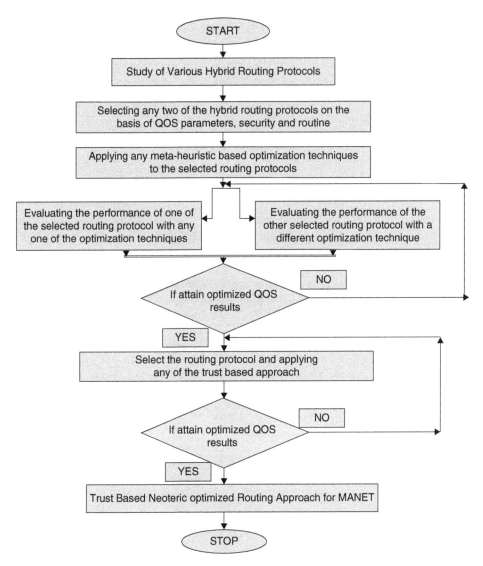

Figure 15.2   Flowchart of the proposed methodology.

of the following metaheuristic-based techniques for QoS optimization of hybrid routing protocols [13, 16]. The various metaheuristic techniques [17, 20] that can be adopted are:

- ACO
- PSO
- ABC
- GSO
- CSO, etc.

After applying different optimization techniques to any two of the hybrid routing protocols, a comparison analysis will be performed in order to choose the best one out of the two selected routing protocols. Furthermore, to enhance the security of the selected routing protocol, any one of the strong trust-based approaches will be applied to the selected routing protocol for building a Trust-Based Neoteric Optimized Routing Approach for MANET [21–23]. The various techniques that can be used to enhance security are:

- Cryptography
- Biometric
- Trust-based approach
- Hash function algorithm, etc.

15.5 Comparison Analysis

Various researches on the above three protocols have revealed various shortcomings and issues faced while using the different types of protocols listed in Table 15.1, which highlights the contrast study of the basic routing protocols of MANET. A few advantages and

Table 15.1 Contrast study of basic routing protocols of MANET [3, 14].

Routing protocol	Advantages	Disadvantages
Proactive protocols (table driven)	Leading and efficient routing information available. Formation of paths is rapid. Latency is abbreviated.	Requires outsized amount of resources and bandwidth, so overheads are high. Most of the preserved routing information remains unused. Loops exists.
Reactive protocols (on- demand)	On demand path formation results in reduced amount of resources. Abridged routing load. Free from loops.	Leading routes not obtainable around the clock. Traffic control overhead costs are relatively high. While discovering routes, latency is comparatively high.
Hybrid protocols	1. Improved control overheads and latency as compared to the other two. 2. Search expenses are confined. 3. Up-to-date routing information in the zones available.	1. Routing overheads for large-sized zones are high so, adding to its complexity. 2. Zone overlapping results in flooding. 3. Large network demands a large memory.

disadvantages [1] of all the three routing protocols have been depicted in the table that support the fact that the Hybrid Routing Protocols have an advantage over the other two types of protocols, which has been proved by other researchers over time.

15.6 Conclusion and Future Scope

This research analysis has covered the concept of MANET along with the advantages of hybrid routing protocols of MANET. The extensive literature review has focussed on various gaps and security apprehensions that exist in the traditional routing methodologies of MANET. This chapter has also analyzed the concept of optimization of the routing approach of MANET. QoS evaluation of hybrid routing protocols of MANET through the metaheuristic approach and further improvement of the security of the selected hybrid routing protocols by application of trust-based mechanism have been proposed. The future scope includes the simulation of the proposed approach.

References

1 Meenakshi, Y. and Nisha, U. (2017). Survey on MANET: routing protocols, advantages, problems and security. *International Journal of Innovative Computer Science & Engineering* 1 (2): 12–17. ISSN: 2393-8528.

2 Abdul, A.H., Mohammad, A.H., and Abdullah, G. (2015). A comparative analysis of energy conservation approaches in hybrid wireless sensor networks data collection protocols. *Telecommunication Systems*, New York: Springer Science+Business Media. doi 10.1007/s11235-015-0092-8

3 Aggarwal, R., Mittal, A., and Kaur, R. (2016). Various optimization techniques used in wireless sensor networks *International Research Journal of Engineering and Technology (IRJET)* 3 (6): 2086–2089. e- ISSN: 2395-0056 p-ISSN: 2395-0072, www.irjet.net

4 Deepshikha, B. and Sharma D.P. (2016). A comparative analysis of proactive, reactive and hybrid routing protocols over open source network simulators in MANET. *International Journal of Applied Engineering Research* 11: 1885–1891. ISSN: 0973-4562

5 Hamdy. E.-S.H. (2016). Shortest paths routing problem in MANETs. *Applied Mathematics & Information Sciences International Journal* 10 (5): 1–5. ISSN: 1885-1891 http://www.naturalspublishing.com/Journals.asp

6 Jayanti, N.V. (2014). Routing protocols in MANET: comparative study. *International Journal of Computer Science and Mobile Computing* 3 (7): 119–125. ISSN 2320–088X

7 Murali, K., Rahul, M., Venkateshwaran, G. et al. (2017). Detecting attacks in MANET using secure zone routing protocol. *International Journal of Engineering Science and Computing* 7 (4): 369–383.

8 Prajkta, N.S. and Vaishali, S.N. (2016). Improving the efficiency of MANET by reducing routing overhead using a NCPR protocol. *International Journal of Advanced Research in Computer Engineering & Technology (IJARCET)* 5 (2): 244–252. ISSN: 2278–1323

9 Pdmalaya, N. and Bhavani, V. (2016). Impact of random mobility models for reactive routing protocols over MANET. *International Journal of Engineering Development and Research* 4 (4): 338–344. ISSN: 1473-804x online

10 Goyal. V. and Arora, G. Review paper on security issues in mobile Adhoc networks, *International Research Journal of Advanced Engineering and Science* 2 (1): 203–207.

11 Jonny, K., Laurence, D.S., and Göran, P. (2012). Routing security in Mobile ad-hoc networks. *Informing Science and Information Technology* 9: 112–115.

12 Rajagopal, A., Somasundaram, S., Sowmya, B. et al. (2015). Soft computing based cluster head selection in wireless sensor network using bacterial foraging optimization algorithm. *International Journal of Electrical, Computer, Energetic, Electronic and Communication Engineering* 9 (3): 85–92.

13 Ramesh, P. and Mathivanan, V. (2017). Bee Inspired Agent Based Routing Protocol-Secondary User (BIABRP-SU). *International Journal of Engineering and Technology (IJET)* 9 (1): 118–127. ISSN (Print):2319-8613, ISSN (Online): 0975-4024

14 Palaniammal, M. and Lalli. M. (2014). Comparative study of routing protocols for MANETS. *International Journal of Computer Science and Mobile Applications* 2 (2): 118–127. ISSN: 2321-8363

15 Vardhini, K.K. and Sitamahalakshmi, T. (2016). A review on nature-based swarm intelligence optimization techniques and its current research directions. *Indian Journal of Science and Technology* 9 (10): 379–384. ISSN (Print) : 0974–6846 ISSN (Online): 0974-5645. doi: 10.17485/ijst/2016/v9i10/81634.

16 Persis, J.D. and Robert, P.T. (2015). Ant based multi-objective routing optimization in Mobile AD-HOC network. *Indian Journal of Science and Technology* 8 (9): 875–888. ISSN (Online):0974-5645, ISSN (Print):0974-6846

17 Sherin, Z., Soni, M.K., and Beg, M.M.S. (2015) QoS optimization in networks through meta-heuristic quartered genetic approach. ICSCTI, IEEE 91–96.

18 Rajneesh, N., Sumeer, K., and Anish. A. (2012). Security issues of routing protocols in MANETs. *International Journal of Computers and Technology* 3 (2): 713–720. ISSN: 2277-3061 www.ijctonline.com

19 Sherin. Z. and Soni. M.K. (2014) Secure routing in MANET through crypt-biometric technique. In: Proceedings of the 3rd International Conference on Frontiers of Intelligent Computing: Theory and Applications (FICTA). 713–720.

20 Heena, R. and Singh, J. (2017). Analysis of swarm intelligence optimization techniques used in MANETS: a survey, *International Journal of Advanced Research in Computer Science* 8 (5): 64–71. ISSN No. 0976-5697

21 Mukesh, G.K. and Singh. N. (2015). Routing and security issues for trust based framework in Mobile Ad Hoc Networks, *IOSR Journal of Computer Engineering (IOSR-JCE)* 17 (3): 01–05. e-ISSN: 2278-0661, p-ISSN: 2278-8727

22 Sherin, Z., Soni, M.K., Beg, M.M.S. (2015) An optimized genetic stowed biometric approach to potent QOS in MANET. *Procedia Computer Science* 62: 410–418.

23 Sherin, Z. and Soni, M.K. (2014) A novel crypt-biometric perception algorithm to protract security in MANET. *International Journal of Computer Network and Information Security* 6 (12): 64–71.

16

Study of Security Issues on Open Channel

Md Mudassir Chaudhary, Siddhartha Sankar Biswas, Md Tabrez Nafis, and Safdar Tanweer

Department of Computer Science and Engineering, School of Engineering Sciences and Technology, Jamia Hamdard, New Delhi, India

16.1 Introduction

The Internet has become an essential part of our daily lives. After stepping into technological innovation, everything is connected to the Internet. Connecting a device to the Internet at one's convenience is the primary purpose many governments are dealing with. Many governments are introducing open channels (Wi-Fi), as setting up wireless broadband is the easiest method of connecting people at high speed without a physical wired connection. [1–4] Wi-Fi uses radio frequencies to send data between your device and the Internet. The implementation of an open channel (Wi-Fi) is done by examining the population density of the location at which it is being set up. Wi-Fi helps to reduce the complexity as well as the mess that is created by wires. When we consider connecting to the Internet via an open channel, we are at massive risk of being attacked anywhere on the globe while using it.

Protecting a user's privacy and confidentiality is the primary responsibility of any organization, whether government or private. When we examine Wi-Fi architecture, we oberve that it consists of basically three elements. Firstly, the wireless router that transmits the radio signals, [4–7] secondly, the Access Point helps to connect us with the Internet through ISP, and finally to the Client or User's device (laptops, mobiles, etc.). Every component can provide an unspecified path opened to attack of our information security policies, i.e., Confidentiality, Integrity, and Availability.

16.2 Wireless Attacks

16.2.1 Reconnaissance Attack

Reconnaissance means the gathering of information. A reconnaissance attack is used to collect information on the targeted network or system. These attacks at first appear harmless, but they can cost severe damage to a company [1, 6]. These attacks are common in society and must be considered as a severe threat to any organization, as attackers use them to steal information to gain access to the organization or to someone's life. Reconnaissance attacks can provide potential information on the privacy of a citizen. It is similar to a thief scouting a neighborhood for unsecure connections using war driving apps. The main goal of a reconnaissance attack is to detect flaws in a network.

16.2.2 Access Attacks

An access attack is used to access another user or network device through improper means. When more than one person connects to a network, there are chances of access attacks within the network. Such an attack allows the attacker access to another users' data present on the same network. An access attack is mainly done via a man-in-the-middle attack (MITM) attack.

16.2.3 Man-in-the-Middle Attack

An MITM is an attack where an attacker creates a spoof AP (Access Point) that allows the user to communicate through that AP. In MITM, two parties communicating together are observed by the attacker as a middle man. An MITM attack is a general term used when an attacker positions himself in a discussion between two users. In MITM, attackers can see all the information sent by users present on that network. The attacker's objective is to acquire individual information such as login ID, credit card numbers, etc.

16.2.4 Denial of Services (DoS)

Denial of Services (DoS) occurs when the attacker stops the traffic movement of the network. This type of attack is done mainly with the intention of denying services to an authorized user. This is achieved by flooding the network with junk traffic requests that clog the network's bandwidth and creates a bottleneck. The creation of a bottleneck is known as a DoS attack. A computer, like us, can only perform one task at a time, so flooding the network with bogus requests causes legitimate users to be unable to connect to the network and can even render the network itself nonfunctional.

16.3 Securing Wireless Transmissions

Open wireless communication creates three basic threats: Interception, Interruption, and Modification.

16.3.1 Protecting the Confidentiality

The best method of protecting the confidentiality of any information transmitted over the wireless network is by encrypting the traffic. If a piece of information is lost while transmitting over the channel, the attacker cannot benefit from it. This is important for each and every user.

16.3.2 Protecting the Modification

Interception and Modification of a wireless network represent the form of a man-in-the-middle attack. Strong encryption and strong authentication of the user can reduce interception and modification of the data on the network.

16.3.3 Preventing Interruption or Denial-of-Service Attack

Wireless networks are also at risk of DoS attack. Organizations can take some steps to reduce the risk of such attacks. These attacks can be resolved by site surveying the locations where signals from other networks are colliding with the organizational network. Proper auditing should be done before setting up the wireless network. Regular audits also help the system work better and find the flaws and perform appropriate remedial action, including removing offending devices or implementing measures to increase signal strength according to the area's coverage.

16.4 Proposed Model for Securing the Client Over the Channel

Before discussing this model, we will see how a person connects to Wi-Fi. In Figure 16.1, a person sends a request with the access key to the AP, and the AP checks the key shared by the person against its saved list. If it matches with the list, it allows the person to connect. So, we need a model that has better security, not in terms of Wi-Fi security, but in terms of data privacy.

While using a simple router, we try to add a VPN on the router's AP that helps us to increase the security of the user by encrypting the user's data, as shown in Figure 16.2. The idea of this operation is to treat the open channel the same as an insecure wired network (the Internet). A secure firewall device can be installed with the router at the point where the wireless network meets the wired network. This will allow only encrypted data to flow in the wired network. All traffic is then tunneled over the wireless network and into a VPN concentrator on the wired network.

Installing a VPN with every router will increase the amount of the setup. VPN creates a tunnel for the secure flow of data, but it decreases the bandwidth and the speed of the data. VPNs require encapsulation, and thus increase overheads and decrease performance.

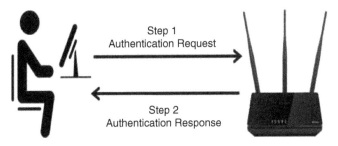

Figure 16.1 Authentication process.

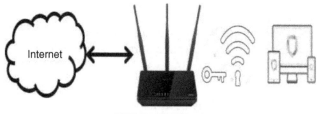

VPN Enabled Router

Figure 16.2 Use of VPN on the router's access point to increase the security of the user data.

16.5 Conclusion

Wireless networking provides countless opportunities to increase the strength and ability to connect to the Web with minimum hardware use. It also raises the risk to users' privacy and confidentiality. However, it is impossible to eliminate risk entirely; instead, it can be reduced by adopting some systematic approaches to managing risk. This chapter discusses the threats on privacy and confidentiality of a user over a wireless network and also discusses some techniques with an architecture that can be used to decrease the risk associated with it.

References

1 Abdulghani, S. (2009). Design and implementation of a network security model for cooperative network. *International Arab Journal of e-Technology* 1: 26–36.

2 Sharma, P.K., Park, J.H., Jeong, Y.S. et al. (2019). SHSec: SDN based secure smart home network architecture for internet of things. *Mobile Networks Applications* 24: 913–924. doi:10.1007/s11036-018-1147-3

3 Mallik, A., Ahsan, A., Shahadat, M. et al. (2019). Man-in-the-middle attack: understanding in simple words. 3: 77–92. 10.5267/j.ijdns.2019.1.001.

4 Sicari, S., Rizzardi, A., Grieco, L. et al. (2015). Security, privacy and trust in Internet of Things: the road ahead. *Computer Networks* 76: 146–164. 10.1016/j.comnet.2014.11.008.

5 Kelly, G. and McKenzie. B. (2002). Security, privacy, and confidentiality issues on the Internet. *Journal of Medical Internet Research* 4 (2): E12. doi:10.2196/jmir.4.2.e12

6 Azad, C. Agrawal, S., and Jha, V.K. (2013). Security, privacy and accountability in wireless network: a review *International Journal of Engineering Research & Technology (IJERT)*. 2 (7): 399–408.

7 Choi, M.-K., Rosslin, J., Robles, R. (2008). Wireless network security: vulnerabilities, threats and countermeasures. *International Journal of Multimedia and Ubiquitous Engineering* 3 (3): 77–86.

Index

Smart and Sustainable Approaches for Optimizing Performance of Wireless Networks: Real-time Applications, First Edition.
Edited by Sherin Zafar, Mohd Abdul Ahad, Syed Imran Ali, Deepa Mehta, and M. Afshar Alam.
© 2022 John Wiley & Sons Ltd. Published 2022 by John Wiley & Sons Ltd.